Innovations in Photogrammetry and Remote Sensing: Modern Sensors, New Processing Strategies and Frontiers in Applications

Innovations in Photogrammetry and Remote Sensing: Modern Sensors, New Processing Strategies and Frontiers in Applications

Editors

Francesco Mancini
Francesco Pirotti

MDPI • Basel • Beijing • Wuhan • Barcelona • Belgrade • Manchester • Tokyo • Cluj • Tianjin

Editors
Francesco Mancini
Department of Engineering
'Enzo Ferrari'
University of Modena and
Reggio Emilia
Modena
Italy

Francesco Pirotti
Department of Land,
Environment, Agriculture
and Forestry
University of Padova
Legnaro
Italy

Editorial Office
MDPI
St. Alban-Anlage 66
4052 Basel, Switzerland

This is a reprint of articles from the Special Issue published online in the open access journal *Sensors* (ISSN 1424-8220) (available at: www.mdpi.com/journal/sensors/special_issues/IPandRS).

For citation purposes, cite each article independently as indicated on the article page online and as indicated below:

LastName, A.A.; LastName, B.B.; LastName, C.C. Article Title. *Journal Name* **Year**, *Volume Number*, Page Range.

ISBN 978-3-0365-3932-4 (Hbk)
ISBN 978-3-0365-3931-7 (PDF)

© 2022 by the authors. Articles in this book are Open Access and distributed under the Creative Commons Attribution (CC BY) license, which allows users to download, copy and build upon published articles, as long as the author and publisher are properly credited, which ensures maximum dissemination and a wider impact of our publications.

The book as a whole is distributed by MDPI under the terms and conditions of the Creative Commons license CC BY-NC-ND.

Contents

About the Editors vii

Preface to "Innovations in Photogrammetry and Remote Sensing: Modern Sensors, New Processing Strategies and Frontiers in Applications" ix

Athos Agapiou
Damage Proxy Map of the Beirut Explosion on 4th of August 2020 as Observed from the Copernicus Sensors
Reprinted from: *Sensors* **2020**, *20*, 6382, doi:10.3390/s20216382 1

Huijiao Qiao, Xue Wan, Youchuan Wan, Shengyang Li and Wanfeng Zhang
A Novel Change Detection Method for Natural Disaster Detection and Segmentation from Video Sequence
Reprinted from: *Sensors* **2020**, *20*, 5076, doi:10.3390/s20185076 23

Ana-Maria Loghin, Johannes Otepka-Schremmer and Norbert Pfeifer
Potential of Pléiades and WorldView-3 Tri-Stereo DSMs to Represent Heights of Small Isolated Objects
Reprinted from: *Sensors* **2020**, *20*, 2695, doi:10.3390/s20092695 43

Qixin Liu, Xingfa Gu, Xinran Chen, Faisal Mumtaz, Yan Liu and Chunmei Wang et al.
Soil Moisture Content Retrieval from Remote Sensing Data by Artificial Neural Network Based on Sample Optimization
Reprinted from: *Sensors* **2022**, *22*, 1611, doi:10.3390/s22041611 63

Ruijin Li, Liming Zhang, Xianhua Wang, Weiwei Xu, Xin Li and Jiawei Li et al.
High-Precision Automatic Calibration Modeling of Point Light Source Tracking Systems
Reprinted from: *Sensors* **2021**, *21*, 2270, doi:10.3390/s21072270 85

Erica Nocerino, Elisavet Konstantina Stathopoulou, Simone Rigon and Fabio Remondino
Surface Reconstruction Assessment in Photogrammetric Applications
Reprinted from: *Sensors* **2020**, *20*, 5863, doi:10.3390/s20205863 105

Ivan Nikolov and Claus Madsen
Rough or Noisy? Metrics for Noise Estimation in SfM Reconstructions
Reprinted from: *Sensors* **2020**, *20*, 5725, doi:10.3390/s20195725 129

Piotr Łabędź, Krzysztof Skabek, Paweł Ozimek and Mateusz Nytko
Histogram Adjustment of Images for Improving Photogrammetric Reconstruction
Reprinted from: *Sensors* **2021**, *21*, 4654, doi:10.3390/s21144654 153

Riccardo Roncella and Gianfranco Forlani
UAV Block Geometry Design and Camera Calibration: A Simulation Study
Reprinted from: *Sensors* **2021**, *21*, 6090, doi:10.3390/s21186090 173

About the Editors

Francesco Mancini

Francesco Mancini (Prof.) received his M.Sc. degrees in Marine Environmental Sciences and a Ph.D. in Geodetic Sciences and Topography from the University of Bologna, Italy. Over 20 years, he has worked in the field of geomatics, as past researcher (University of Bologna and Technical University of Bari, Italy) and currently associate professor (University of Modena and Reggio Emilia, Italy). He is involved as surveyor in application to ground deformation, natural hazard assessment and landscape archaeology. He is professor in Geomatics technology, Geographical Information System, Cartography and Photogrammetry at schools of Civil and Environmental Engineering, Geology and Geography in diversified academic institutions. His current interests include, precision surveys, spatial data analysis, ground subsidence, monitoring of coastal areas, SAR interferometry, GNSS positioning and photogrammetric surveys by unmanned aerial vehicle (UAV). He has published a number of papers on international journals focused on surveying for environmental researches and modeling of natural/anthropogenic phenomena.

Francesco Pirotti

Francesco Pirotti is associate professor at University of Padova teaching at the Department of Land Environment and Agriculture and Forestry (TESAF). His research touches on geospatial frameworks (geodatabases and GIS), surveying, photogrammetry and remote sensing for environmental applications. He focuses primarily on methods for processing data from imagery –optical and radar –and lidar/photogrammetric point clouds for understanding forest structure dynamics. He is author of 100+ scientific publications in national and international peer reviewed journals. He is active in several national and international scientific societies. He has been involved in the scientific and organizing committee of 10+ international conferences. He is also involved in several international and national projects with different roles, as well as having been called to evaluate European-funded projects (FP7, EU Structural and Cohesion Fund 2014-2020, ERC Starting Grants) and academic evaluations both in Italy and abroad.

Preface to "Innovations in Photogrammetry and Remote Sensing: Modern Sensors, New Processing Strategies and Frontiers in Applications"

The recent development and rapid evolution of modern sensors and new processing strategies of collected data has paved the way for innovations in photogrammetry and remote sensing. The observation of the natural and built environment from aerial and satellite platforms benefits from improvements in data processing strategies thanks to the introduction of approaches based on multi-sensor data fusion, near real-time mapping and monitoring, the introduction of a new generation of algorithms from the artificial intelligence for data processing and increased ability in the handling of big datasets. The new generation of sensors and methods in the fields of photogrammetry and remote sensing has also been complemented by an ever-greater level of automation and capabilities in data processing. A very wide range of professional and research activities, previously limited to geomatics engineering and Earth observation, now have the opportunity to exploit the aforementioned tools in timely investigation of natural or anthropogenic features and phenomena. More flexible and automated approaches are also required by markets related to the space economies promoted by several countries around the world and the new opportunities arising from the use of sensors designed for photo or video-monitoring applications from low-altitude unmanned aerial vehicles (UAVs) or fixed locations in terrestrial surveys. In this direction, much has been achieved in the academic and professional world in terms of accurate reconstruction of three-dimensional features at very high spatial and temporal resolutions using close-range photogrammetry and structure from motion (SfM) technology. The interest of researchers is now moving towards the use of very dense point clouds to derive surface properties at desired spatial scales. However, a limited number of papers and experiences focused on time-related issues, monitoring, dynamic assessments, quasi-real time investigations of phenomena from repeated surveys and automated processing. This still represents a challenging approach to surveying. This Special Issue aimed at papers related to novel research and advances in these key areas of photogrammetry and remote sensing. Emphasis was placed on new algorithms, modern and/or forthcoming sensors, improvements in data fusion and processing strategies, as well as the assessment of their reliability with careful consideration of quality assurance/quality control and error budget.

After the revision procedure, nine papers that focused on satellite and aerial/terrestrial images and videos processing were selected and published. Five of them referred to innovative applications from satellite data and to the development of satellite sensors. In particular, Agapiou [1] used backscattered signal and interferometric synthetic aperture radar (InSAR) analysis from Sentinel-1 radar in addition to medium-resolution Sentinel-2 optical data images and a cloud-based facility to produce a damage proxy map of the Beirut explosion, which occurred on the 4th of August 2020. Qiao et al. [2] presented a novel method for change detections after a natural disaster using an optical flow-based method and adaptive thresholding segmentation able to detect the pixel-based motion tracking at fast speed from video sequences composed of successive passages of high-resolution satellites. Papers based on satellite data include the work by Loghin et al. [3]. The authors discuss the use of high-resolution stereo and multi-view imagery for the derivation of a digital surface model (DSM) over large areas for numerous applications in topography, cartography, geomorphology, and 3D surface modelling. In particular, the paper was focused on applications based on Pléiades and WorldView-3 imagery, that have, respectively, 0.70 m and 0.31 m ground sampling distance (GSD), for the reconstruction of small isolated objects with the assessment of their detectability, by estimating heights as a function of object type and size, and the successive validation of measurements. Liu et al. [4] developed a methodology based on remote sensing data for soil moisture content (SMC) retrieval using an artificial neural network (ANN). In this paper, the quantity and quality of samples for ANN training and testing were tested in addition to a sparse sample exploitation (SSE) method developed to solve problems arising from sample scarcity, resultant from cloud obstruction in optical images and/or the malfunction of in situ SMC-measuring instruments. To this aim, Sentinel-1 SAR and Landsat-8 images were adopted. The series of papers focused on satellite methods includes the contribution from Li et al. [5], where the authors propose a high-precision automatic calibration model based on a novel point light source tracking system for mirror arrays. Such a method can satisfy the requirements of high-resolution, high-precision, high-frequency on-orbit satellite radiometric calibration and modulation transfer function detection.

Four papers focused on photogrammetric data processing were included in this Special Issue. The paper by Nocerino et al. [6] evaluated whether the integration of visibility information (image orientation) and photo-consistency could potentially lead to an improvement of the mesh quality (and successive products) with tests carried out on diverse datasets of varying nature, scale, shape, image resolution and network designs. Metrics were introduced and considered to present qualitative and quantitative assessment of the results. Nikolov and Madsen [7] introduced a metric for noise estimation in SfM highly detailed 3D reconstructions from close-range photogrammetry. In particular, the authors discussed a possible approach to distinguishing real surface roughness from reconstruction noise and proposed a number of geometrical and statistical metrics for noise assessment based on both the reconstructed object and the capturing camera setup. Łabędź et al. [8] proposed a methodology based on the enhancement image histograms for the improvement of raster images and the final increasing of the accuracy level achieved in photogrammetric reconstruction. Various types of objects were reconstructed starting from photogrammetric dataset acquired by close-range and aerial photogrammetry. Finally, the paper by Roncella and Forlani [9] introduce a simulation study where the accuracy estimation of camera parameters and tie points' ground coordinates is evaluated as a function of various project parameters in the designing of photogrammetric fights with aerial unmanned vehicle in varying terrain shape. Several scenarios were investigated and conclusions show that the accuracy between different block configurations reach an order of magnitude of difference while oblique imaging is confirmed as a key requisite to the accurate reconstruction of flat terrain, while ground control point (GCP) density is not. Moreover, the accurate camera positioning constitutes a benefit in the accurate 3D reconstruction rather than an increasing number of GCPs in flat terrain. This study completes the collection of papers included in the Special Issue that testify to the growing interest in the key areas of photogrammetry and remote sensing.

We hope that the scientific community involved in photogrammetry and remote sensing will find the papers published in this Special Issue useful for their future investigations.

References

1. Agapiou, A. Damage Proxy Map of the Beirut Explosion on 4th of August 2020 as Observed from the Copernicus Sensors. Sensors 2020, 20, 6382, doi:10.3390/s20216382.

2. Qiao, H.; Wan, X.; Wan, Y.; Li, S.; Zhang, W. A Novel Change Detection Method for Natural Disaster Detection and Segmentation from Video Sequence. Sensors 2020, 20, 5076, doi:10.3390/s20185076.

3. Loghin, A.-M.; Otepka-Schremmer, J.; Pfeifer, N. Potential of Pléiades and WorldView-3 Tri-Stereo DSMs to Represent Heights of Small Isolated Objects. Sensors 2020, 20, 2695. doi:10.3390/s20092695.

4. Liu, Q.; Gu, X.; Chen, X.; Mumtaz, F.; Liu, Y.; Wang, C.; Yu, T.; Zhang, Y.; Wang, D.; Zhan, Y. Soil Moisture Content Retrieval from Remote Sensing Data by Artificial Neural Network Based on Sample Optimization. Sensors 2022, 22, 1611. doi.org/10.3390/s22041611.

5. Li, R.; Zhang, L.; Wang, X.; Xu, W.; Li, X.; Li, J.; Hu, C. High-Precision Automatic Calibration Modeling of Point Light Source Tracking Systems. Sensors 2021, 21, 2270. doi.org/10.3390/s21072270.

6. Nocerino, E.; Stathopoulou, E.K.; Rigon, S.; Remondino, F. Surface Reconstruction Assessment in Photogrammetric Applications. Sensors 2020, 20, 5863. doi:10.3390/s20205863.

7. Nikolov, I.; Madsen, C. Rough or Noisy? Metrics for Noise Estimation in SfM Reconstructions. Sensors 2020, 20, 5725. doi:10.3390/s20195725.

8. Łabedź, P.; Skabek, K.; Ozimek, P.; Nytko, M. Histogram Adjustment of Images for Improving Photogrammetric Reconstruction. Sensors 2021, 21, 4654. doi.org/10.3390/s21144654.

9. Roncella, R.; Forlani, G. UAV Block Geometry Design and Camera Calibration: A Simulation Study. Sensors 2021, 21, 6090. doi.org/10.3390/s21186090.

Francesco Mancini and Francesco Pirotti
Editors

Communication

Damage Proxy Map of the Beirut Explosion on 4th of August 2020 as Observed from the Copernicus Sensors

Athos Agapiou [1,2]

[1] Department of Civil Engineering and Geomatics, Faculty of Engineering and Technology, Cyprus University of Technology, Saripolou 2-8, Limassol 3036, Cyprus; athos.agapiou@cut.ac.cy

[2] Eratosthenes Centre of Excellence, Saripolou 2-8, Limassol 3036, Cyprus

Received: 15 September 2020; Accepted: 7 November 2020; Published: 9 November 2020

Abstract: On the 4th of August 2020, a massive explosion occurred in the harbor area of Beirut, Lebanon, killing more than 100 people and damaging numerous buildings in its proximity. The current article aims to showcase how open access and freely distributed satellite data, such as those of the Copernicus radar and optical sensors, can deliver a damage proxy map of this devastating event. Sentinel-1 radar images acquired just prior (the 24th of July 2020) and after the event (5th of August 2020) were processed and analyzed, indicating areas with significant changes of the VV (vertical transmit, vertical receive) and VH (vertical transmit, horizontal receive) backscattering signal. In addition, an Interferometric Synthetic Aperture Radar (InSAR) analysis was performed for both descending (31st of July 2020 and 6th of August 2020) and ascending (29th of July 2020 and 10th of August 2020) orbits of Sentinel-1 images, indicating relative small ground displacements in the area near the harbor. Moreover, low coherence for these images is mapped around the blast zone. The current study uses the Hybrid Pluggable Processing Pipeline (HyP3) cloud-based system provided by the Alaska Satellite Facility (ASF) for the processing of the radar datasets. In addition, medium-resolution Sentinel-2 optical data were used to support thorough visual inspection and Principal Component Analysis (PCA) the damage in the area. While the overall findings are well aligned with other official reports found on the World Wide Web, which were mainly delivered by international space agencies, those reports were generated after the processing of either optical or radar datasets. In contrast, the current communication showcases how both optical and radar satellite data can be parallel used to map other devastating events. The use of open access and freely distributed Sentinel mission data was found very promising for delivering damage proxies maps after devastating events worldwide.

Keywords: Copernicus; Sentinel-1; Sentinel-2; InSAR; change detection; damage proxy map; Beirut; Lebanon; explosion

1. Introduction

In emerging situations such as industrial and technological accidences, earth observation sensors can provide near-real-time information to local stakeholders and international organizations. This has been already demonstrated in the past in several examples such as the case study of the blast at Cyprus Naval Base (11th of July 2011, at Mari area, Cyprus), at the Fukushima Daiichi nuclear disaster (11th of March 2011, Ōkuma, Japan), etc. High-resolution satellite sensors are mainly used for rapid mapping and recording of the damage over large extents. However, the use of freely distributed sensors such as those of the Copernicus Programme did not thoroughly investigate in the past. This is in contrast with their wide use in monitoring other hazards all over the world, which has been demonstrated in several articles [1–4]. This communication article aims to investigate whether the use of these sensors, namely the Sentinel-1 and Sentinel-2, can provide reliable information, in case of emerging situations.

For this reason, a recent catastrophic event was examined in Lebanon, aiming to produce Damage Proxy Maps (DPM) over the area.

The event took place in the evening of the 4th of August 2020 in the area near the main harbor of Beirut, the capital of Lebanon. According to the media, a fire near the port ignited a large nearby store of ammonium nitrate, which is a highly explosive chemical often used in fertilizer. More than 100 people died, while more than 5000 were wounded. It was estimated that at least 300,000 people were left homeless from this event, while the destruction was estimated to cost between 10 and 15 billion USA dollars [5]. The Beirut explosion generated seismic waves equivalent to a magnitude 3.3 earthquake, according to the United States Geological Survey (USGS) Earthquake Hazard Program [6].

From the early beginning, several agencies have tried to support ground rescue investigations. Examples of this effort are by the Center for Satellite Based Crisis Information (ZKI) of the German Aerospace Centre (DLR) [7] that used high-resolution WorldView-2 multispectral images acquired just a day after the event. The Advanced Rapid Imaging and Analysis (ARIA) team at NASA's Jet Propulsion Laboratory and California Institute of Technology in Pasadena, California, in collaboration with the Earth Observatory of Singapore (EOS), have also processed Sentinel-1 radar images indicating areas that are likely damaged caused by the explosion in Lebanon [8].

Figure 1 shows high-resolution red-green-blue (RGB) WorldView-2 images over the area of the harbor (Figure 1, top) and the broader area of the explosion (Figure 1, bottom) before (Figure 1a,c) and after (Figure 1b,d) the event. These images were released some hours after the explosion by the European Space Imaging [9], thus allowing visualization of the destruction.

Figure 1. Top: (**a**) WorldView-2 high-resolution optical image over the Beirut harbor area before the explosion and (**b**) after the event. Bottom: (**c**) WorldView-2 high-resolution optical image over the broader area of the Beirut harbor before the explosion and (**d**) after the event (copyrights European Space Imaging [9]).

While the above-mentioned results are very important for local stakeholders as they provide damage proxy maps in a short time, the detection of damaged areas is usually based on the processing on either optical or radar products. In this paper, we aim to investigate the Beirut explosion that took place on 4th of August 2020 by using in parallel both optical and radar Copernicus Sentinel data. The detection of damage areas in different wavelengths of the spectrum can further strengthen the final damage proxy maps.

The paper is organized as follow: initially, the methodology and the data used are presented, while the results from the change detection and InSAR analysis from the Sentinel-1 datasets are shown. These results are also compared with optical Sentinel-2 images as well as other published—in the media—high-resolution results. Finally, a general discussion for the potential of space-based applications for monitoring areas under threat is presented.

2. Study Area

The study area was focused on the harbor area of Beirut in Lebanon, where the explosion occurred, and its surrounding area (Figure 2). While much damage was observed in the harbor area (see also Figure 1a,b), other damages have been also reported and mapped from high resolution satellite images in an area 2000 m away from the blast zone, by the DLR team (see [7]). In this study, we extended the area of interest to cover a circular area of a radius of 3000 m around the harbor area. Four main zones were defined as follows: Zone A is an area up to 500 m from the blast site, Zone B is an area from 500 m to 1000 m from the site of the explosion, and Zone C covers areas up to 2000 m away from the harbor. Finally, Zone D is defined as the area from 2000 to 3000 m away from the harbor site. The northern part of the study area is the Mediterranean Sea; therefore, no images analysis was carried out at this part.

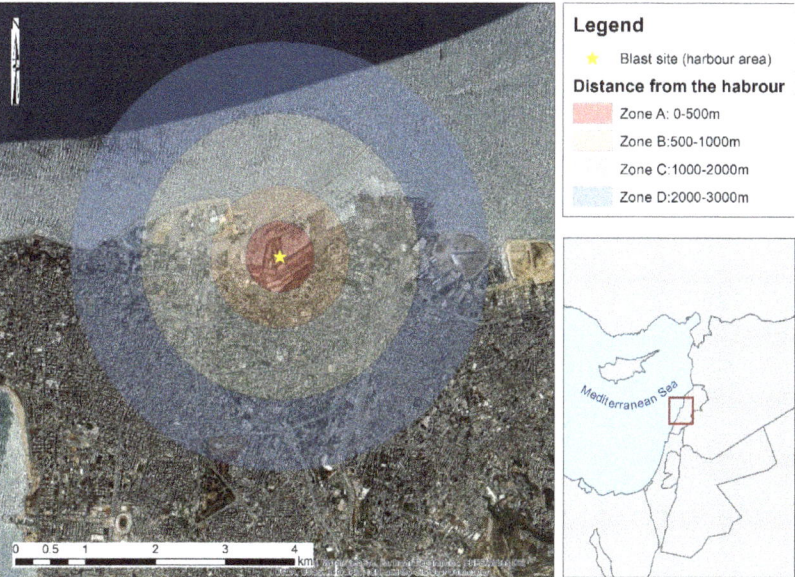

Figure 2. Study area indicating the harbor area (blast site) with a yellow star and the various zones created for further consideration covering distances from 0 to 3000 m away from the blast site.

3. Materials and Methods

3.1. Methodology

For the needs of the study, Copernicus radar freely distributed datasets were used. Once the available images were selected from the Alaska Satellite Facility (ASF) services [10], these were further processed by the Hybrid Pluggable Processing Pipeline (HyP3) cloud platform [11] (see more in Section 3.3).

The radar processing at the HyP3 cloud platform included two products as follows. A change detection map based on a pair of Radiometrically Terrain-Correct (RTC) Sentinel-1 data and an Interferometric Synthetic Aperture Radar (InSAR) analysis map, using both ascending and descending Sentinel-1 images, was used to map changes in the area. For the first product, the log difference for both VV (vertical transmit, vertical receive) and VH (vertical transmit, horizontal receive) backscattering gamma 0 amplitude polarizations of the RTC Sentinel-1 images was estimated based on the following equation for each image pixel:

$$\text{Change detection} = \text{Log}10\,(D2/D1), \tag{1}$$

where D1 refers to the Sentinel-1 with the earlier acquisition date (before the explosion) and D2 refers to the Sentinel-1 image taken after the explosion. Due to the small temporal window of the images, any changes can be linked to the blast event. A significant change of the gamma-0 amplitude can indicate areas of damage. Positive values indicate an increase in radar backscatter from the first date to the second, while negative values indicate a decrease. It should be mentioned that Sentinel-1 images are first filtered using the Enhanced Lee speckle filter, while the images are corrected based on the input Digital Elevation Model (DEM) of the area of interest. Then, the images are co-registered and radiometrically corrected (removal of radiometric distortions). The log difference of the gamma-0 amplitude for the VV and VH polarizations input images is estimated following Equation (1).

In addition, the coherence values were also mapped over the area. The degree of coherence is defined as the normalized complex correlation coefficient of the complex backscatter intensities S_1 and S_2 [12]. Coherence values can be estimated using Equation (2), where the * denotes the complex conjugate.

$$\gamma = |\langle S_2 \times S_1{}^* \rangle| / \sqrt{(\langle S_1 \times S_1{}^* \rangle \langle S_2 \times S_2{}^* \rangle)} | \tag{2}$$

Coherence values may range from 0 to 1; the larger the number, the higher the coherence. A low coherence value may indicate areas that have changed between the two overpasses of the Sentinel-1 radar sensors.

The InSAR analysis of an ascending and a descending pair of Sentinel-1 images was executed—as mentioned before in the HyP3 platform—using the Gamma software. In short, the InSAR Gamma algorithm for the Sentinel-1 images comprises of eleven (11) steps as follows. (1) determine the overlapping area using the two Sentinel-1 images. (2) Download the European Union Digital Elevation Model (EU-DEM) v1.1, a hybrid Shuttle Radar Topography Mission digital elevation model (SRTM), and ASTER Global Digital Elevation Map (ASTER GDEM) data fused by a weighted averaging approach over the area of interest [13]. (3) Create a lookup table between the DEM and Sentinel-1 imagery. A lookup table for SLC co-registration, considering terrain heights, is created. The data are filtered with adaptive data filtering (ADF). (4) Create a differential interferogram using the DEM height along with the co-registration with the DEM. (5–6) Removal of the flat earth phase (5) and the topographic phase (6). (7) Refinement of the slave image (prior to the event) with the master image (after the event). A check for convergence is performed using the azimuth offset as a limit (less than 0.02 pixels). (8) Resampling of the slave to match the master image. (9) Create the final interferogram. (10) Unwrap the phase using the Minimum Cost Flow (MCF) algorithm. The MCF algorithm permits a global automatic optimization robust phase unwrapping, taking into consideration disconnected areas of high coherence. (11) Geocode the results. The end products, as well as the sub-products developed through this processing chain of the InSAR analysis, are also available for downloading by the end-user (see more in [11]).

Then, the final products from the change detection and the InSAR analysis were imported into a Geographical Information System (GIS), namely the ArcGIS v.10 software, whereas zonal statistics per each zone, namely Zone A to Zone D, were estimated. The overall results generated a damage detection proxy map that was also compared with available high-resolution RGB WorldView-2 images as well as other reports from international agencies, namely the German Aerospace Center (DLR) and the ARIA teams. In addition, Sentinel-2 optical images, taken before and after the explosion, were downloaded from the Sentinel Data Hub [14] in order to compare the results from the Sentinel-1 analysis.

3.2. Datasets

As mentioned earlier, two pairs of Sentinel-1 images were used for the change detection and the InSAR analysis. Table 1 indicates the characteristics of the two Sentinel-1B Interferometric Wide (IW) images used for the change detection analysis (see Equation (1)). An image with an acquisition date of the 24th of July 2020 and an image taken on the 5th of August 2020 were used. The images were taken from the same sensor (S1B) as well as with the same pass direction (ascending orbit) to minimize any noise.

Table 1. Sentinel-1 data used for the change detection analysis.

Name	Characteristics Prior to the Event (Reference–D1)	Characteristics After the Event (Secondary–D2)
Date	The 24th of July 2020	The 5th of August 2020
Mode	Interferometric Wide swath (IW)	Interferometric Wide swath (IW)
Satellite	Sentinel-1B	Sentinel-1B
Absolute Orbit number	22,615	22,790
Pass direction	ASCENDING	ASCENDING
Polarization	VV + VH	VV + VH
Product type	ground range (GRD)	ground range (GRD)
Path	14	14
Frame	107	107

In addition, two pairs of Sentinel-1B images in descending and ascending orbits were processed for the InSAR analysis. The images were taken on the 31st of July 2020 and the 6th of August 2020 with a descending orbit (see Table 2), while another pair of Sentinel-1B images in ascending orbit, with an overpass at the 29th of July 2020 and the 10th of August 2020, was used (see Table 3).

Table 2. Sentinel-1 data used for the Interferometric Synthetic Aperture Radar (InSAR) analysis (descending orbit).

Name	Characteristics Prior to the Event (Reference—D1)	Characteristics After the Event (Secondary—D2)
Date	The 31st of July 2020	The 6th of August 2020
Mode	Interferometric Wide swath (IW)	Interferometric Wide swath (IW)
Satellite	Sentinel-1B	Sentinel-1B
Absolute Orbit number	33,693	22,797
Pass direction	DESCENDING	DESCENDING
Polarization	VV + VH	VV + VH
Product type	Single Look Complex (SLC)	Single Look Complex (SLC)
Path	21	21
Frame	480	480

Table 3. Sentinel-1 data used for the InSAR analysis (ascending orbit).

Name	Characteristics Prior to the Event (Reference—D1)	Characteristics After the Event (Secondary—D2)
Date	The 29th of July 2020	The 10th of August 2020
Mode	Interferometric Wide swath (IW)	Interferometric Wide swath (IW)
Satellite	Sentinel-1B	Sentinel-1B
Absolute Orbit number	22,688	22,863
Pass direction	ASCENDING	ASCENDING
Polarization	VV + VH	VV + VH
Product type	Single Look Complex (SLC)	Single Look Complex (SLC)
Path	87	87
Frame	107	107

Finally, Sentinel-2 images of Level 2A (Bottom of Atmosphere (BOA) reflectance images) with a spatial resolution of 10/20 m acquired before (24th of July 2020) and after (8th of August 2020) the explosion, with limited cloud coverage over the area of interest were downloaded from the Sentinel Data Hub. The RGB optical composite of the Sentinel-2 image taken after the explosion and the NIR-R-G pseudo color composite of the same image were used for comparison purposes through interpretation with the Sentinel-1 image analysis.

In addition, Principal Component Analysis (PCA) was applied to the integrated Sentinel-2 dataset (images of 24th of July 2020 and 8th of August 2020 together) to detect any significant spectral changes near the harbor area. The PCA is a well-known statistical analysis process that takes into account the spectral variations within the image [15]. While this type of analysis is implemented in single image processing, it can also be tested for a multi-temporal dataset, whereas the temporal variance will be taken into consideration. Therefore, the PCA can be used as a fast change detection method, in cases where the radiometric noise and the time span between the images is minimum [16,17]. In our example, we used Sentinel 2 optical images of Level 2A processing, meaning that the images are geometric, radiometric, and atmospherically corrected. Therefore, the PCA can explain the changes due to the explosion over the area.

3.3. Big-Data Cloud Platform

The HyP3 big data cloud platform provided by the ASF was used to process the radar Sentinel-1 datasets. The platform is designed on Amazon cloud services, and upon submitting the request form, the end users may select several products and processing algorithms. Further details for the potentials of this platform can be found in [11]. The final products are distributed through Amazon's simple storage service (S3) and are available to the end users for downloading. The HyP3 platform currently operates in a beta version, and its access is limited to restricted users. It runs a series of different radar processing chains such as Interferometric SAR (InSAR), Radiometric Terrain Correction (RTC), and change detection. The processing of the radar images is based on either the Sentinel Toolbox [18] or Gamma software [19].

4. Results

This section provides an overview of the results generated for this analysis of the radar images. These results are also compared with optical Sentinel-2 data as well as other high-resolution images such as those of WorldView-2 and the public products from the DLR and ARIA teams.

4.1. Sentinel-1 Analysis

Below, the results from the Sentinel-1 radar images is provided, following the change detection image analysis in Section 4.1.1 and the InSAR analysis in Section 4.1.2.

4.1.1. Sentinel-1 Change Detection Analysis

Figure 3 presents the results of the change detection analysis. The two Sentinel-1B images with acquisition dates of 24th of July 2020 and 5th of August 2020 (Table 1) have been processed based on Equation (1). Two threshold values were selected to classify the change detection results for both VV and VH polarizations of the Sentinel-1 images. The default value of −0.25 and +0.25 from the HyP3 platform was kept, while another lower threshold of −0.15 and +0.15 was also applied to visualize any minor changes in the magnitude of the backscattered signals of the two images. Areas that undergo a negative change between the two images, and therefore recorded a decrease in the backscatter returns, are displayed in red, while those displaying a positive change that indicates an increase in backscatter returns are displayed in blue color. In Figure 3, the four zones under study (Zone A to Zone D) are also presented. Figure 3 is a focus of the results around the harbor area (blast site) indicating the same results. The extent of Figure 4 is indicated in Figure 3, with a white dash line around the blast site. As the individual products from the VV and the VH change, detection polarizations tend to give different accuracy levels; these can be found in Appendix A. However, although the VH polarization signals (Figures A1 and A3) tend to have much lower coherence than the VV polarization signals (Figures A2 and A4), both of them were able to indicate areas of changes near the blast zone.

Integrating both these polarization results, the majority of the differences of the backscattered signal (VV and VH polarizations) are around the blast zone mainly in Zone A and Zone B (Figures 3 and 4). This is also evident in the focus area in Figure 4. However, damages have also been recorded on a broader area mainly to the east and south of the blast site, extended in Zones C and D, and in some areas even beyond Zone D (>3000 m). It is also important to highlight that the majority of these changes are considered quite significant (>−0.25 or >0.25, indicated with red color in Figures 3 and 4), which are the result from the sudden catastrophe around the harbor zone. Details for the individual results of the VV and the VH polarizations can be found in Appendix A (see Figures A1–A4).

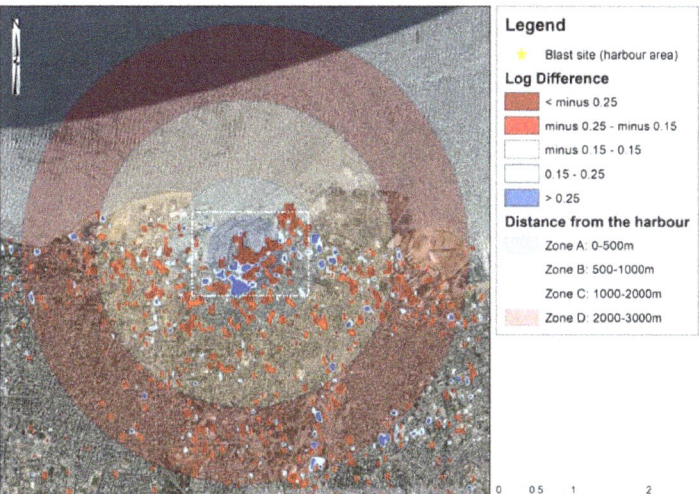

Figure 3. Change detection results using the log difference of the VV and the VH backscattering amplitude. Significant changes of the pair of images are presented with dark red and blue colors (>−0.25 or >0.25 differences), while other minor changes in the range of −0.25 to −0.15 and 0.15 to 0.25 are also presented in light red and blue colors, respectively. The four zones under study (Zone A to Zone D) are also given. The location of the blast size is shown with the yellow star at the center of the figure. The white dashed rectangle around the blast site is indicating the zoom area presented in Figure 4.

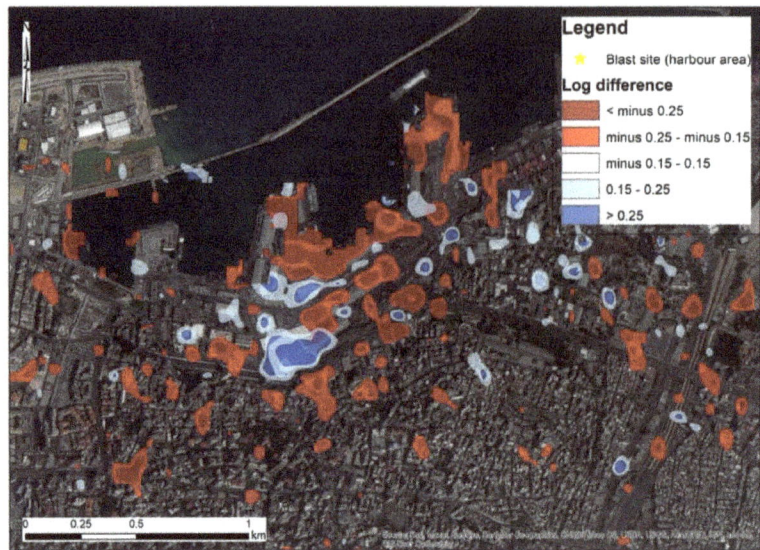

Figure 4. Change detection results—as in Figure 3, see before—for the area around the blast site near the harbor of Beirut (indicated with a white dashed line in Figure 3).

Table 4 summarizes the damaged areas for each zone (Zone A to Zone D) as estimated from the GIS environment using the log difference of the VH polarizations (see also Figures A1 and A3). The area of destruction for each zone is estimated in hectares units (10,000 m^2). As shown, the majority of the damaged area is found in Zone A with an area of 17.9 hectares, which equals 37.2% of the total damaged area, while another 24.5% of the total damage area is reported in Zone B that extends up to 1000 m away from the blast site. Zonal statistics analysis indicate a similar pattern for Zone C and Zone D with an approximately average damage area of 9 hectares (18.5%); however, the majority from these areas are in the medium log difference threshold units (−0.25 to −0.15 and 0.15 to 0.25). Zone A also reports the highest percentage in terms of areas of having a log difference more than >−0.25 and >0.25 (17.9 hectares or 62.0% and 11.8 hectares or 76.9%, respectively).

Table 4. Zonal statistics using the VH polarization for each log difference threshold and zone (Zone A to Zone D). Areas are measured in hectares (10,000 m^2).

Log Difference	Zone A	Zone B	Zone C	Zone D	Total
>−0.25	4.4	0.5	0.6	1.6	7.1
−0.25 to −0.15	4.4	7.2	6.2	5.6	23.4
0.15 to 0.25	5.5	3.8	1.8	1.9	13.0
>0.25	3.6	0.4	0.3	0.5	4.7
Total	17.9	11.8	8.9	9.5	48.2

The next table (Table 5) shows the zonal statistics for the log difference of the VV polarizations (see also Figures A2 and A4). The VV polarizations reported a larger area of damage (99.0 hectares) against the 48.2 hectares of the VH polarization analysis (Table 4). Zone A reports the highest percentage of the top threshold value (>−0.25) with an area of 14.4 hectares out of the total 25.8 hectares of Zone A. It is worth noting also that statistics from all zones (Zone A to Zone D) tend to give similar total areas of damage ranging from 20.9 hectares (Zone D) to 26.9 hectares (Zone C).

Table 5. Zonal statistics using the VV polarization for each log difference threshold and zone (Zone A to Zone D). Areas are measured in hectares (10,000 m^2).

Log Difference	Zone A	Zone B	Zone C	Zone D	Total
>−0.25	14.4	6.0	3.8	2.5	26.7
−0.25 to −0.15	6.0	12.6	16.1	10.7	45.5
0.15 to 0.25	3.1	4.7	5.3	5.9	19.0
>0.25	2.3	2.1	1.7	1.7	7.8
Total	25.8	25.4	26.9	20.9	99.0

Synthesizing the results from Tables 4 and 5, we can see the overall changes both for the VV and the VH polarizations. It should be mentioned that the new outcome that corresponds to the total damaged areas (shown in Table 6) is not the sum of Tables 4 and 5, but the statistics after the spatial union of these tables. This affects the overlapping areas, whereas the higher-ranking threshold was kept. Based on the findings of Table 6, we can estimate the total damaged areas within 3000 m away from the blast site to be around 117 hectares. In detail, an area of 29.0 hectares was mapped in Zone A (0–500 m from the blast site), while an area of approximately 32.0 hectares was mapped for Zones B and C (500–1000 m and 1000–2000 m from the blast site, respectively). Finally, an area of 24.2 hectares was estimated in Zone D (2000–3000 m).

Table 6. Zonal statistics using the synthesis of the VH and VV polarizations for each log difference threshold and zone (Zone A to Zone D). Areas are measured in hectares (10,000 m^2).

Log Difference	Zone A	Zone B	Zone C	Zone D	Total
>−0.25	11.0	5.2	4.1	2.9	23.3
−0.25 to −0.15	8.3	17.0	19.6	12.6	57.4
0.15 to 0.25	5.9	7.8	6.4	6.9	27.0
>0.25	3.7	2.1	1.6	1.9	9.2
Total	29.0	32.0	31.7	24.2	116.9

The majority of the damaged areas are found between a medium threshold value (from −0.25 to −0.15 and 0.15 to 0.25 log difference of the VV and VH polarization) with a total area of 84.4 hectares (57.4 and 27.0 hectares, respectively) that equals 72.1% of the total damaged area. The remaining 27.9% of the damaged area (32.5 hectares) is found at the higher threshold values (>−0.25 and >0.25).

4.1.2. Sentinel-1 InSAR Image Analysis

Following the change detection analysis, the InSAR analysis for mapping the micromovements of the area because of the generated seismic waves from the explosion was performed. As mentioned earlier for the needs of the InSAR analysis, a pair of Sentinel-1B Single Look Complex (SLC) images was used, with acquisition dates of 31st July 2020 and the 6th of August 2020.

Figure 5 shows the wrapped interferogram around the harbor site for the descending orbit (see Table 2). The wrapped interferogram is proportional to the difference in path lengths for the SAR Sentinel-1 image pair [12]. Interferometric fringes, visible in Figure 5, represent a full 2π cycle of phase change. The deformation fringes, visible around the area of the harbor in Figure 5, can be linked to the ground movement of that area due to the explosion. It should be mentioned that this observation is only visible in the area around the harbor site. The relative ground movement between the blast site and any other point of the area can be calculated by counting the fringes and multiplying them by half of the wavelength. The closer together the fringes, the higher the deformation on the ground.

Figure 5. Wrapped interferogram showing the deformation fringes as a result of the Beirut explosion as derived from the Sentinel-1 SAR images. The color shows a full 2π cycle of phase change.

Based on this wrapped interferogram, which is mainly useful for visualization purposes, the unwrapped interferogram was calculated for the descending orbit Sentinel-1 images (Figure 6a), which corresponds to the change in the distance along the line of sight of the sensor. An unwrapped interferogram converts the wrapped 2-π scale into a continuous scale (of multiples of pi). The Minimum Cost Flow (MCF) and triangulation methods (see more in [20]) are applied through the HyP3 platform for the phase unwrapping process.

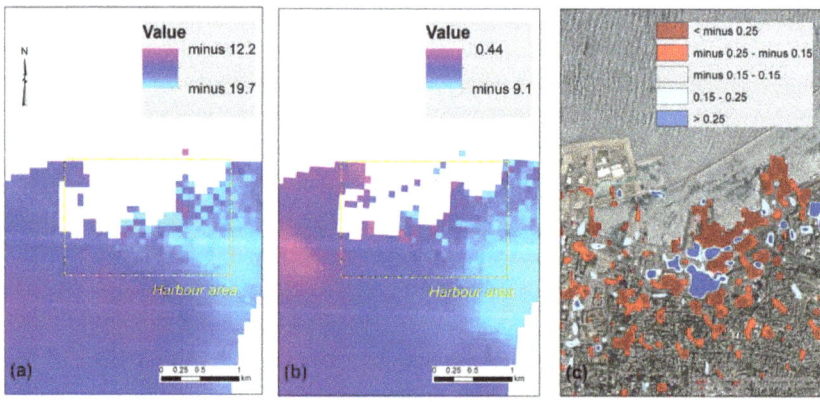

Figure 6. (**a**) Unwrapped interferogram as derived from the Sentinel-1 SAR images in descending orbit (see Table 2). (**b**) Unwrapped interferogram as derived from the Sentinel-1 SAR images in ascending orbit (see Table 3), and (**c**) VV and VH log difference polarizations of the same area (as in Figure 3). Harbor area is indicated in a yellow square.

Figure 6b shows the results from the ascending orbit Sentinel-1 images. Values greater than zero (positive) indicate relative movement away from the sensor (subsidence), while negative values indicate movement toward the sensor (uplift). Figure 6c shows the same area with the results from the change detection analysis. The ground relative displacements around the area of the harbor (see the yellow rectangle in Figure 6a,b) were estimated between −15 mm relative displacement along with the line of sight of the sensor for the descending orbit and −5 mm for the ascending orbit.

The line of sight (LOS) displacement map was estimated for both the ascending and the descending pairs of SLC Sentinel-1 (Figure 7a,b respectively). For estimating the LOS, we convert the unwrapped differential phase (Figure 6) into measurements of ground movement along the look vector (line of sight). Positive values indicate movement toward the sensor (such as uplift), while negative values indicate movement away from the sensor (such as subsidence).

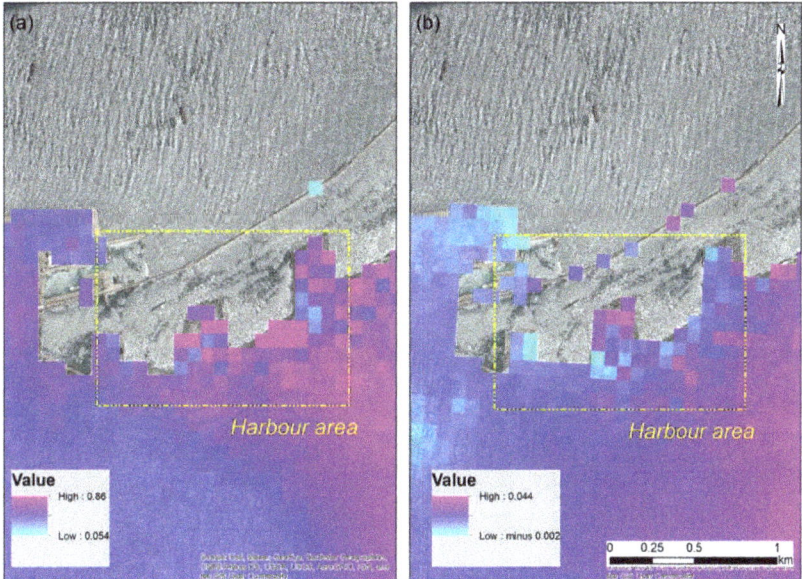

Figure 7. (**a**) Line of sight (LOS) displacements as derived from the Sentinel-1 SAR images (descending orbit). Harbor area, as indicated in a yellow square. (**b**) Line of sight (LOS) displacements as derived from the Sentinel-1 SAR images (ascending orbit). Harbor area, as indicated in a yellow square.

In the area of interest, and based on the descending orbit Sentinel-1 images (see Table 2), we can observe only positive movements (toward the sensor) that range between 5 and 9 cm; however, the more significant movement is found in the area over the harbor area. Similar findings were also retrieved from the analysis of the ascending orbit Sentinel-1 images of Table 3. These results are shown in Figure 7b. Once again, the positive values indicate uplift, while the negative values indicate subsidence. Specific pixels within the harbor area indicate the higher relative displacements.

Of course, these examples should be taken with great caution, as a low coherence is reported at this site due to the VV and VH changes observed from the explosion, as well as the sensitivity of the Sentinel-1 sensors (in relation to their wavelength that corresponds to ≈5.54 cm). However, the findings here are quite noteworthy, as even with low coherence, the InSAR analysis was able to detect the blast site and locate significant changes, such as those reported from the change detection analysis.

The coherence map for both descending and ascending orbits are shown in Figure 8a,b, respectively. Areas highlighted with purple color indicate high coherence values, while the blue color indicates regions with low coherence. The latest (regions with low coherence), as shown in both Figure 8a,b, are around the blast zone at the Beirut harbor, indicating a significant degree of change between the two overpasses of the Sentinel-1 (both in the ascending as well as for the descending orbits).

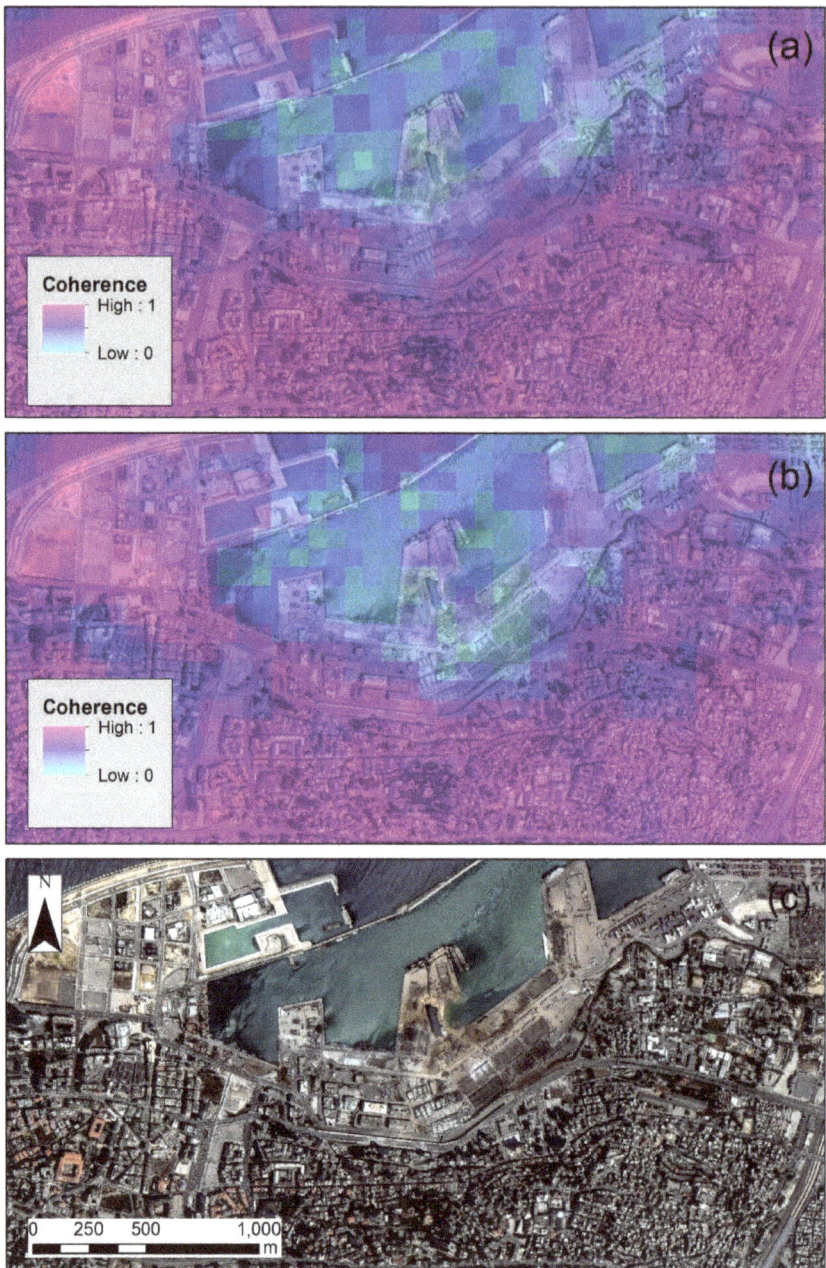

Figure 8. (a) Coherence map generated from the descending orbit images (see Table 2). (b) Coherence map generated from the ascending orbit images (see Table 3). (c) High-resolution WorldView-2 image.

4.2. Sentinel-2 Image Analysis

The results from the Sentinel-1 images have also been compared with the Sentinel-2 optical image taken after the 4th of August 2020. The RGB composite of the image is shown in Figure 9a. Destroyed buildings can be observed in the area near the exposition site; however, these are more evident in the NIR-R-G pseudo color composite shown in Figure 9b. Vegetated areas are shown with red color in this figure. The correlation between the visual inspection of the Sentinel-2 and the Sentinel-1 change detection results can be seen in Figure 9c.

Figure 9. (a) Sentinel-2 optical image taken over the area of interest on the 8th of August 2020 (RGB composite). (b) The NIR-R-G pseudo color composite of the same image and (c) the change detection results from the Sentinel-1 image analysis. Black arrows indicate destroyed areas from image interpretation of the Sentinel-2 image, while yellow arrows indicate the non-detectable destroyed areas from this analysis. The location of the blast size is shown with the yellow star.

Visual interpretation of the optical image can depict some of the most apparent damages over the area (see black arrows in Figure 9).

To further process the optical Sentinel-2 images, PCA was implemented in the integrated datasets of the Sentinel-2 images taken on 24th of July 2020 and 8th of August 2020. The results are shown in Figure 10. Figure 10a shows the results from the first principal component (PC1) while Figure 10b shows an RGB pseudo color composite of PC1–PC3. Figure 10c shows the high-resolution optical WorldView-2 image over the area. The red color in Figure 10 shows higher PC1 values, thus indicating changes in the integrated temporal Sentinel-2 dataset (i.e., 24th of July 2020 and 8th of August 2020 together). As evident, higher PC1 values are located around the blast area as well as in the western part of the harbor. Similar findings can be also reported from the pseudo color composite where the first three principal components, namely PC1 to PC3, have been used.

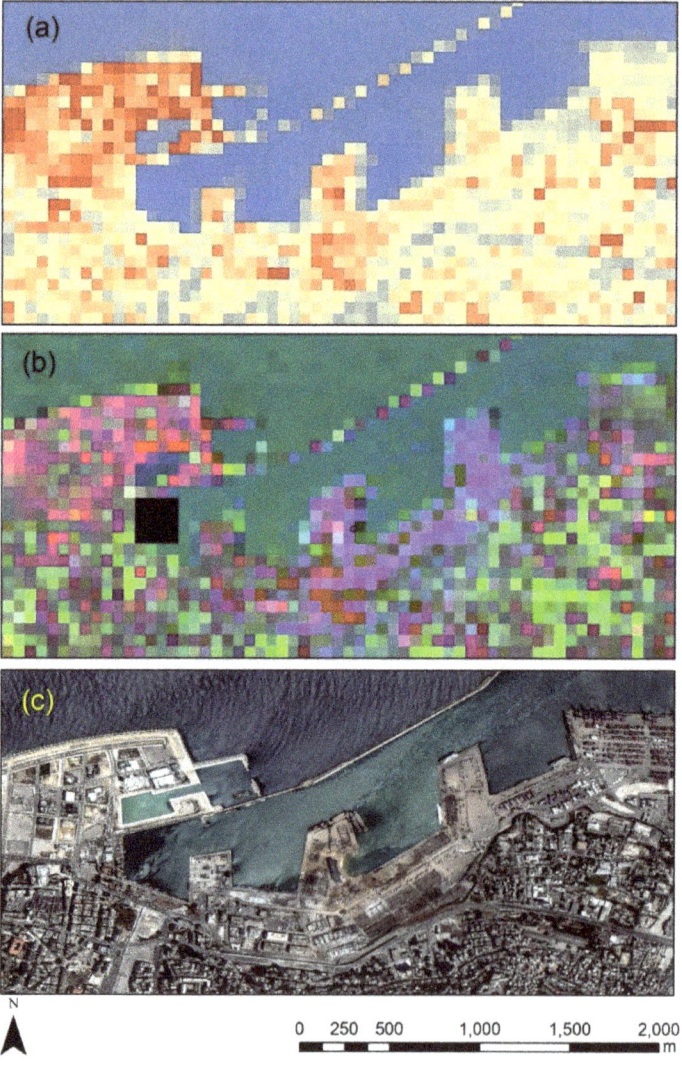

Figure 10. (a) The first principal component (PC1) results over the area around the harbor after the Principal Component Analysis (PCA) of the integrated Sentinel-2 optical images of 24th of July 2020 and 8th of August. (b) The PC1–PC3 pseudo color composite of the same integrated dataset and (c) the high-resolution WorldView-2 image.

5. Comparison with Other Results

The explosion at Beirut harbor has left significant and extensive damage in the surrounding buildings but also in other areas far away from the blast site. In the previous section, we have presented how this destruction could be detected by the Copernicus Sentinel-1 sensor via two difference analyses: (a) a change detection approach and (b) an InSAR analysis. The overall results were also confirmed through other studies that have been reported only some hours after the event, such as the case of the ARIA and the DLR teams. In addition, the availability of a high-resolution WorldView-2 image taken over the area of interest has confirmed the results generated from the analysis of the Sentinel-1 sensors.

To evaluate the overall performance of this study, we have compared the results presented above with other published material (maps) and available high-resolution WorldView-2 images. Initially, we have compared our change detection results (Damage Proxy Map) with the one of the ARIA team [8], as these have been generated with the Sentinel-1 sensor (as here). The comparison is shown in Figure 11. Figure 11a shows the change detection results with the threshold used in our study, while Figure 11b shows the damaged area as reported from the ARIA team. Although there is no threshold unit for this classification, it was evident that a similar pattern is reported from both studies, while adjusting the thresholds values would generate similar results. Once again, the area with the most damages is the one around the blast site.

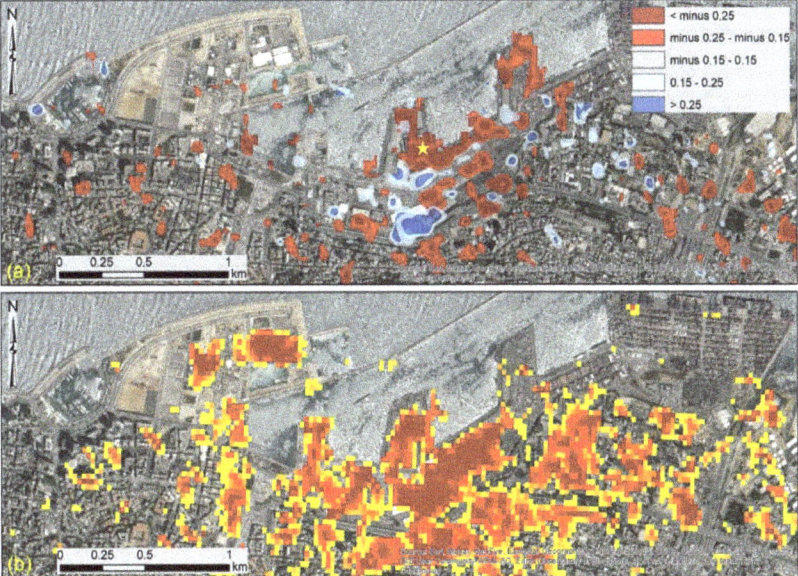

Figure 11. (**a**) Change detection results as generated from the VV and VH log differences of the Sentinel-1 images, while (**b**) indicates the Damage Proxy Map generated from the Advanced Rapid Imaging and Analysis (ARIA) team [8]. The location of the blast size is shown with the yellow star.

Within the area around the harbor of Beirut, digitization of the buildings that have been damaged has been reported by the MapAction platform [21]. Similar findings with this report are also found in the maps generated by the DLR team, which are not shown here. Figure 12 shows the focus area, whereas the VV and the VH log difference polarizations (as found from this study) are presented, and the digitized results from the MapAction platform. Comparing these two results, we can see a high correspondence between the two products; however, these were generated from different resolutions.

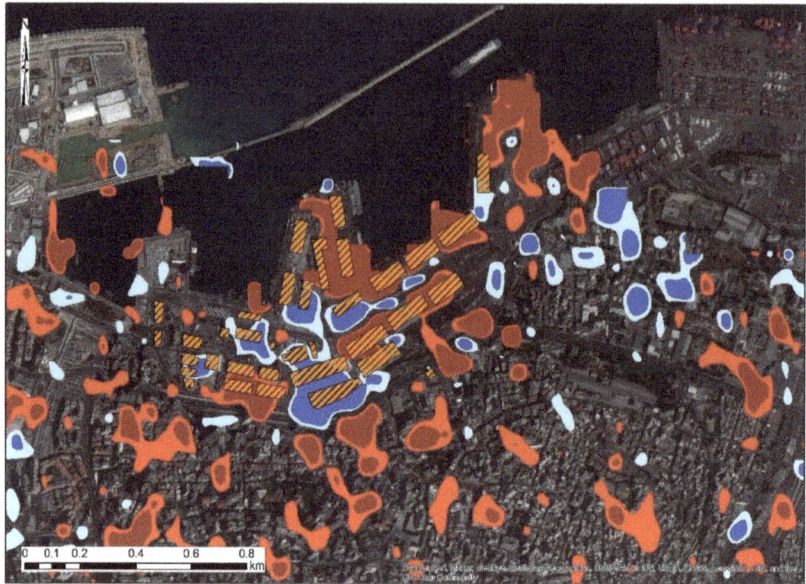

Figure 12. Change detection results as generated from the VV and VH log differences of the Sentinel-1 images around the blast site, while orange polygons indicate the buildings that have been damaged as reported by the MapAction platform [21] (digitized by the author).

A direct comparison of the high-resolution WorldView-2 image that was taken just after the explosion over the harbor of Beirut was carried out. This image is shown in Figure 13a, whereas the destroyed buildings can be observed in the image. The blast site, indicated with a yellow star in Figure 13a, has now been vanished due to the highly explosive chemicals that were kept in the site. Both the change detection analysis (Figure 13b), as well as the InSAR analysis (Figure 13c), shows a good correlation with the visual inspection of the high-resolution WorldView-2 image despite the difference in their spatial resolution (0.30 m of the WorldView-2 image against the 10m resolution of the Sentinel-1 results).

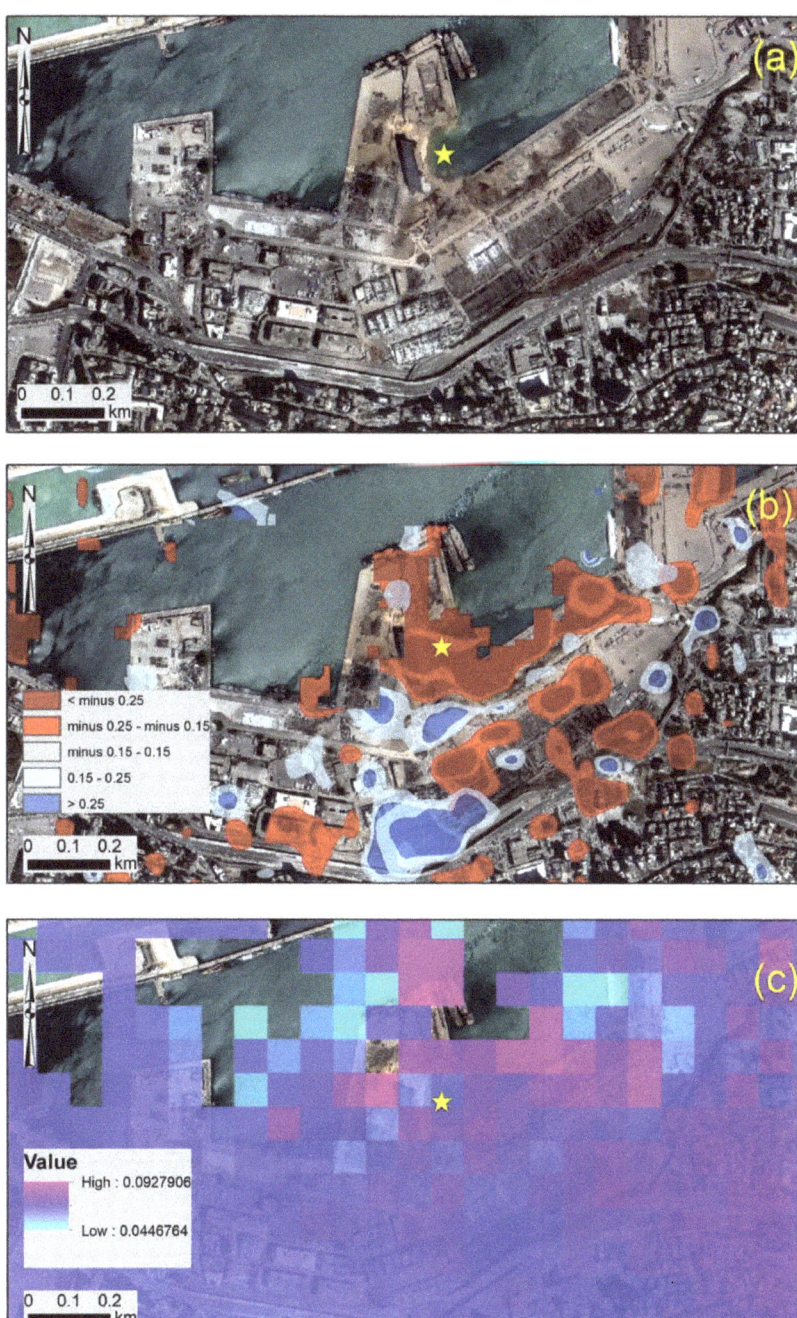

Figure 13. (**a**) High-resolution WorldView-2 image taken some hours after the exposition over the harbor of Beirut. The blast site is indicated with a yellow star. (**b**) Change detection as generated from the VV and VH log differences of the Sentinel-1 image and (**c**) the InSAR analysis (LOS of the descending orbit images) over the area. The location of the blast size is shown with the yellow star.

6. Conclusions

A damage proxy map based on the processing of Sentinel-1 images was carried out in this study over the harbor area of Beirut, Lebanon after the high explosion occurred in the area on the 4th of August 2020. The analysis of this study was achieved through a change detection analysis of Sentinel-1 images, while the seismic waves results were also mapped using an InSAR Sentinel-1 image analysis.

The analysis was able to be carried out due to the systematic observations of the Copernicus Sentinel-1 sensors operated by the European Space Agency (ESA), while the processing chain was carried out by the big-data cloud HyP3 platform operated by ASF. This combination allows end users to process in a short time (less than 1-h computational time) a series of radar processing chains in almost global coverage. Of course, this is feasible due to the free and fully open policy (FFO) of the Sentinel datasets.

The findings of this study are well aligned with other products delivered by specialized space centers such as the DLR and the ARIA teams, while a comparison was also performed using a high-resolution WorldView-2 image. While the medium-resolution Sentinel-1 images and their products do not allow us to detect individual destruction in a building level, the 10-m resolution is enough for estimating with high accuracy the damaged area. The use of Sentinel-2 images could support the detection analysis through visual interpretation of significant changes in the landscape as well as through a PCA temporal analysis. This was more obvious using the NIR-R-G pseudo color composite; however, due to the spatial resolution of the image (20 m), the detection of smaller areas was quite challenging.

The use of both sensors, namely the Sentinel-1 and Sentinel-2, can be further utilized in the future to support emergency situations, with an almost global coverage. In this case, other events not so well known can be monitored by the scientific local community to support specific emergency needs. Of course, these maps can be only indicative of the destruction, as ground verifications are needed. In addition, the analysis of the results from this series of analyses should be seen with great caution, as the images might suffer from other factors that influence the final outcomes. For instance, the relative micromovements can be due to other sources, such as the atmospheric component and even DEM accuracy, while the change detection of the VV and the VH polarizations need to be linked also with their coherence.

Overall, the use of Sentinel data was found very promising for supporting ground investigations after an event (such as the one presented here, or even a natural hazard). Future steps can include considering the automation of the whole procedure, minimizing the time between the data processing and the event itself.

Funding: This communication is submitted under the NAVIGATOR project. The project is being co-funded by the Republic of Cyprus and the Structural Funds of the European Union in Cyprus under the Research and Innovation Foundation grant agreement EXCELLENCE/0918/0052 (Copernicus Earth Observation Big Data for Cultural Heritage).

Acknowledgments: The author would like to acknowledge the ASF DAAC 2020 using GAMMA software. The articles contain modified Copernicus Sentinel data 2020, processed by ESA. Acknowledgements are also given to the European Space Imaging for the WorldView-2 images.

Conflicts of Interest: The author declares no conflict of interest.

Appendix A

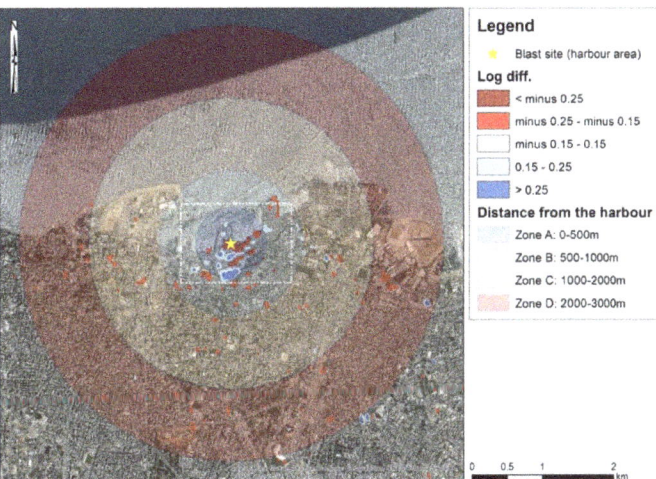

Figure A1. Change detection results using the log difference of the VH backscattering amplitude. Significant changes of the pair of images are presented with dark red and blue colors (>−0.25 or >0.25 differences), while other minor changes in the range of −0.25 to −0.15 and 0.15 to 0.25 are also presented in light red and blue colors, respectively. The four zones under study (Zone A to Zone D) are also given. The location of the blast size is shown with the yellow star. The white dashed rectangle around the blast site is indicating the zoomed-in area presented in Figure A3.

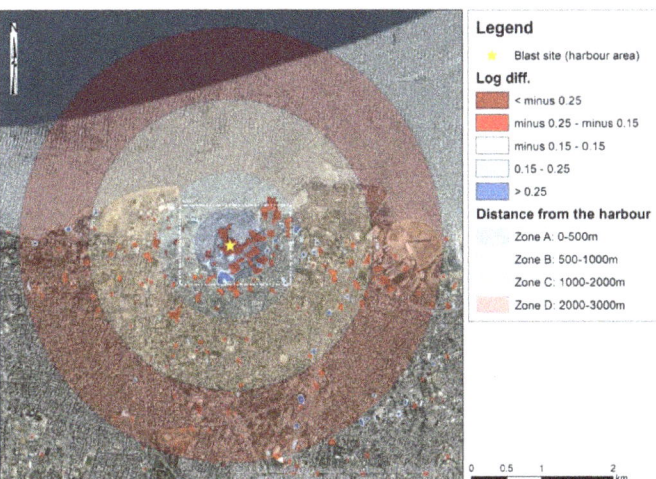

Figure A2. Change detection results using the log difference of the VV backscattering amplitude. Significant changes of the pair of images are presented with dark red and blue colors (>−0.25 or >0.25 differences), while other minor changes in the range of −0.25 to −0.15 and 0.15 to 0.25 are also presented in light red and blue colors, respectively. The four zones under study (Zone A to Zone D) are also given. The location of the blast size is shown with the yellow star. The white dashed rectangle around the blast site is indicating the zoomed-in area presented in Figure A4.

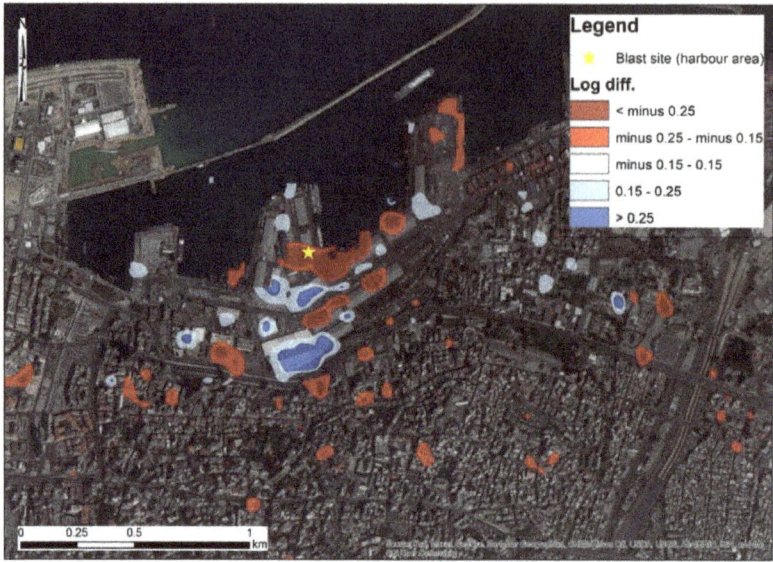

Figure A3. Change detection results as shown in Figure A1 (see before) for the area around the blast site near the harbor of Beirut (indicated with a white dashed line in Figure A1). The location of the blast size is shown with the yellow star.

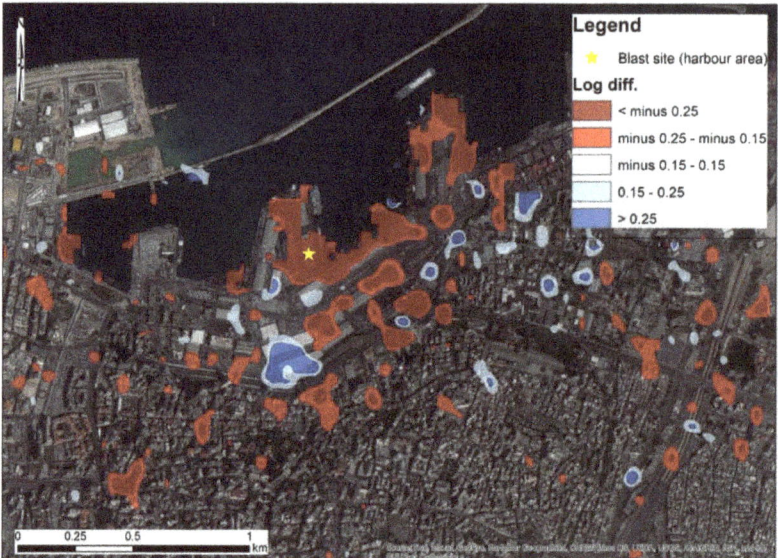

Figure A4. Change detection results—as in Figure A2, see before—for the area around the blast site near the harbor of Beirut (indicated with a white dashed line in Figure A2). The location of the blast size is shown with the yellow star.

References

1. Lazecký, M.; Hatton, E.; González, P.J.; Hlaváčová, I.; Jiránková, E.; Dvořák, F.; Šustr, Z.; Martinovič, J. Displacements monitoring over Czechia by IT4S1 System for Automatised Interferometric measurements using sentinel-1 data. *Remote Sens.* **2020**, *12*, 2960. [CrossRef]
2. Pawluszek-Filipiak, K.; Oreńczak, N.; Pasternak, M. Investigating the effect of cross-modeling in landslide susceptibility mapping. *Appl. Sci.* **2020**, *10*, 6335. [CrossRef]
3. Agapiou, A.; Lysandrou, V.; Hadjimitsis, D.G. Earth observation contribution to cultural heritage disaster risk management: Case study of Eastern Mediterranean open air archaeological monuments and sites. *Remote Sens.* **2020**, *12*, 1330. [CrossRef]
4. Zhuo, G.; Dai, K.; Huang, H.; Li, S.; Shi, X.; Feng, Y.; Li, T.; Dong, X.; Deng, J. Evaluating potential ground subsidence geo-hazard of Xiamen Xiang'an new airport on reclaimed land by SAR interferometry. *Sustainability* **2020**, *12*, 6991. [CrossRef]
5. Scientists Map Beirut Blast Damage. Available online: https://earthobservatory.nasa.gov/images/147098/scientists-map-beirut-blast-damage (accessed on 16 August 2020).
6. United States Geological Survey (USGS) Earthquake Hazard Program. Available online: https://earthquake.usgs.gov/earthquakes/eventpage/us6000b9bx/executive (accessed on 16 August 2020).
7. Center for Satellite Based Crisis Information (ZKI) of the German Aerospace Centre (DLR). Available online: https://activations.zki.dlr.de/en/activations/items/ACT148.html (accessed on 16 August 2020).
8. ARIA-SG Damage Proxy Map (Copernicus Sentinel-1) on 8/5/2020 for the Explosion in Lebanon 2020. Available online: https://maps.disasters.nasa.gov/arcgis/home/item.html?id=d9f8c773bfd04042819b646d8f2980c0 (accessed on 16 August 2020).
9. European Space Imaging. Available online: https://www.euspaceimaging.com/beirut-lebanon-satellite-images-show-explosion-damage/ (accessed on 16 August 2020).
10. Alaska Satellite Facility (ASF). Available online: https://asf.alaska.edu (accessed on 16 August 2020).
11. About HyP3. Available online: https://hyp3.asf.alaska.edu/about (accessed on 16 August 2020).
12. Werner, C.L.; Wegmüller, U.; Strozzi, T. Processing Strategies for Phase Unwrapping for InSar Applications. 2002. Available online: https://www.gamma-rs.ch/uploads/media/2002-4_PhaseUnwrapping.pdf (accessed on 16 August 2020).
13. EU-DEM. Available online: https://www.eea.europa.eu/data-and-maps/data/copernicus-land-monitoring-service-eu-dem (accessed on 30 October 2020).
14. Sentinel Data Hub. Available online: https://scihub.copernicus.eu (accessed on 30 October 2020).
15. Campbell, J.B. *Introduction to Remote Sensing*; The Guilford Press: New York, NY, USA, 2007.
16. Agapiou, A. Detecting looting activity through earth observation multi-temporal analysis over the archaeological site of Apamea (Syria) during 2011–2012. *J. Comput. Appl. Archaeol.* **2020**, *3*, 219–237. [CrossRef]
17. Bustos, C.; Campanella, O.; Kpalma, K.; Magnago, F.; Ronsin, J. A method for change detection with multi-temporal satellite images based on Principal Component Analysis. In Proceedings of the 6th International Workshop on the Analysis of Multi-temporal Remote Sensing Images (Multi-Temp), Trento, Italy, 12–14 July 2011; pp. 197–200. [CrossRef]
18. Sentinel Toolbox. Available online: https://sentinel.esa.int/web/sentinel/toolboxes (accessed on 13 August 2020).
19. Gamma Software. Available online: https://www.gamma-rs.ch (accessed on 13 August 2020).
20. Sun, X.; Zimmer, A.; Mukherjee, S.; Kottayil, N.K.; Ghuman, P.; Cheng, I. DeepInSAR—A deep learning framework for SAR Interferometric phase restoration and coherence estimation. *Remote Sens.* **2020**, *12*, 2340. [CrossRef]
21. MapAction Platform. Available online: https://maps.mapaction.org/event/2020-lbn-001 (accessed on 16 August 2020).

Publisher's Note: MDPI stays neutral with regard to jurisdictional claims in published maps and institutional affiliations.

© 2020 by the author. Licensee MDPI, Basel, Switzerland. This article is an open access article distributed under the terms and conditions of the Creative Commons Attribution (CC BY) license (http://creativecommons.org/licenses/by/4.0/).

Article

A Novel Change Detection Method for Natural Disaster Detection and Segmentation from Video Sequence

Huijiao Qiao [1,2], Xue Wan [2,3], Youchuan Wan [1,*], Shengyang Li [2,3] and Wanfeng Zhang [2,3]

1. School of Remote Sensing and Information Engineering, Wuhan University, Wuhan 430079, China; qiaohj@whu.edu.cn
2. Technology and Engineering Center for Space Utilization, Chinese Academy of Science, Beijing 100094, China; wanxue@csu.ac.cn (X.W.); shyli@csu.ac.cn (S.L.); wfzhang@csu.ac.cn (W.Z.)
3. Key Laboratory of Space Utilization, Chinese Academy of Sciences, Beijing 100094, China
* Correspondence: ychwan@whu.edu.cn; Tel.: +86-15802797355

Received: 22 July 2020; Accepted: 2 September 2020; Published: 7 September 2020

Abstract: Change detection (CD) is critical for natural disaster detection, monitoring and evaluation. Video satellites, new types of satellites being launched recently, are able to record the motion change during natural disasters. This raises a new problem for traditional CD methods, as they can only detect areas with highly changed radiometric and geometric information. Optical flow-based methods are able to detect the pixel-based motion tracking at fast speed; however, they are difficult to determine an optimal threshold for separating the changed from the unchanged part for CD problems. To overcome the above problems, this paper proposed a novel automatic change detection framework: OFATS (optical flow-based adaptive thresholding segmentation). Combining the characteristics of optical flow data, a new objective function based on the ratio of maximum between-class variance and minimum within-class variance has been constructed and two key steps are motion detection based on optical flow estimation using deep learning (DL) method and changed area segmentation based on an adaptive threshold selection. Experiments are carried out using two groups of video sequences, which demonstrated that the proposed method is able to achieve high accuracy with F1 value of 0.98 and 0.94, respectively.

Keywords: change detection; natural disasters; deep learning; threshold selection; optical flow estimation

1. Introduction

Natural disasters, such as earthquakes, tsunamis, floods and landslides, have shown a dramatically and globally increasing trend, both in frequency and intensity [1–3]. Accurate determination of changes on ground features associated with destructive disaster events is crucial to quick disaster response, post-disaster reconstruction and financial planning [4]. Change detection (CD) using remote sensing data can effectively capture changes before and after disasters [5–7], which has been widely used in various fields of natural disasters such as flood monitoring [8], landslide displacement tracking [9,10] and earthquake damage assessment [11,12], as well as relief priority mapping [13,14].

With the continuing growth of earth observation techniques and computer technology, massive amounts of remote sensing data for natural disaster with different spectral-spatial-temporal resolution are available for surveying and assessing changes in natural disaster, which greatly promotes the development of change detection methodologies. Many change detection approaches for natural disaster scenes have been proposed and they can be broadly divided into traditional and deep learning (DL)-based [15]. For traditional CD methods, the simplest approaches are algebra-based methods. Hall and Hay [16] firstly segmented two panchromatic SPOT data observed at different times and then detected changes through an image differencing method. Matsuoka et al. [17], on the basis of the difference between the backscattering coefficient and correlation coefficient achieved in an earthquake, applied supervised classification of the pre- and post-event optical images to present the distribution of damaged areas in Bam. These directly algebraic operations were easy to be implemented but always generated noisy outputs, such as isolated pixels or holes in the changed objects; thus, some transformation and models were used in CD researches. Sharma et al. [18] finished a damage assessment of landslides in a minimum time by pseudo color transformation and extracting the landslide affected area based on the pre- and post-earthquake Landsat-8 images. Lee et al. [19] first proposed an optimization algorithm based on Stepwise Weight Assessment Ratio Analysis (SWARA) model and geographic information system to assess seismic vulnerability. In order to overcome the limitation of the sole band and improve the identification of change detection, fusing datasets acquired from various remote sensors and geographical data are paramount to monitoring the environmental impacts of natural disasters. ElGharbawi et al. [20] estimated the crustal deformation affected by the 2011 Tohoku earthquake combined with two deformation patterns using Synthetic Aperture Radar (SAR) and GPS data. With the purpose of determining the changed buildings in forested areas, Du et al. [21] adopted the graph cuts method taking account of spatial relationships and took grey-scale similarity from old aerial images and height difference based on Digital Surface Model (DSM) generated from LiDAR data as two change detection indexes to optimize building detection.

Due to the rapid development of computer technology, the research of traditional change detection approaches has tuned into integrating deep learning techniques in recent years. Deep learning-based methods have presented promising potentials based on the extraction of high-level features. Saha et al. [22] detected collapsed buildings from SAR images and Ji et al. [23] further put them into a random forest classifier to detect post-seismic destroyed buildings using pre- and post-disaster remote sensing images. In order to achieve higher accuracy, some new neural networks have been introduced into disaster monitoring researches. Ci et al. [24] proposed a novel Convolutional Neural Network (CNN) model in combination with a CNN feature extractor, a new loss function and an ordinal regression classifier to evaluate the degree of building damage caused by earthquakes using aerial imagery. Peng et al. [25] utilized an end-to-end CD method named UNet++ to fuse multiple feature maps from different semantic segmentation levels to generate a final change map with high accuracy. Yavariabdi et al. [26] proposed a new change detection method based on multiobjective evolutionary algorithm (MOEA), which is robust to multispectral Landsat images with atmospheric changes. In this method, the similarity index measure (SSIM) is used to generate the difference image. After that, MOEA is applied to obtain a set of multiple binary change masks by iteratively minimizing two objective functions for changed and unchanged regions and the final binary mask is optimally fused by MRF. With the purpose of improving efficiency, in Ghaffarian et al. [27], extended U-net based on deep residual (ResUnet) followed a Conditional Random Field (CRF) implementation was proposed to update the post-disaster buildings from very high resolution imagery. Alizadeh et al. [12] established a new hybrid framework of Analytic Network Process (ANP) and Artificial Neural Network (ANN) models for earthquake vulnerability assessment. To avoid labeling a massive number of data for the training network, transfer learning has received increased attention. Pi et al. [28] employed transfer learning to train eight CNN models based on You-Only-Look-Once (YOLO), so as to recognize undamaged building roofs in disaster-affected areas. Transferring learning was used by Kung et al. [29] to manage disaster by combination of data augmentation, reference model and augmented model. Li et al. [30] proposed SDPCANet by combining PCANet and saliency detection to make change detection based on SAR images, which effectively reduced the number of training samples but kept higher change detection performance.

Recently, the development of commercial video satellites and the spread of mobile devices makes it possible to thoroughly monitor the changing process in natural disaster. For example, high resolution video sequences from video satellites, such as SkySat and Jilin_1, can provide valuable data during different disaster phases [31,32]. Thus, change detection can now move from pre- and post-image analysis to almost real-time disaster monitoring using video sequences. Although change detection for natural disaster has been researched for years in the society of remote sensing, the main studies are focused on pre- and post-disaster satellite imagery, while the change detection based on video satellite for natural disaster monitoring has rarely been studied. These video sequences, which capture disaster motion change, bring a challenge for existing CD methods because the color and texture of objects usually remain the same while only the positions have been changed.

The aim of this paper is to explore an effective method that detects the motion change in disaster from video sequences. Optical flow in the field of computer vision, is very likely to detect the pixel in this video sequence owing to its fast speed and pixel-based motion tracking. However, to fuse the result of optical flow into the final change detection map is a challenge. The generation of change detection map required the empirical threshold for optical flow, which may vary from case to case. Thus, this paper first presents the investigation of the motion detection property of the optical flow estimation algorithm based on deep learning and then proposes a novel change detection framework, OFATS, based on a new objective function for video sequence in natural disaster which combines the optical flow results and an adaptive thresholding segmentation algorithm based on the ratio of maximum between-class variance and minimum within-class variance.

The rest of this paper is organized as follows: Section 2 briefly reviews the optical flow estimation methods. The proposed change detection method is then introduced in Section 3. In Section 4, the effective of the proposed method is tested and compared with some most commonly used CD methods using two different natural disaster datasets. Finally, the paper is concluded in Section 5.

2. Optical Flow Estimation Methods

Optical flow, which represents change of the pixels' displacement vectors between image frames, is most widely used in motion tracking [33]. For example, optical flow has been used to detect human/animal movements [34,35] and medical organ lesions [36,37], robots or vehicle navigation [38,39], measure flow motion [40], airfoil deformation and surface strain [41]. With the assumption of instantaneous pixel value invariance over a short displacement, optical flow can be separated into two categories: local computation method on the basis of Lucas–Kanade (LK) method and global computation method based on Horn and Schunck (HS) formulation [42]. LK method supposes that the adjacent pixels in a sliding window share the same motion and keep locally constant [43]. However, the size of a sliding window is difficult to be determined and further affect the final accuracy [33]. HS assumes that the velocity field varies globally smoothly, which is more fit for real scenes [44]. Horn and Schunck introduced the optical constraint equation based on the combination of velocity field and gray value to build a basic algorithm of optical flow estimation [45]. However, these traditional optical flow computation methods often provide blurred boundaries and are hard to be used in real time [46,47]. Convolutional neural networks (CNNs) have a strong ability of feature extraction and speckle noise suppressing [15,48,49], which has attracted more attention to numerous computer vision tasks.

FlowNet is the first end-to-end optical flow estimation model with CNN in 2015 and it uses an encoder-decoder structure making up of convolutional and deconvolutional layer with additional crosslinks between these contracting and expanding networks [50]. For the encoder module, it is made up of nine convolutional layers and Rectified Linear Unit (ReLU), and mainly used to compute abstract features from respective fields of increasing seize, but the latter reestablishes the original resolution by an expanding upconvolutional architecture using four deconvolutional layers and ReLU active function layer. It turned out to be an achievable training network and can directly compute optical flow from two input images, but it is not competitive with fine-tuned traditional methods at accuracy and the running speed is also slower [51]. On the basis of FlowNet, FlowNet 2.0, a novel end-to-end optical flow estimation network was proposed in the winter of 2016, which can effectively solve the above-mentioned problems in close accuracy with the state-of-the-art methods while running orders of magnitude faster and be marked as a milestone for optical flow estimation based on CNN [52]. This success benefits from the following three aspects: new adding training dataset including tiny motion and real-word data, stacking numerous networks by warping operations and a novel leaning schedule of multiple datasets fusion. The schematic view of FlowNet 2.0 is shown in Figure 1. The network is separated into two parts: large displacement and small displacement optical flow network. For the computation of large displacement optical flow, two FlowNetS is combined and the warping layers are introduced as a refinement. To cope with small displacements, a smaller network, FlowNet-SD is added. Then the former stacked network and the small network are fused into FlowNet 2.0 in an optimal manner, which can achieve optimal performance on arbitrary displacements.

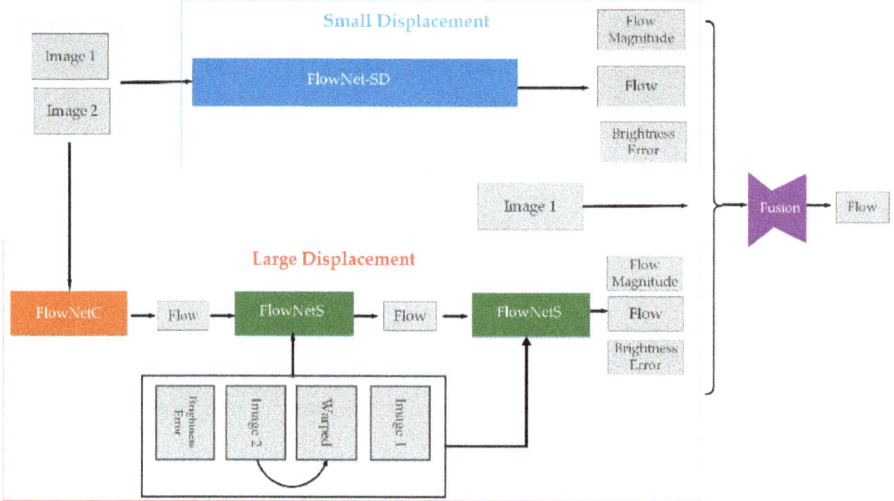

Figure 1. Scheme of FlowNet 2.0.

Optical flow estimation method based on FlowNet 2.0 has achieved considerable progress. Nevertheless, it has rarely been used to make change detection in natural disasters, as far as we know. In fact, FlowNet 2.0 based on HS generates a dense velocity field, that is, each pixel has the corresponding optical flow field [43]. In order to accurately divide pixels into changed and unchanged part based on the optical flow field, the selection of an appropriate threshold is critical. However, the threshold selection for image segmentation needs to consider the data characteristics with expert knowledge. Thus, in this paper, a novel CD framework, OFATS, for disaster detection has been proposed by combing motion detection based on FlowNet 2.0 and the adaptive threshold determination method based on a novel objective function.

3. Proposed OFATS Method

In this section, OFATS, the automated CD framework for natural disaster detection from video sequence is proposed and the workflow is as shown in Figure 2.

It consists of two main steps: motion detection where FlowNet 2.0, the optical flow estimation method based on deep learning, is introduced to compute the displacement and change boundary extraction based on an adaptive threshold determination algorithm which takes the ratio of maximum between-class variance and minimum within-class variance as the new objective function. Specially, two optimization strategies are proposed: narrowing the searching range of potential thresholds and dynamic normalization of motion information.

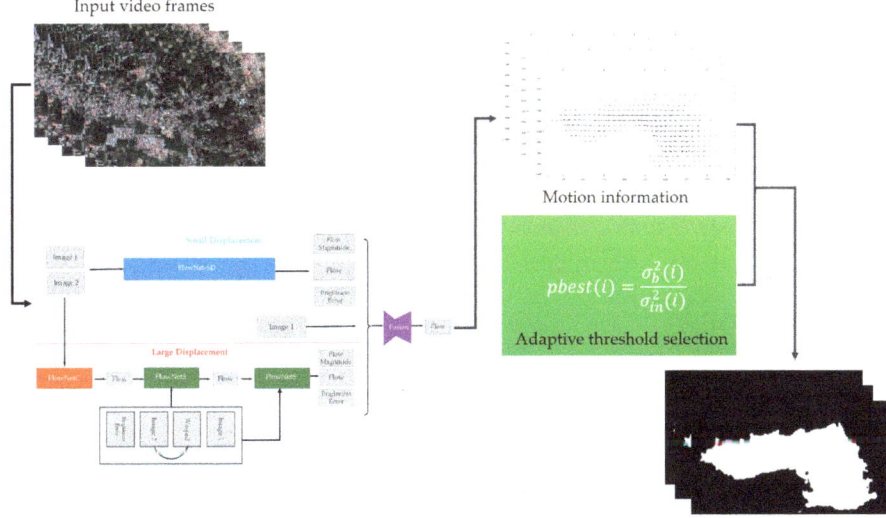

Figure 2. The workflow of OFATS (optical flow-based adaptive thresholding segmentation).

3.1. Motion Detection

In this paper, the pixel displacements in horizontal and vertical are calculated by FlowNet 2.0, denoted as $u(x, y)$ and $v(x, y)$, respectively, as shown in Figure 3.

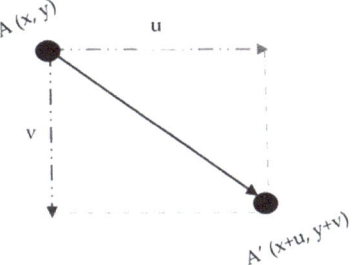

Figure 3. Illustration of displacements based on FlowNet 2.0.

The displacement can be calculated as follow:

$$r(x, y) = \sqrt{u(x, y)^2 + v(x, y)^2} \qquad (1)$$

Figure 4 shows an example of displacements' distribution based on sequence "RubberWhale" [53]. The sample frame and the corresponding optical flow field visualization result are shown in Figure 4a,b, respectively. In Figure 4b, the angles of arrows represent directions of each pixel's displacement $r(x, y)$ and the lengths show the magnitudes of displacements. Four boxes of different colors in Figure 4a,b representing various objects with different types of motions. In order to demonstrate the detailed difference of four boxes in Figure 4b, they are zoomed in Figure 4c–f. Overall, these figures show that the changed area can be roughly determined by the magnitudes and directions of optical flow. Given that magnitude changes are more obvious to detection, a proper segmentation threshold based on magnitude should be taken into consideration, separating the changed and unchanged part in the next step.

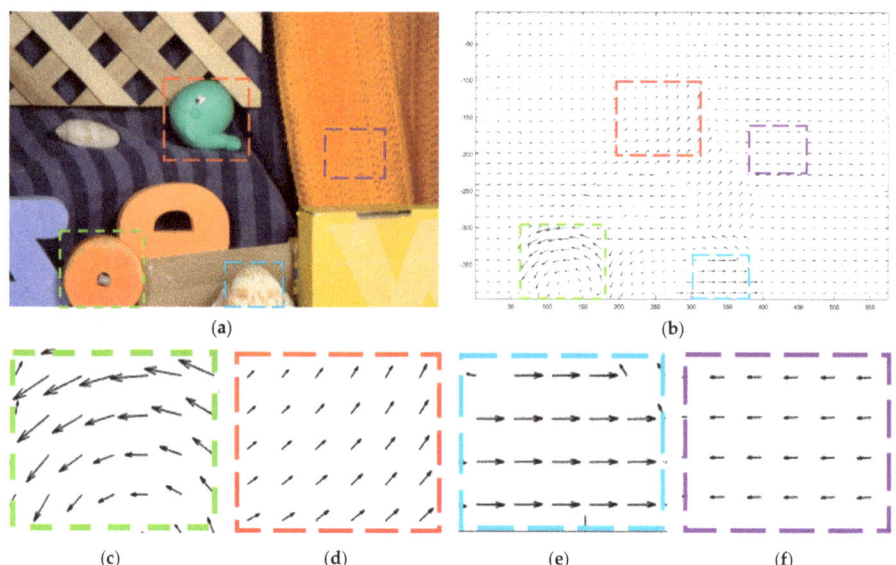

Figure 4. An example of displacements' distribution (**a**) sample frame on sequence "RubberWhale" [53]; (**b**) the corresponding $r(x,y)$ visualization based on FlowNet 2.0; (**c**–**f**) zoom-in boxes.

3.2. Change Boundary Extraction

After the motion detection, the next step is to determine the global optimal threshold of the displacements so as to divide the changed and unchanged part which can be viewed as a two-class classification problem. Based on displacements characteristics, the novel objective function and two optimizing strategies for the optimal threshold selection are proposed, respectively.

3.2.1. The Objective Function

For any CD algorithm, the key factor differentiating change from non-change is the objective function. The goal of setting the objective function is to find the global optimal threshold which can maximize the between-class variance and minimize the within-class variance at the same time by exploring a finite set of the possible displacement values as the possible threshold. Otsu is widely used for global thresholding selection [54], but it does not work when the target and background vary widely and two classes are very unequal [55,56]. Thus, in this paper, the objective function is set as the ratio of maximum between-class variance and minimum within-class variance:

$$pbest(i) = \frac{\sigma_b^2(i)}{\sigma_{in}^2(i)} \qquad (2)$$

where i is the iteration number and the value range is from 0 to Num, the number of unique value of motion detection results. $pbest(i)$ is the fitness value of ith iteration, $\sigma_b^2(i)$ and $\sigma_{in}^2(i)$ are the between-class and within-class variance, and the calculations are based on Equations (3) and (4), respectively.

$$\sigma_b^2 = P(C_1)P(C_2)(\mu_1 - \mu_2)^2 \qquad (3)$$

$$\sigma_{in}^2 = P(C_1)\sigma_1^2 + P(C_2)\sigma_2^2 \qquad (4)$$

where $P(C_1), P(C_2), \mu_1, \mu_2, \sigma_1^2, \sigma_2^2$ represent probabilities of class occurrence, class mean levels and the class variances of unchanged class C_1 and changed class C_2, respectively. They are defined as following:

$$P(C_1) = \sum_{i=1}^{m} p_i \tag{5}$$

$$P(C_2) = \sum_{i=m+1}^{n} p_i \tag{6}$$

$$\mu_1 = \frac{1}{P(C_1)} \sum_{i=1}^{m} w_i p_i \tag{7}$$

$$\mu_2 = \frac{1}{P(C_2)} \sum_{i=m+1}^{n} w_i p_i \tag{8}$$

$$\sigma_1^2 = \frac{1}{m} \sum_{i=1}^{m} \left(w_i - \frac{1}{m} \sum_{i=1}^{m} w_i \right)^2 \tag{9}$$

$$\sigma_2^2 = \frac{1}{n-m} \sum_{i=m+1}^{n} \left(w_i - \frac{1}{n-m} \sum_{i=m+1}^{n} w_i \right)^2 \tag{10}$$

The displacements can be represented in n levels $[1, 2, \ldots, n]$ and C_1 denotes pixels with levels $[1, \ldots, m]$, and C_2 denotes pixels with levels $[m+1, \ldots, n]$. The pixel value and the corresponding percentage at level i are denoted by w_i and p_i.

To concluded, $\sigma_b^2(i)$ and $\sigma_{in}^2(i)$ are determined by the iteration threshold value based on motion detection results and the iteration threshold w_i corresponding to the maximal fitness value $pbest(i)$ is considered as the optimal threshold t_{best}. The displacement of each pixel which is larger than the optimal threshold could be classified into the changed part, while the smaller ones are unchanged.

$$g(x, y) = \begin{cases} 1, & r(x, y) > t_{best} \\ 0, & r(x, y) \leq t_{best} \end{cases} \tag{11}$$

However, it requires large numbers of iteration and costs much time of too many unique values in motion detection results because the changeable range of different pixels' displacements being really wide. Thus, the optimizing strategies are further proposed for optimal threshold selection.

3.2.2. Optimizing Strategies for Threshold Selection

According to the distribution of displacement data, we proposed two strategies to optimize the threshold selection criterion: narrowing the searching range of iterations and dynamic normalization of displacements which are greater than the currently selected threshold for each iteration.

Narrowing searching range is to efficiently reduce the scope of potential thresholds determined by the wide range of experimental data. Comparing with the change of pre- and post- disasters, the video data with 30 FPS during disasters can record the whole minor change of each frame. Thus, it should be labeled as 'change' when pixels with displacements are larger than 1 pixel. Then, the searching range can be narrowed from 0 to 1 and the corresponding iteration number is reduced to $[0, N]$ and N is the number of unique values of displacements with the value from 0 to 1. According to this, a large number of the useless pixel values are excluded and speed can be enhanced greatly.

In order to reduce the influence of large range of displacement on the calculation of objective function, quite a lot of pixels with displacements exceeding 1 pixel have to be normalized for the whole image. Generally, pixels with larger displacements and great variations in magnitudes of pixels' movements must be classified as change class; therefore, normalization is executed only for the pixels in changed class whose displacements are more than 1 pixel. To be more elaborate, the partial normalization dynamically changes with each iteration. For ith iteration, the threshold is w_i and pixels with displacements' values $w_1, \ldots, w_i$ are automatically labeled as unchanged class C_1 and other pixels with displacements $w_{i+1}, \ldots, w_n$ which are larger than w_i are classed as changed part C_2. The pixels in C_2 whose displacements $w_j, \ldots, w_n$ ($j \geq i+1$), are greater than 1 need to be normalized to $[w_{i+1}, w_{end}]$ but the corresponding percentages remain unchanged. The formulas are as follows:

$$k_j = \frac{w_{end} - w_{i+1}}{w_{max} - w_{min}} \tag{12}$$

$$w_j = w_{end} + k_j(w_j - w_{min}) \tag{13}$$

w_{end} is the maximum displacement value which is most close to 1; w_{max}, w_{min} are the maximum and minimum pixel value of changed class C_2 which need to be normalized.

Based on the two strategies, the threshold calculation can be more efficient and lay a foundation for the selection of the optimal threshold.

3.3. The Proposed CD Method

The whole flowchart of proposed OFATS is as Figure 5 shown and the details of change detection process are implemented in Algorithm 1. The essential steps of OFATS are motion detection based on FlowNet 2.0 and segmentation based on adaptive threshold selection criteria. For motion detection, the selected frames are input into FlowNet 2.0 to compute the magnitude in horizontal and vertical directions, based on which the displacements are calculated by Equation (1). After that, the next steps are the iterative process for optimal threshold selection. Following Algorithm 1, Equations (2)–(13) are repeatedly applied to calculate the fitness value for a fixed number to enable the iterative optimization. Based on the optimal threshold, the displacement result can be segmented into changed and unchanged parts.

The details of OFATS are implemented in Algorithm 1:

Algorithm 1. The proposed OFATS for change detection in natural disaster

---

Input: The two frames extracted from the video sequence.
Output: The change detection result.
1: Input the two frame images and calculate the movement in horizontal and vertical directions based on FlowNet 2.0;
2: Calculate the displacements reserving a decimal fraction based on Equation (1);
3: Generate initial global fitness value *gbest* and iteration value *i*;
4: **while** the algorithm does not reach the termination condition **do**
5:    I = I + 1;
6:    Divide into unchanged class C_1 and changed class C_2 threshold w_i and normalize displacements which are larger than 1 in C_2 according to Equation (12) and (13) and then involve in arithmetic by using Equations (5)–(10);
7:    Calculate between- and within-class variance by using Equations (3) and (4);
8:    Calculate fitness value by using Equation (2);
9:    **if** The solution is better **then**
10:      Replace the current individual;
11:    **else**
12:      The individual does not change;
13:    **End if**
14:    Find out the current global best agent;
15: end while
16: **return** The optimal threshold.
17: Divide the image into two parts by optimal threshold value by using Equation (11).

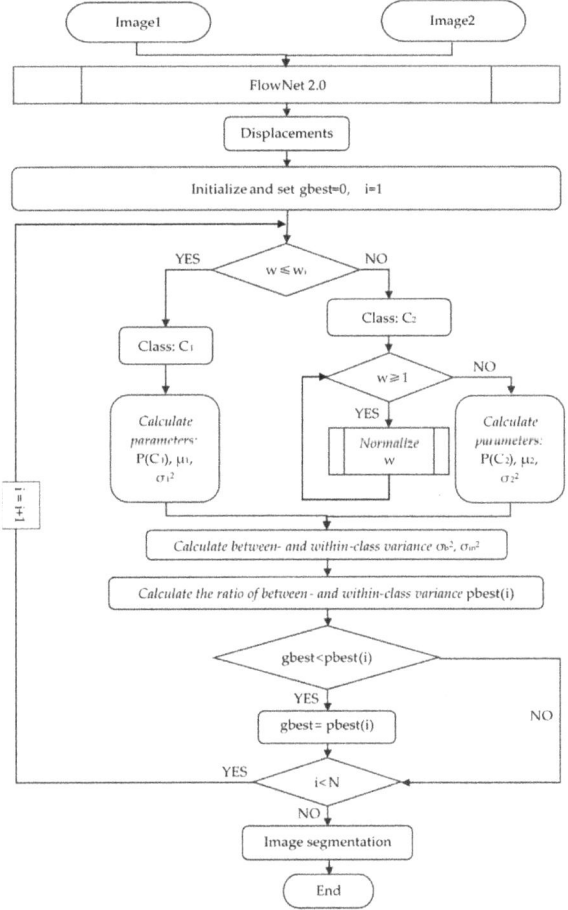

Figure 5. Flowchart for the proposed OFATS.

4. Data and Experiments

In this section, the proposed OFATS is applied to the detection of motion in two real video datasets including tsunami dataset and landslide dataset. The experimental results can be divided into two parts. Firstly, we verify the performance of the proposed OFATS method using video frame images with different input parameters. Secondly, the proposed method is compared with state-of-the-art CD methods using the two datasets.

4.1. Study Datasets

The proposed CD method is evaluated using two video frame datasets representing different natural disasters. The first video data gives a glimpse at tsunami in Petobo, Indonesia, where a 7.5 magnitude earthquake trigged a tsunami on 28 September 2018. Digital Globe's WorldView captured these change progress by satellite images and transformed into a video consisting of 301 effective video clips [57]. Another example is about the slow-moving landslide produced by a subject named massive landslides caught on camera 2 and a video clip with 172 effective frames is selected [58]. In this research, we both select six frames of two video datasets but the quantities of alternate frame are different (at frame 160 and 165, 162 and 163, and 175 and 180 for the tsunami scene and at frame

4960 and 4970, frame 4970 and 4980, frame 4980 and 4985 for landslide scene, respectively) as the input image sequences for change detection in order to test OFATS's robust to arbitrary movements. The experimental data together with the ground truth generated by visual interpretation are shown in Figure 6.

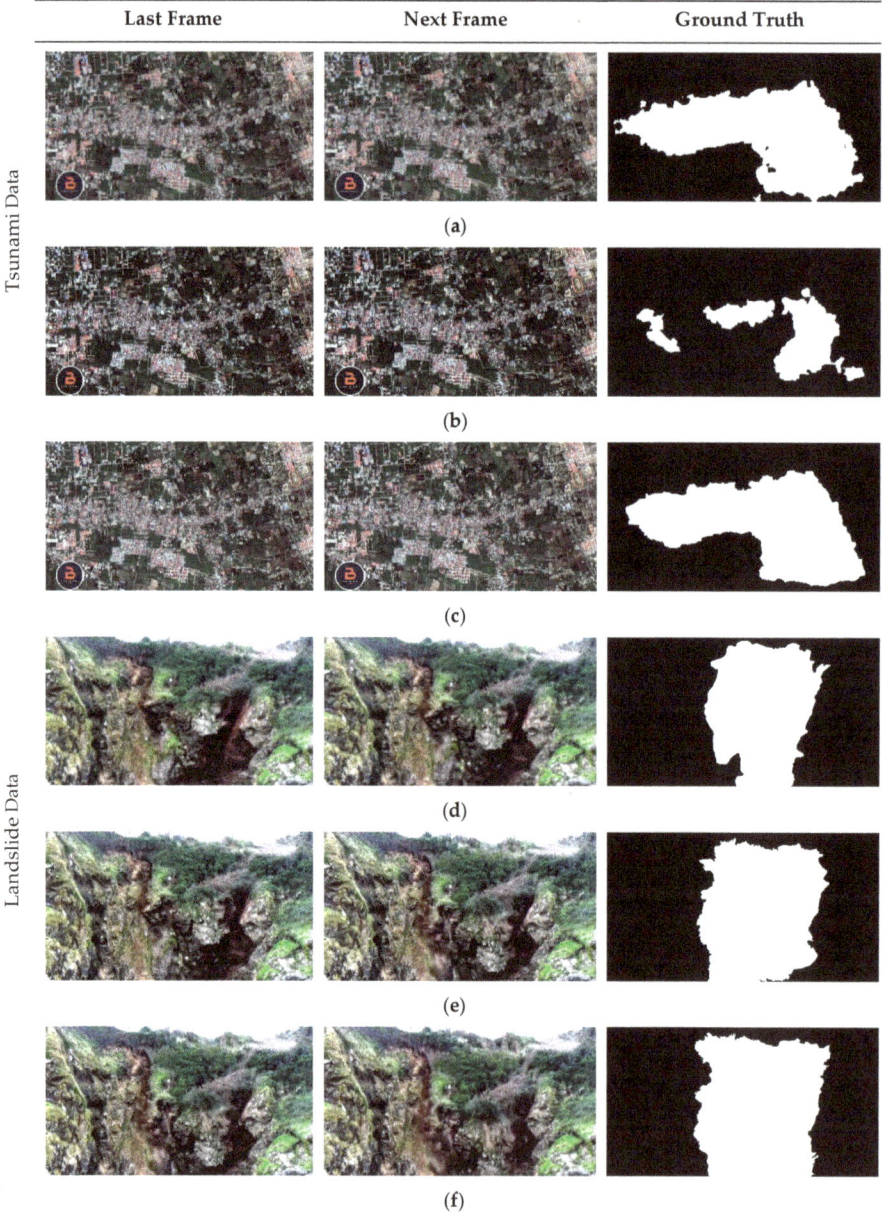

Figure 6. Experimental video frames and the corresponding ground truth: (**a–c**) Frame 160 and 165, Frame 162 and 163, and Frame 175 and 180 of tsunami video, respectively; (**d–f**) Frame 4960 and 4970, Frame 4970 and 4980, and Frame 4980 and 4985 of landslide video, respectively.

4.2. Evaluation of the Proposed Threshold Selection Method

In this section, the aim is to verify whether the proposed algorithm is able to automatically determine the optimal threshold for CD. Considering the change detection as a binary classification problem, the F1-measure is often used to test the selection of the optimal threshold. F1, which can synthetically consider precision (P) and recall (R) in binary classification, is shown in Table 1. The threshold that has the highest F1 will be considered as the optimum threshold. The value of F1 indicates the accuracy of change detection, and the closer to 1 means more accurate. This verification is executed on two sides: the correspondence of the optimal threshold and the maximum F1-measure value, and the performance to determine the optimal threshold value based on adaptive threshold selection proposed in OFATS and Otsu, a classic thresholding way for binarization in image processing.

Table 1. Formulas related to calculating F1.

Parameter Name	Formula	Explanation of Abbreviations
P	$\frac{tp}{tp+fp}$	tp (true positive): detects that are correctly identified as changed
R	$\frac{tp}{tp+fn}$	tn (true negative): detects that are correctly identified as unchanged fp (false positive): detects that are falsely identified as changed
F1	$\frac{2*P*R}{P+R}$	fn (false negative): detects that are falsely identified as unchanged

In the first experiment, the frames 160 and 165 in the tsunami video are taken as an example to test whether the proposed OFATS can generate the optimal threshold with the corresponding maximum F1-measure value. If the threshold determined by the proposed objective function is in accordance with the peak value of F1, it means that the generated threshold is optimum and OFATS has the best performance. The variations of objective function value with respect to threshold value are demonstrated in Figure 7, as well as the corresponding F1 value. The blue bars represent the objective function values and the red line represents F1 values with iterated threshold values. The optimal threshold based on OFATS and the corresponding F1 are labeled with green circle. According to Figure 7, the maximum objective function value (6.04×10^4) corresponds to the optimum threshold (0.3) based on which can achieve the highest peak of F1 during the whole iterations. This indicates that OFATS can automated produce the optimum threshold with the highest F1-measure value.

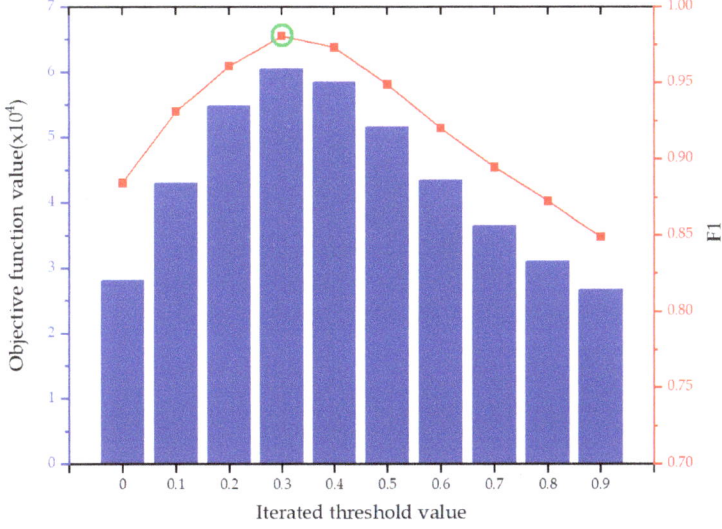

Figure 7. The objective function and F1 values with the variation of iterated threshold values.

In order to illustrate the robustness of adaptive threshold selection in OFATS, Otsu is introduced as a comparison of the optimal threshold selection based on the same displacements' results in the second experiment. There are three groups of tsunami data and three groups of landslide data and the comparison results based on different threshold selection methods for experimental data have been shown in Table 2.

Table 2. The optimum thresholds and the corresponding F1 values based on adaptive threshold selection in OFATS and Otsu for the experimental data.

Study Data	Experimental Frame	Adaptive Threshold Selection in OFATS		Threshold Selection Based on Otsu	
		Optimum Threshold	F1	Optimum Threshold	F1
Tsunami data	Frame 160–165	0.3	0.98	0.4	0.97
	Frame 162–163	0.2	0.97	0.38	0.90
	Frame 175–180	0.3	0.99	0.48	0.96
Landslide data	Frame 4960–4970	0.4	0.94	0.4	0.94
	Frame 4970–4980	0.3	0.92	0.5	0.91
	Frame 4980–4985	0.3	0.92	0.5	0.91

According to Table 2, the corresponding F1 values based on the adaptive threshold selection in OFATS are higher than Otsu in most cases, except the case in frames 4960–4970, which OFATS has the same threshold with Otsu. For the majority experimental frames, average F1 values based on OFATS are 0.02 higher than Otsu. This comparison indicates that the proposed OFATS is more robust to generate the optimum threshold thus accurately detecting the natural disaster change between video sequences.

4.3. Comparing the Proposed Method with Other CD Methods

This section takes frames 160–165 of the tsunami video and frames 4960–4970 of the landslide video as examples and compares the proposed algorithm with state-of-the-art CD algorithms, including Image Differencing, Image Rationing, Change Vector Analysis (CVA), Post-classification comparison (PCC), Kullback–Leibler divergence (KL), and Classic optical flow (COF) based on HS, a traditional optical flow estimation method. The aim is to test the superiority but to verify the robustness of OFATS to different range of movements.

The evaluation methods in this experiment are Producer's Accuracy (PA), and User's Accuracy (UA), Overall Accuracy (OA), Kappa coefficient (K). PA and UA are local indexes, where PA is obtained by dividing the number of correctly classified pixels in each class by the number of ground truth pixels in the corresponding class and UA is obtained by dividing the number of the total correctly classified pixels in the same class. Thus, there are four related indexes, that are PA_c, PA_{un}, PA for changes and unchanges, and UA_c, UA_{un}, UA for changes and unchanges, as shown by Equations (14)–(17). OA and K are global indexes, where OA is the proportion of number of correctly identified pixels, both changed and unchanged, to the number of total pixels, and K builds on OA by taking into account both the omission and commission of pixels. As OA, K, and F1 increase and approach 100%, 1, and 1, respectively, so too does the accuracy of the CD method in differentiating changes from non-changes.

$$PA_c = \frac{tp}{tp+fn} \quad (14)$$

$$PA_{un} = \frac{fp}{fp+tn} \quad (15)$$

$$UA_c = \frac{tp}{tp+fp} \quad (16)$$

$$UA_{un} = \frac{fn}{fn+tn} \quad (17)$$

OA and K are calculated by Equations (17) and (18):

$$OA = \frac{tp + tn}{tp + tn + fp + fn} \tag{18}$$

$$k = \frac{k_1 - k_2}{1 - k_2} \tag{19}$$

where k_1 and k_2 are computed as follows:

$$k_1 = \frac{tp + tn}{tp + tn + fp + fn} \tag{20}$$

$$k_2 = \frac{(tp + fn)*(tp + fp) + (fp + tn)*(fn + tn)}{(tp + tn + fp + fn)^2} \tag{21}$$

All of the previously mentioned CD methods were used to analyze tsunami and landslide data, and CD maps are shown in Figures 8 and 10 and the accuracy results were tabulated in Tables 3 and 4, respectively, and demonstrated in Figures 9 and 11, respectively. The results indicate that the proposed OFATS method has K and F1 values closing to 1 and also has the highest OA values, which shows that it is capable of accurately distinguishing between changed and unchanged pixels. The values of K, F1, and OA, according to Tables 3 and 4, are 0.98%, 0.97% and 98.5%, respectively, for the tsunami dataset, and 0.94%, 0.91%, and 96.3%, respectively, for the landside dataset.

Table 3. Confusion matrices along with indexes of the tsunami data.

Method		Ground Truth			F1	K	OA (%)
		C	U	UA (%)			
Image differencing	C	507,934	37,899	93.0			
	U	237,043	1,183,204	83.3	0.79	0.69	86.0
	PA (%)	68.2	96.7				
Image rationing	C	624,451	841,520	42.6			
	U	120,526	379,583	76.0	0.57	0.13	51.1
	PA (%)	83.8	31.1				
CVA	C	420,863	229,199	64.7			
	U	324,114	991,904	75.4	0.60	0.39	71.9
	PA (%)	56.5	81.2				
PCC	C	294,325	401,789	42.2			
	U	450,652	819,314	64.5	0.41	0.07	56.6
	PA (%)	39.5	67.1				
KL	C	548,695	750,222	42.2			
	U	196,282	470,881	70.6	0.54	0.11	51.9
	PA (%)	73.7	38.6				
COF	C	688,338	36,873	94.9			
	U	56,639	1,184,230	95.4	0.94	0.90	95.2
	PA (%)	92.4	97.0				
OFATS	C	737,237	21,164	97.2			
	U	7740	1,199,939	99.3	0.98	0.97	98.5
	PA (%)	98.9	98.2				

Figure 8a is the ground truth in which white represents changes and black means unchanges. Figure 8b–g are the results from comparative CD methods, from which most of them have difficulty to provide a clear boundary and complete changed area, especially for the CD methods of PCC (Figure 8e) and KL (Figure 8f). Many changed areas are wrongly detected as non-changed areas of image differencing (Figure 8b); however, the results of image rationing and CVA (Figure 8c,d) are the opposite. Figure 8g from COF is better than the traditional CD methods, and the result is very similar to OFATS (Figure 8h); however, the change detection accuracy of COF is less than the proposed method with several tiny false alarm areas.

Table 4. Confusion matrices along with indexes of landslide data.

Method		Ground Truth			F1	K	OA (%)
		C	U	UA (%)			
Image differencing	C	512,244	218,623	70.1	0.76	0.64	83.8
	U	100,309	1,134,904	91.9			
	PA (%)	83.6	83.8				
Image rationing	C	540,840	205,471	72.5	0.80	0.69	85.9
	U	71,713	1,148,056	94.1			
	PA (%)	88.3	84.8				
CVA	C	577,780	90,087	86.5	0.90	0.86	93.6
	U	34,773	1,263,440	97.3			
	PA (%)	94.3	93.3				
PCC	C	380,536	613,376	38.3	0.47	0.14	57.0
	U	232,017	740,151	76.1			
	PA (%)	62.1	54.7				
KL	C	525,254	208,691	71.6	0.78	0.67	84.9
	U	87,299	1,144,836	92.9			
	PA (%)	85.7	84.6				
COF	C	611,643	140,762	81.3	0.90	0.84	92.8
	U	910	1,212,765	99.9			
	PA (%)	99.8	89.6				
OFATS	C	584,927	44,777	92.9	0.94	0.91	96.3
	U	27,626	1,308,750	97.9			
	PA (%)	95.4	96.7				

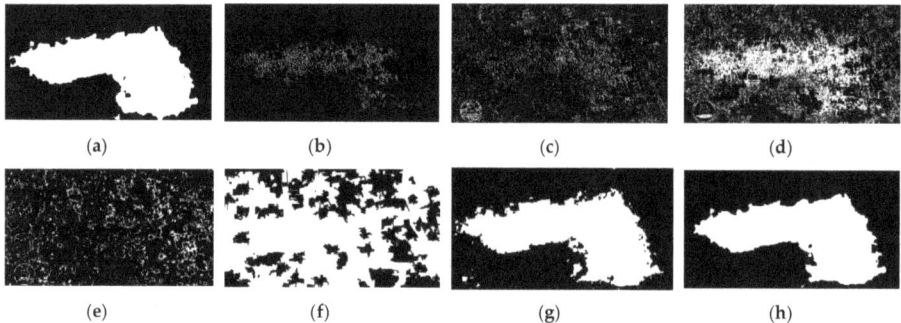

Figure 8. The change detection results for tsunami dataset: (a) Ground Truth; (b) Image differencing; (c) Image rationing; (d) CVA; (e) PCC; (f) KL; (g) COF and (h) OFATS.

To further analyze the experimental results, Table 3 presents the quantitative performance indexes of these different CD methods for the tsunami dataset and Figure 9 visually displays this. For image rationing, PCC and KL CD methods, the F1 values are around 0.5, OA is less than 60% and K is smaller than 0.15, which shows these methods fail to identify the changed area. Image differencing and CVA are slightly better than them but the results are still unsatisfying. This is because most of these CD methods only use one band of the RGB images to directly do algebraic operation or some transformation computing, which is unable to digging deep features to detect the change information of the corresponding pixel with small changes. Thus, some originally changed land cover types are hard to be tested. Moreover, the CD results are extremely fragmented and have to be implemented by morphologic erosion and dilation. The selection of the optimum threshold for the final binary image also produces errors. KL detects changes based on the spectral similarity of two single band, but the changed pixels in our research are only with displacements and no obvious spectral change in appearance. Thus, KL is less sensitive to this kind of changes that occurring when the displacements are small. PCC concurrently obtains the changed boundary and the "from-to" change information; however, PCC is the least accurate of the algorithms that were studied in this paper. Only small displacements or slight deformations occur rather than land cover type changes, resulting in PCC's ineffectiveness in the experimental data. Both COF and the proposed OFATS method produce reasonable CD accuracy with values of F1, OA and K higher than 0.9. Compared to COF, the proposed OFATS can achieve higher accuracy because high-level features can be extracted based on deep learning and the corresponding motion results can keep a higher precision [59]. Thus, the final CD accuracy based on OFATS are higher than that based on COF even if they use the same optimal threshold value.

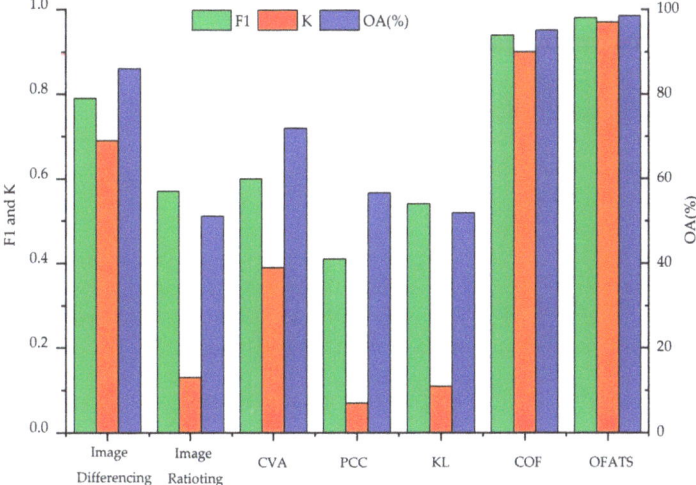

Figure 9. Comparison between other change detection (CD) methods and OFATS using Tsunami images.

Figure 10 shows the results for landslide dataset. Table 4 and Figure 11 present the quantitative analysis indexes of different CD methods. By observing the CD results in Figure 10, we can still find that the proposed OFATS is the most similar one to the change ground truth.

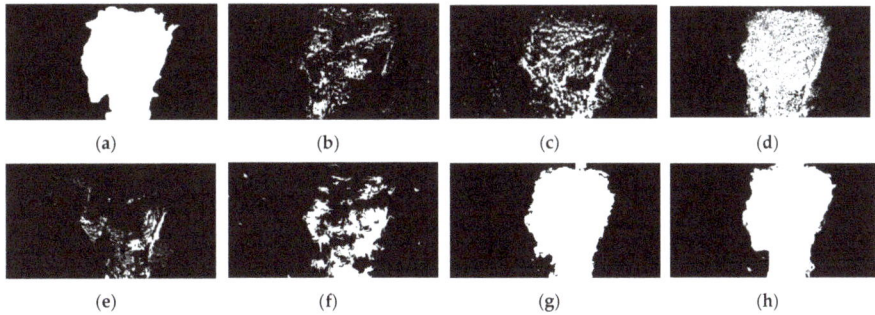

Figure 10. The change detection results for landslide dataset: (**a**) Ground Truth; (**b**) Image differencing; (**c**) Image rationing; (**d**) CVA; (**e**) PCC; (**f**) KL; (**g**) COF and (**h**) OFATS.

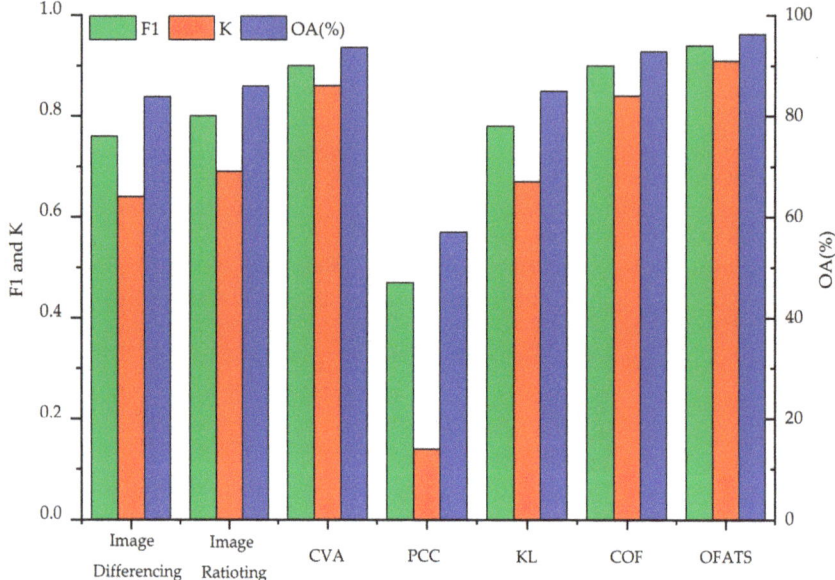

Figure 11. Comparison between other CD methods and OFATS using Landslide images.

Similar to Table 3, Table 4 also demonstrates the superior performance of OFATS with the highest F1, OA and K indexes, reaching 0.94, 96.3% and 0.91, respectively. Figure 11 also visually displayed the comparison of different CD algorithms, where the three accuracy indexes of OFATS are all the maximum but the corresponding values of PCC are minimal. The similar accuracies of the two datasets show that the proposed OFATS method maintains excellent performance and is robust to different motion changes. It should be noticed that nearly all CD methods perform better than in the Indonesia tsunami case.

Although the two experimental datasets are both from natural disaster scenes, the mean pixel displacement in the landslide dataset is larger than that of the tsunami dataset. The difference can be used to test the robustness of OFATS. The comparative analysis will be based on K values for all of the algorithms for both the landslide and tsunami datasets, as shown in Figure 12. The K values based on CD methods for the two datasets have varied but the change trends are basically identical. The K values for the traditional CD methods, other than image differencing, all follow the trend of the value obtained from the landslide dataset being higher than that of the tsunami dataset. It is worthwhile to be mentioned that the K values obtained from the landslide dataset are significantly higher than that of the tsunami dataset for image rationing, CVA and KL. Specially, the K values are greater than 0.6 and less than 0.4 for the landslide and tsunami datasets, respectively, which illustrates that these CD methods should only be fit for large displacements. The different performances of these CD methods indicate that only when the differences between corresponding pixels on bi-temporal images are large, like in the case of landslides, can these CD algorithms detect the change. The K values that were obtained using PCC are less than 0.2, therefore these results further indicate that PCC is not applicable for these two types of situations regardless of the magnitude of displacement.

Despite the variations of K values, the two CD methods based on optical flow estimation algorithms achieved excellent results for both experimental datasets and K values were all greater than 0.8. However, the K values obtained using OFATS are 10% higher than COF for both datasets and their absolute values are both greater than 0.9. The superiority of OFATS is not only the accuracy, but it is also significantly more efficient in terms of computing time. Therefore, our proposed OFATS is more practical in these actual circumstances.

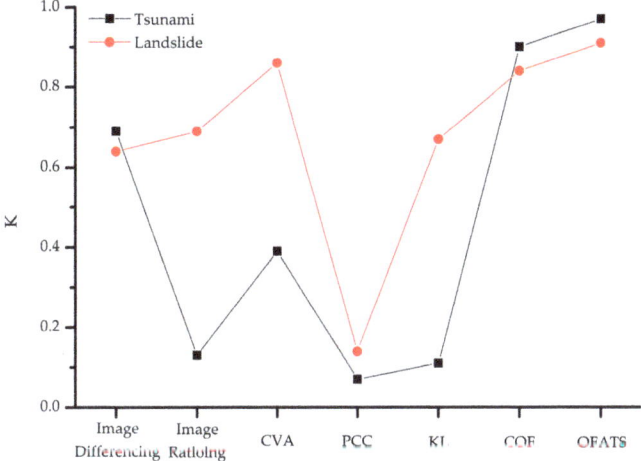

Figure 12. The comparison of K for experimental datasets based on CD methods.

5. Conclusions

The challenging problems in natural disaster detection are how to detect the motion change and how to determine an adaptive threshold that can automatically and rapidly produce accurate change detection results. To solve this problem, an automatic change framework, termed as OFATS, is proposed in this paper. First, the displacement was computed from two frames using optical flow estimation algorithm based on deep learning. Then, the optimal threshold for rapidly separating changed from unchanged parts was automatically generated using an adaptive threshold selection based on a new objective function by narrowing the threshold searching range and dynamic normalization.

The proposed OFATS has been applied to two different natural disaster videos. The CD results have been compared with seven other state-of-the-art CD methods, visually and quantitatively. The quantitative evaluation demonstrated that the accuracies of proposed method are greater than 95% for the two experimental datasets and it surpasses the most excellent CD algorithms by almost 4% for tsunami data and 5% for landslide data. Experiments showed three advantages of the proposed method: (1) it can detect the change using video datasets for natural disasters in an automatic way, which have rarely been studied before; (2) it is highly efficient to conduct natural disaster change detection, even for small motion; (3) it can automatically generate the optimum threshold for the following image segmentation.

Author Contributions: H.Q., X.W. provided the original idea for this study; Y.W., S.L. and W.Z. contributed to the discussion of the design; H.Q., X.W. designed and performed the experiments, supervised the research and contributed to the article's organization; X.W. and Y.W. edited the manuscript, which was revised by all authors. All authors have read and agreed to the published version of the manuscript.

Funding: This work was supported in part by the National Key Research and Development Program of China (No. 2018YFD1100405) and the National Natural Science Foundation of China (No. 41701468).

Conflicts of Interest: The authors declare no conflict of interest.

References

1. Milly, P.C.D.; Wetherald, R.T.; Dunne, K.A.; Delworth, T.L. Increasing risk of great floods in a changing climate. *Nature* **2002**, *415*, 514. [CrossRef] [PubMed]
2. Sublime, J.; Kalinicheva, E. Automatic post-disaster damage mapping using deep-learning techniques for change detection: Case study of the Tohoku Tsunami. *Remote Sens.* **2019**, *11*, 1123. [CrossRef]
3. Crooks, A.T.; Wise, S. GIS and agent-based models for humanitarian assistance. *Comput. Environ. Urban Syst.* **2013**, *41*, 100–111. [CrossRef]

4. Lu, C.; Ying, K.; Chen, H. Real-time relief distribution in the aftermath of disasters—A rolling horizon approach. *Transp. Res. Part E Logist. Transp. Rev.* **2016**, *93*, 1–20. [CrossRef]
5. Asokan, A.; Anitha, J. Change detection techniques for remote sensing applications: A survey. *Earth Sci. Inform.* **2019**, *12*, 143–160. [CrossRef]
6. Klomp, J. Economic development and natural disasters: A satellite data analysis. *Global Environ. Chang.* **2016**, *36*, 67–88. [CrossRef]
7. Yu, H.; Wen, Y.; Guang, H.; Ru, H.; Huang, P. Change detection using high resolution remote sensing images based on active learning and Markov random fields. *Remote Sens.* **2017**, *9*, 1233. [CrossRef]
8. Pulvirenti, L.; Chini, M.; Pierdicca, N.; Guerriero, L.; Ferrazzoli, P. Flood monitoring using multi-temporal COSMO-SkyMed data: Image segmentation and signature interpretation. *Remote Sens. Environ.* **2011**, *115*, 990–1002. [CrossRef]
9. Lacroix, P.; Bièvre, G.; Pathier, E.; Kniess, U.; Jongmans, D. Use of Sentinel-2 images for the detection of precursory motions before landslide failures. *Remote Sens. Environ.* **2018**, *215*, 507–516. [CrossRef]
10. Cai, J.; Wang, C.; Mao, X.; Wang, Q. An adaptive offset tracking method with SAR images for landslide displacement monitoring. *Remote Sens.* **2017**, *9*, 830. [CrossRef]
11. Gautam, D.; Dong, Y. Multi-hazard vulnerability of structures and lifelines due to the 2015 Gorkha earthquake and 2017 central Nepal flash flood. *J. Build. Eng.* **2018**, *17*, 196–201. [CrossRef]
12. Alizadeh, M.; Ngah, I.; Hashim, M.; Pradhan, B.; Pour, A. A hybrid analytic network process and artificial neural network (ANP-ANN) model for urban earthquake vulnerability assessment. *Remote Sens.* **2018**, *10*, 975. [CrossRef]
13. Carlotto, M.J. Detection and analysis of change in remotely sensed imagery with application to wide area surveillance. *IEEE T. Image Process.* **1997**, *6*, 189–202. [CrossRef] [PubMed]
14. Bejiga, M.; Zeggada, A.; Nouffidj, A.; Melgani, F. A convolutional neural network approach for assisting avalanche search and rescue operations with UAV imagery. *Remote Sens.* **2017**, *9*, 100. [CrossRef]
15. Shi, W.; Zhang, M.; Zhang, R.; Chen, S.; Zhan, Z. Change detection based on artificial intelligence state-of-the-art and challenges. *Remote Sens.* **2020**, *12*, 1688. [CrossRef]
16. Hall, O.; Hay, G.J. A multiscale object-specific approach to digital change detection. *Int. J. Appl. Earth Obs.* **2003**, *4*, 311–327. [CrossRef]
17. Matsuoka, M.; Yamazaki, F. Building damage mapping of the 2003 Bam, Iran, earthquake using Envisat/ASAR intensity imagery. *Earthq. Spectra* **2005**, *21*, 285–294. [CrossRef]
18. Sharma, K.; Saraf, A.K.; Das, J.; Baral, S.S.; Borgohain, S.; Singh, G. Mapping and change detection study of Nepal-2015 earthquake induced landslides. *J. Indian Soc. Remote* **2018**, *46*, 605–615. [CrossRef]
19. Alizadeh, M.; Shirzadi, A.; Khosravi, K.; Melesse, A.M.; Yekrangnia, M.; Rezaie, F.; Al, E. SEVUCAS a novel GIS-based machine learning software for seismic vulnerability assessment. *Appl. Sci.* **2019**, *9*, 3495.
20. ElGharbawi, T.; Tamura, M. Coseismic and postseismic deformation estimation of the 2011 Tohoku earthquake in Kanto Region, Japan, using InSAR time series analysis and GPS. *Remote Sens. Environ.* **2015**, *168*, 374–387. [CrossRef]
21. Du, S.; Zhang, Y.; Qin, R.; Yang, Z.; Zou, Z.; Tang, Y.; Fan, C. Building change detection using old aerial images and new LiDAR data. *Remote Sens.* **2016**, *8*, 1030. [CrossRef]
22. Sudipan, S.; Francesca, B.; Lorenzo, B. Destroyed-buildings detection from VHR SAR images using deep features. In Proceedings of the Image and Signal Processing for Remote Sensing XXIV, Berlin, Germany, 10–12 September 2018.
23. Ji, M.; Liu, L.; Du, R.; Buchroithner, M.F. A comparative study of texture and convolutional neural network features for detecting collapsed buildings after earthquakes using pre- and post-event satellite imagery. *Remote Sens.* **2019**, *11*, 1202. [CrossRef]
24. Ci, T.; Liu, Z.; Wang, Y. Assessment of the degree of building damage caused by disaster using convolutional neural networks in combination with ordinal regression. *Remote Sens.* **2019**, *11*, 2858. [CrossRef]
25. Peng, D.; Zhang, Y.; Guan, H. End-to-end change detection for high resolution satellite images using improved UNet++. *Remote Sens.* **2019**, *11*, 1382. [CrossRef]
26. Yavariabdi, A.; Kusetogullari, H. Change detection in multispectral landsat images using multiobjective evolutionary algorithm. *IEEE Geosci. Remote Sens.* **2017**, *14*, 414–418. [CrossRef]
27. Ghaffarian, S.; Kerle, N.; Pasolli, E.; Jokar Arsanjani, J. Post-disaster building database updating using automated deep learning: An integration of pre-disaster OpenStreetMap and multi-temporal satellite data. *Remote Sens.* **2019**, *11*, 2427. [CrossRef]

28. Pi, Y.; Nath, N.D.; Behzadan, A.H. Convolutional neural networks for object detection in aerial imagery for disaster response and recovery. *Adv. Eng. Inform.* **2020**, *43*, 101009. [CrossRef]
29. Kung, H.; Hsieh, C.; Ho, C.; Tsai, Y.; Chan, H.; Tsai, M. Data-augmented hybrid named entity recognition for disaster management by transfer learning. *Appl. Sci.* **2020**, *10*, 4234. [CrossRef]
30. Li, M.; Li, M.; Zhang, P.; Wu, Y.; Song, W.; An, L. SAR image change detection using PCANet guided by saliency detection. *IEEE Geosci. Remote Sens.* **2018**, *16*, 402–406. [CrossRef]
31. Curtis, A.; Mills, J.W. Spatial video data collection in a post-disaster landscape: The Tuscaloosa Tornado of 27 April 2011. *Appl. Geogr.* **2012**, *32*, 393–400. [CrossRef]
32. Curtis, A.J.; Mills, J.W.; McCarthy, T.; Fotheringham, A.S.; Fagan, W.F. *Space and Time Changes in Neighborhood Recovery after a Disaster Using a Spatial Video Acquisition System*; Springer: Berlin, Germany, 2009; pp. 373–392.
33. Tu, Z.; Xie, W.; Zhang, D.; Poppe, R.; Veltkamp, R.C.; Li, B.; Yuan, J. A survey of variational and CNN-based optical flow techniques. *Signal Process. Image Commun.* **2019**, *72*, 9–24. [CrossRef]
34. Guo, Y.; Zhang, Z.; He, D.; Niu, J.; Tan, Y. Detection of cow mounting behavior using region geometry and optical flow characteristics. *Comput. Electron. Agric.* **2019**, *163*, 104828. [CrossRef]
35. Gronskyte, R.; Clemmensen, L.H.; Hviid, M.S.; Kulahci, M. Monitoring pig movement at the slaughterhouse using optical flow and modified angular histograms. *Biosyst. Eng.* **2016**, *141*, 19–30. [CrossRef]
36. Yan, W.; Wang, Y.; van der Geest, R.J.; Tao, Q. Cine MRI analysis by deep learning of optical flow: Adding the temporal dimension. *Comput. Biol. Med.* **2019**, *111*, 103356. [CrossRef] [PubMed]
37. Wang, L.; Clarysse, P.; Liu, Z.; Gao, B.; Liu, W.; Croisille, P.; Delachartre, P. A gradient-based optical-flow cardiac motion estimation method for cine and tagged MR images. *Med. Image Anal.* **2019**, *57*, 136–148. [CrossRef]
38. Cao, Y.; Renfrew, A.; Cook, P. Comprehensive vehicle motion analysis using optical flow optimization based on pulse-coupled neural network. *IFAC Proc. Vol.* **2008**, *41*, 158–163. [CrossRef]
39. Tchernykh, V.; Beck, M.; Janschek, K. Optical flow navigation for an outdoor UVA using a wide angle mono camera and dem matching. *IFAC Proc. Vol.* **2006**, *39*, 590–595. [CrossRef]
40. Liu, Y.; Xi, D.; Li, Z.; Hong, Y. A new methodology for pixel-quantitative precipitation nowcasting using a pyramid Lucas Kanade optical flow approach. *J. Hydrol.* **2015**, *529*, 354–364. [CrossRef]
41. Zhao, R.; Sun, P. Deformation-phase measurement by optical flow method. *Opt. Commun.* **2016**, *371*, 144–149. [CrossRef]
42. Osman, A.B.; Ovinis, M. A review of in-situ optical flow measurement techniques in the Deepwater Horizon oil spill. *Measurement* **2020**, *153*, 107396.
43. Yuan, W.; Yuan, X.; Xu, S.; Gong, J.; Shibasaki, R. Dense Image-Matching via Optical Flow Field Estimation and Fast-Guided Filter Refinement. *Remote Sens.* **2019**, *11*, 2410. [CrossRef]
44. Sun, D.; Roth, S.; Black, M.J. Secrets of optical flow estimation and their principles. In Proceedings of the 2010 IEEE Computer Society Conference on Computer Vision and Pattern Recognition, San Francisco, CA, USA, 13–18 June 2010; pp. 2432–2439.
45. Horn, B.K.; Schunck, B.G. Determining optical flow. *Artif. Intell.* **1981**, *17*, 185–203. [CrossRef]
46. Prajapati, D.; Galiyawala, H.J. *A Review on Moving Object Detection and Tracking*; Department of Electronics and Communication Engineering, UKA Tarsadia University: Bardoli, India, 2015.
47. Wei, S.; Yang, L.; Chen, Z.; Liu, Z. Motion detection based on optical flow and self-adaptive threshold segmentation. *Procedia Eng.* **2011**, *15*, 3471–3476. [CrossRef]
48. Hou, B.; Wang, Y.; Liu, Q. Change detection based on deep features and low rank. *IEEE Geosci. Remote Sens.* **2017**, *14*, 2418–2422. [CrossRef]
49. Yuan, Q.; Shen, H.; Li, T.; Li, Z.; Li, S.; Jiang, Y.; Xu, H.; Tan, W.; Yang, Q.; Wang, J.; et al. Deep learning in environmental remote sensing: Achievements and challenges. *Remote Sens. Environ.* **2020**, *241*, 111716. [CrossRef]
50. Dosovitskiy, A.; Fischer, P.; Ilg, E.; Hausser, P.; Hazirbas, C.; Golkov, V.; Van Der Smagt, P.; Cremers, D.; Brox, T. Flownet: Learning optical flow with convolutional networks. In Proceedings of the IEEE International Conference on Computer Vision, Nice, France, 13–16 October 2003; pp. 2758–2766.
51. Hui, T.W.; Tang, X. LiteFlowNet: A lightweight convolutional neural network for optical flow estimation. In Proceedings of the 2018 IEEE/CVF Conference on Computer Vision and Pattern Recognition, Salt Lake City, UT, USA, 18–23 June 2018; pp. 8981–8989.

52. Ilg, E.; Mayer, N.; Saikia, T.; Keuper, M.; Dosovitskiy, A.; Brox, T. Flownet 2.0: Evolution of optical flow estimation with deep networks. In Proceedings of the IEEE conference on computer vision and pattern recognition, San Francisco, CA, USA, 18–20 June 1996; pp. 2462–2470.
53. Baker, S.; Scharstein, D.; Lewis, J.P.; Roth, S.; Black, M.J.; Szeliski, R. A Database and evaluation methodology for optical flow. *Int. J. Comput. Vis.* **2011**, *92*, 1–31. [CrossRef]
54. Vala, H.J.; Baxi, A. A review on Otsu image segmentation algorithm. *Int. J. Adv. Res. Comput. Eng. Technol.* **2013**, *2*, 387–389.
55. Pal, N.R.; Pal, S.K. A review on image segmentation techniques. *Pattern Recogn.* **1993**, *26*, 1277–1294. [CrossRef]
56. Waseem Khan, M. A Survey: Image segmentation techniques. *Int. J. Future Comput. Commun.* **2014**, *3*, 89–93. [CrossRef]
57. Digital Globe Data in Indonesia Earthquake. Available online: https://www.youtube.com/watch?v=-41ENJF0wVwx (accessed on 24 October 2018).
58. Slow-Moving Landslide Des Caught on Camera 2. Available online: https://www.youtube.com/watch?v=PmLHg-mLrMU (accessed on 10 July 2019).
59. Qiao, H.J.; Wan, X.; Xu, J.Z.; Li, S.Y.; He, P.P. Deep learning based optical flow estimation for change detection: A case study in Indonesia earthquake. *ISPRS Ann. Photogramm. Remote Sens. Spat. Inf. Sci.* **2020**, *3*, 317–322. [CrossRef]

© 2020 by the authors. Licensee MDPI, Basel, Switzerland. This article is an open access article distributed under the terms and conditions of the Creative Commons Attribution (CC BY) license (http://creativecommons.org/licenses/by/4.0/).

Article

Potential of Pléiades and WorldView-3 Tri-Stereo DSMs to Represent Heights of Small Isolated Objects

Ana-Maria Loghin *, Johannes Otepka-Schremmer and Norbert Pfeifer

Department of Geodesy and Geoinformation, Technische Universität Wien, Wiedner Hauptstraße 8-10, 1040 Vienna, Austria; Johannes.Otepka@geo.tuwien.ac.at (J.O.-S.); Norbert.Pfeifer@geo.tuwien.ac.at (N.P.)
* Correspondence: ana-maria.loghin@geo.tuwien.ac.at; Tel.: +43-(1)-58801-12259

Received: 30 March 2020; Accepted: 3 May 2020; Published: 9 May 2020

Abstract: High-resolution stereo and multi-view imagery are used for digital surface model (DSM) derivation over large areas for numerous applications in topography, cartography, geomorphology, and 3D surface modelling. Dense image matching is a key component in 3D reconstruction and mapping, although the 3D reconstruction process encounters difficulties for water surfaces, areas with no texture or with a repetitive pattern appearance in the images, and for very small objects. This study investigates the capabilities and limitations of space-borne very high resolution imagery, specifically Pléiades (0.70 m) and WorldView-3 (0.31 m) imagery, with respect to the automatic point cloud reconstruction of small isolated objects. For this purpose, single buildings, vehicles, and trees were analyzed. The main focus is to quantify their detectability in the photogrammetrically-derived DSMs by estimating their heights as a function of object type and size. The estimated height was investigated with respect to the following parameters: building length and width, vehicle length and width, and tree crown diameter. Manually measured object heights from the oriented images were used as a reference. We demonstrate that the DSM-based estimated height of a single object strongly depends on its size, and we quantify this effect. Starting from very small objects, which are not elevated against their surroundings, and ending with large objects, we obtained a gradual increase of the relative heights. For small vehicles, buildings, and trees (lengths <7 pixels, crown diameters <4 pixels), the Pléiades-derived DSM showed less than 20% or none of the actual object's height. For large vehicles, buildings, and trees (lengths >14 pixels, crown diameters >7 pixels), the estimated heights were higher than 60% of the real values. In the case of the WorldView-3 derived DSM, the estimated height of small vehicles, buildings, and trees (lengths <16 pixels, crown diameters <8 pixels) was less than 50% of their actual height, whereas larger objects (lengths >33 pixels, crown diameters >16 pixels) were reconstructed at more than 90% in height.

Keywords: VHR tri-stereo satellite imagery; digital elevation model; isolated objects; dense image matching

1. Introduction

For more than thirty years, civilian satellite sensors have been used for digital elevation model (DEM) extraction over large areas in a timely and cost-effective manner for a wide range of applications in engineering, land planning, and environmental management. Beginning with the year 1986, when SPOT—the first satellite providing stereo-images, with a panchromatic Ground Sampling Distance (GSD) of 10 m—was launched [1], the optical satellite industry has been experiencing continuous development. Today more and more space sensors are available that acquire not only stereo but also tri-stereo satellite imagery. The generation of high and very high resolution commercial space imaging systems for DEM generation started in September 1999, with the launch of IKONOS [2]. Among the Very High Resolution (VHR) optical satellites providing along- and across-track stereo, the following systems can be mentioned: Ziyuan-3 (2.1 m), KOMPSAT-2 (1 m), Gaofen-2 (0.8 m), TripleSat (0.8 m), EROS B (0.7 m), KOMPSAT-3 (0.7 m), Pléiades 1A/1B (0.7 m), SuperView 1-4 (0.5 m), GeoEye-1 (0.46 m), WorldView-1/2 (0.46 m) and WorldView 3 (0.31 m).

The new generation of Earth observation satellites are characterized by an increased acquisition capacity and the possibility of collecting multiple images of the same area from different viewing angles during a single pass [3,4]. This multi-view aspect is essential for extracting 3D information. In recent years, the potential of tri-stereo acquisition from high-resolution satellite images for surface modelling has become an interesting research topic that has been addressed in various publications. The capacity of the Pléiades system in performing 3D mapping was analyzed by Bernard et al. [5], where 17 images acquired from different points of view were used. They showed that by means of "triple stereo" configurations reliable digital surface models can be generated in urban areas. From their best image combination, a root mean square error (RMSE) of 0.49 m was obtained at 295 ground control points (GCPs). The radiometric and geometric characteristics of Pléiades imagery with a focus on digital surface modelling are analyzed by Poli et al. [6]. The model derived from a "triple stereo" scene (nadir, forward and backward) showed median values close to zero and an RMSE in the range of 6–7 m, when compared with a reference light detection and ranging (LiDAR) DSM. An accurate 3D map with tri-stereo images can be obtained by optimizing the sensor models with GCPs, leading to accuracies in the range of 0.5 m in planimetry and of 1 m in height as demonstrated in [7].

Much of the previous research using WorldView-3 satellite images focused on their high resolution multi-spectral information, with applications in topographic mapping, land planning, land use, land cover classification, feature extraction, change detection, and so on [8–11]. The 3D potential of WorldView-3 data is assessed by Hu et al. [12], where the reconstructed DEM shows height differences of less than 0.5 m for 82.7% of 7256 ground LiDAR checkpoints located along road axes. A new algorithm for generating high quality digital surface models is proposed by Rupnik et al. [13], where a dense image matching method is applied for multi-view satellite images from Pléiades and WorldView-3.

The capability of satellite images regarding a rapid evaluation of urban environments is addressed in Abduelmula et al. [14]. They compared 3D data extracted from tri-stereo and dual stereo satellite images combined with pixel-based matching and Wallis filter to improve the accuracy of 3D models, especially in urban areas. The result showed that 3D models achieved by Pleiades tri-stereo outperformed the result obtained from a GeoEye pair, in terms of both accuracy and detail. This could mean that tri-stereo images can be successfully used for urban change analyses. The potential of VHR optical sensors for 3D city model generation has been addressed in [15–17], showing promising results for automatic building extraction when compared to a LiDAR elevation model, although highlighting some difficulties in the case of small individual house reconstruction. A quantitative and qualitative evaluation of 3D building models from different data sources was presented in [18], where a DSM at 1 m resolution derived from a GeoEye-1 stereo-pair, a DSM from an aerial block at 50 cm GSD, and a LiDAR-based DSM at 1 m resolution were used. Their results show that the percentage of correctly reconstructed models is very similar for airborne and LiDAR data (59% and 67%, respectively), while for GeoEye data it is lower (only 41%). The real dimensions of the 17 buildings surveyed were used as ground truth-reference for the 3D building model's quality assessment, obtaining a mean value for the residual heights of 1.94 m for the photogrammetric DSM. In [19] the authors analyze and validate the potential of high-resolution DSMs produced from stereo and tri-stereo Pléiades-1B satellite imagery acquired over the Athens Metropolitan Area. From their tests, the tri-stereo model shows the best performance in height accuracy, with an RMSE of 1.17 m when compared with elevations measured by a differential global positioning system.

The advantages of using tri-stereo instead of stereo image pairs are described by Piermattei et al. [20], where the nadir image increases the DSM completeness, reducing the occlusions usually caused by larger convergence angles on the ground. Additionally, they investigate in detail the influence of the acquisition geometry (viewing and incidence angles) of VHR imagery on DSM accuracy.

The cited literature concentrates on the accuracy assessments either on open areas or on (large) buildings within city areas, but not on smaller objects like cars.

The standard method of DSM generation from stereo-pairs or triples is dense image matching using global or semi-global optimization. Because of the smoothness constraint of dense image matching [21], the heights of small individual objects may not be reconstructed. Hence, the corresponding 3D points will not have higher elevations compared to their surroundings. Based on this hypothesis, we investigated the capability of dense image matching when evaluating the height of small individual, i.e., detached, objects. While previous studies addressed the general 3D capabilities of VHR sensors, our investigation concentrates on small, isolated object detectability by height estimation. To the best of our knowledge, our study is the first to analyze and quantify the capability of the tri-stereo Pléiades and WorldView-3 reconstructed DSMs for single object detection by height estimation, with focus on vehicles, buildings, and trees. The object's height compared with its surrounding terrain (in the following simply referred to as height) was investigated with respect to the following parameters: building length and width, vehicle length and width, and tree crown diameter. We investigate DSMs from different sensors with very small GSD, Pléiades and WorldView-3, but the focus is not a comparison of the two satellite systems. Specifically, our research investigation's purpose is to answer the following questions: (1) which geometric signature and minimum size must individual objects have to be detected in the DSM-based on their reconstructed heights; and (2) what are the influences of different acquisition times, geometries, and GSDs on dense image matching quality for single objects. In the following, we first describe the tri-stereo satellite imagery used together with the study site (Section 2), followed by image block orientation (in Section 3.1) using Rational Polynomial Coefficients (RPCs) and a set of GCPs. Once the orientation is completed, the 3D reconstruction follows. The 3D coordinates of the specific points corresponding to buildings, vehicles and trees are monoscopically measured in all three images: forward, nadir, and backward. The elevations obtained were used for computing the reference individual object's height, by subtracting the correspondent elevations from a LiDAR digital terrain model (DTM) (in Section 3.2). Subsequently, the accuracy of image orientation and the procedure of dense image matching are detailed in Sections 4.1 and 4.2. After that, individual objects are grouped into different classes depending on their corresponding sizes. Their automatically-reconstructed heights are then compared with reference values (in Sections 4.3 and 4.4). Finally, the major findings of the current work are summarized in Section 5.

2. Study Area and Data Analysis

The study area is located in Lower Austria, the north-eastern state of the country (48°30′30″ N; 15°08′34″ E; WGS84). With elevations ranging from 537 to 846 m above sea level, the region contains different land cover types such as: urban, suburban, rural, arable, grassland, and forested areas. Analysis was conducted based on tri-stereo satellite images acquired with both Pléiades-1B and WorldView-3 optical satellite sensors. Each triplet consists of images that were collected during the same pass from different sensor-viewing angles (along-track stereo): forward, close to nadir, and backward. The location of the study area acquired with the two sensors, with an overlapping surface of 44.5 km², is shown in Figure 1.

For the current analyses, the reference data contains 43 GCPs measured by means of real time kinematic (RTK) GPS with an accuracy of approximately 1 cm. A DTM generated from a LiDAR flight campaign conducted in December 2015 is available, too. The raster DTM is in UTM 33 North map projection, datum WGS 84 with a grid spacing of 1 m. Its height accuracy was checked against the RTK GCPs yielding a σ_Z of 0.12 m. We used a digital orthophoto from 2017 at 0.20 m resolution for defining the positions of check points (CP). The planimetric accuracy of the digital orthophoto was verified by computing the differences between the RTK point coordinates and their corresponding position on the orthophoto. The result showed that no shifts larger than one pixel were observed.

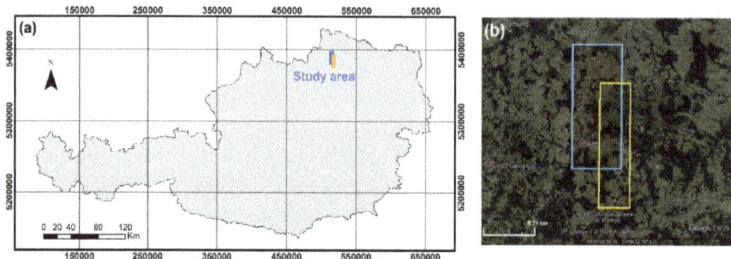

Figure 1. Study area: (**a**) overview map of Austria with location of the study area (coordinates in UTM zone 33N); (**b**) satellite imagery–blue: Pléiades tri-stereo pair and orange: WorldView-3 tri-stereo pair.

Each satellite image was delivered with RPCs that allow the transformation between object and image space [22]. Table 1 summarizes the main acquisition parameters like along, across and overall incidence angles, together with corresponding times and covered areas. Additionally, using the equations found in [23] we computed the stereo intersection angles (also called convergence angles in [24]) between each scene and their base-to-height (B/H) ratios. For WorldView-3 satellite imagery the resulting values for the B/H ratios were twice as large as those of the Pléiades imagery.

Table 1. Acquisition parameters of Pléiades and WorldView-3 data.

Sensor Type & Acquisition Date	View	Acquisition Time (hh:hm:ss.s)	Incidence Angles (°)			Area (km²)	B/H Ratio	Convergence Angle (°)
			Across	Along	Overall			
Pléiades 13-06-2017	Forward (F)	10:09:51.5	−2.23	−6.75	6.75	158.73	0.10 (FN)	5.71 (FN)
	Nadir (N)	10:10:03.7	−3.31	−1.13	3.50	158.49	0.11 (NB)	6.30 (NB)
	Backward (B)	10:10:14.0	−5.00	4.95	7.02	158.78	0.21 (FB)	12.0 (FB)
WorldView-3 08-04-2018	Forward (F)	10:22:07.0	7.71	11.00	13.57	100.00	0.20 (FN)	11.52 (FN)
	Nadir (N)	10:22:25.5	7.23	−0.62	7.36	100.00	0.20 (NB)	11.51 (NB)
	Backward (B)	10:22:44.1	6.72	−12.20	13.97	100.00	0.40 (FB)	23.04 (FB)

(FN): Forward-Nadir, (NB): Nadir-Backward and (FB): Forward-Backward image pairs.

The tri-stereo Pléiades images were provided at sensor processing level, corrected only from on-board distortions such as viewing directions and high frequency attitude variations [25]. For all three images, we performed an optical radiometric calibration using the open source software Orfeo Tool Box [26]. The pixel values were calibrated by the influence of the following parameters: sensor gain, spectral response, solar illumination, optical thickness of the atmosphere, atmospheric pressure, water vapor, and ozone amount, as well as the composition and amount of aerosol gasses.

In contrast, the WorldView-3 images were delivered as tri-stereo with relative radiometrically-corrected image pixels. The relative radiometric calibration included a dark offset subtraction and a non-uniformity correction (e.g., detector-to-detector relative gain), which is performed on the raw data during the early stages of product generation. We computed the absolute radiometric calibration for each WorldView-3 image. This was done in two steps: the conversion from image pixels to top-of-atmosphere spectral radiance; and the conversion from top-of-atmosphere spectral radiance to top-of-atmosphere reflectance. The calculations were performed independently for each band, using the equations found in the technical sensor description [27]. Usually, the optical radiometric calibration step is necessary before making any physical interpretation of the pixel values. In particular, this processing is mandatory before performing any comparison of pixel spectrum between several images from the same sensor. In our case, this pre-processing step could be omitted, since it did not change the geometric quality of the images.

An important characteristic of both satellites—Pléiades-1B and WorldView-3—is the fast rotation to enable recording three images within the same orbit in less than 1 min. Therefore, the images have the same illumination conditions and shadow changes are not significant. With radiometric resolutions of 16 bit/pixel, the images provide a higher dynamic range than the traditional 8- or 12-bit/pixel images.

From a visual inspection, some radiometric effects were observed in the WorldView-3 forward image (Figure 2). Here, the reflective roof surface, in combination with the imaging incidence angle and sun elevation, caused saturation and spilling effects. In contrast, no radiometric artefacts were observed in the Pléiades satellite images.

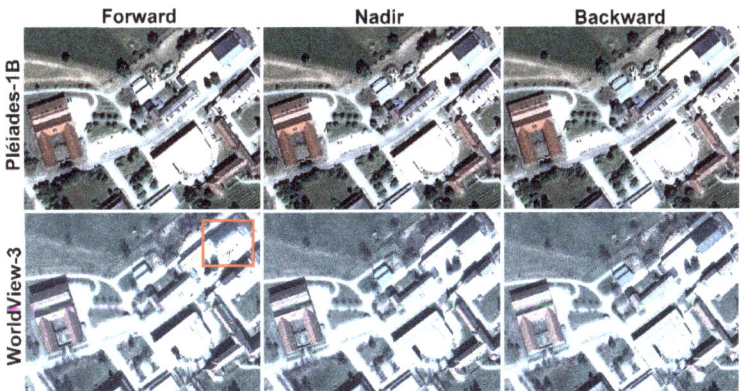

Figure 2. Detail of Pléiades (first row) and WorldView-3 (second row) tri-stereo satellite images on the same built-up area, acquired with forward-, nadir- and backward-looking.

2.1. Pléiades-1B Triplet

The Pléiades satellite system is composed of two identical satellites, Pléiades 1A/1B, which were launched by the French space agency (CNES) in December 2011 and December 2012, respectively. Both satellites are flying on the same sun-synchronous low-Earth orbit at an altitude of 694 km with a phase of 18° and an orbital period of 98.8 min. The sensors are able to collect both panchromatic and multispectral images [3].

The Pléiades-1B tri-stereo images used in the current investigations were acquired in the late morning of 13 June 2017 within 23 s. Within this time, the satellite travelled a total distance (Base) of 167.87 km, leading to an intersection angle of rays on the ground of 12°. The B/H ratios are of 0.10, 0.11, and 0.21 for forward-nadir, nadir-backward and forward-backward image combinations. The pansharpened images composing the triplet are available in 16 bit, each of them with four spectral bands, i.e., red, green, blue, and near-infrared. Depending on the viewing angle, the mean values for the GSD vary between 0.70 and 0.71 m.

2.2. WorldView-3 Triplet

The WorldView-3 Digital Globe's very high-resolution optical sensor was launched in August 2014. Operating at an altitude of 617 km with an orbital period of 97 min, the sensor provides 0.31 m panchromatic resolution.

According to the metadata, the tri-stereo WorldView-3 images for our study were acquired in spring 2018, on 8 April, within 37 s. The corresponding base has a value of 279.69 km and the intersection angle of rays on the ground is 23°. Even though both sensor platforms fly at approximately the same speed (~7.5 km/s), the WorldView-3 B/H ratios have higher values, because of the lower altitude and increased acquisition time, compared to the Pléiades triplet. Hence, the B/H ratio is 0.20 for both forward-nadir and nadir-backward images, whereas for forward-backward it is 0.40. Each image is pan-sharpened, with four spectral bands, i.e., red, green, blue, and near-infrared, with 16-bit depth and zero cloud coverage. Depending on the viewing direction, the mean GSD values vary from 0.31 m to 0.32 m. The images were delivered as eight, two, and two tiles for close-to-nadir, forward, and backward, respectively. As a pre-processing step, the tiles for each image were mosaicked accordingly. For both image triplets, Pléiades and WorldView-3, auxiliary data including the third-order rational function model (RFM) coefficients are provided as separate files.

Figure 3 shows a visual comparison between the pan-sharpened (near) nadir images acquired with Pléiades (left) and WorldView-3 (right) over a small area, in order to highlight the effects of the two different resolutions of 0.70 m and 0.31 m. The same objects can be distinguished in both images: streets, buildings, cars, and trees. The higher resolution of WorldView-3 provides a more detailed image with clear visible cars, tree branches, and building roof edges.

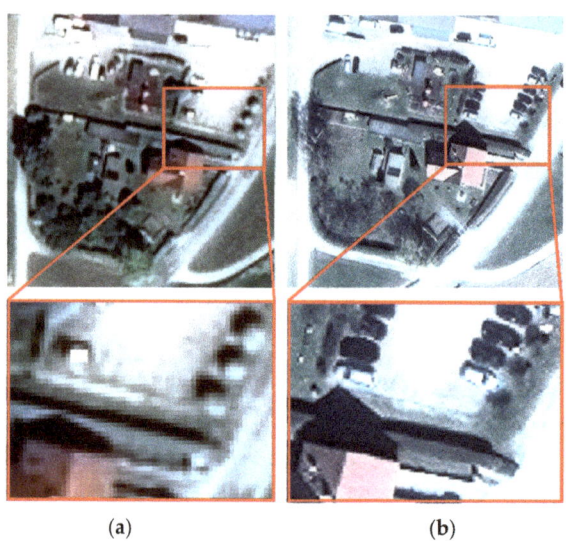

Figure 3. Comparative view of the same area from Pléiades (**a**) and World View-3 images (**b**) with a magnified detail.

3. Data Processing and Analyses

3.1. Image Orientation

The prerequisite for a successful photogrammetric 3D reconstruction is that both interior and exterior orientations are correctly known. The physical sensor model based on the collinearity condition, describing the geometric relation between image points and their corresponding ground points, yields high accuracies (typically a fraction of one pixel), but is complex, and varies depending on different sensors types. Moreover, information about the camera model, ephemerides, and satellite attitude may not be available to users, since they are kept confidential by some commercial satellite image providers [28]. Usually, in practice, the RFM containing eighty RPCs is used for replacing the rigorous physical model of a given sensor [29]. In our research work, we have used the RFM for the photogrammetric mapping.

The workflow for DEM extraction from tri-stereo images begins with image block triangulation and geo-referencing, based on provided RPCs and available GCPs. The main steps for the satellite triangulation are:

1. Image point measurement of GCPs. The number of GCPs is different for the two image sets because of the different data extent and visibility in the scenes. Therefore, 43 GCPs and 22 GCPs were manually measured in each Pléiades and WorldView-3 tri-stereo pair using the multi-aerial viewer, which allows a simultaneous display of the images. This step is performed in order to stabilize the image block and to achieve higher accuracies by improving the given RPCs' values.

2. Tie points (TPs) extraction and RPC refinement. The orientation of the satellite imagery includes the automatic TPs extraction as well as an automatic and robust block adjustment. During the adjustment, a maximum number of six parameters (affine transformation in image space) can be computed: two offsets, two drift values, and two shear values (for each image). Depending on their significance, only a subset of these corrections could be computed by the software: two shifts (on X and Y) and a scale on Y. TPs were automatically extracted by applying Feature Based Matching using the Förstner operator and refining them with Least Squares Matching [30]. TPs with residuals (in image space) larger than one pixel are considered mistakes (gross errors) and rejected. The RPCs are refined through a subsequent adjustment procedure where the differences between image- and backprojected- (with the RFM) coordinates of the GCPs and TPs are minimized.

3. Geo-positioning accuracy analysis. To evaluate the accuracy of the georeferenced images, 50 CPs were used. They were manually acquired from the available orthophoto at 0.2 m GSD and their heights were extracted at the same locations from the LiDAR DTM (1 m resolution). For CP selection, stable details on the ground such as road marks (e.g., pedestrian crossing lines), road surface changes, corners of paved areas and corners of parking lots were selected. Considering the horizontal accuracy of the orthophoto (0.10 m) and the vertical accuracy of the DTM (0.12 m), these points are less accurate than the RTK point measurements.

For the current work, the entire photogrammetric workflow was implemented in the Inpho 8.0 software from Trimble [31], designed to perform precise image block triangulation through bundle block adjustment and 3D point cloud reconstruction using dense image matching techniques for push-broom cameras. The same processing line was followed for Pléiades and WorldView-3 tri-stereo images.

3.2. Manual Reference Measurements

After the image orientation was completed, the manual measurement of the 3D points was performed. For each object, the highest point was selected, such as points on a building's ridge, on a car's roof and tree crown centre (approximation of the tree top). For the 3D restitution, we manually measured the points monoscopically in all three oriented images (forward, nadir and backward), in a multi-aerial view mode. The mean-square error of the Z object coordinate is given by the following formula [32]:

$$\sigma_Z = \sqrt{2} \cdot m_B \cdot \sigma_B \cdot \frac{Z}{B}, \quad (1)$$

with σ_Z the object elevation accuracy, σ_B the accuracy of image measurement (1/3rd of a pixel), Z the altitude of the satellite orbit, B the base and m_B the satellite image-scale number given by the Z/c ratio, where c is the focal length of the optical system. Due to the large heights, we considered the special case of parallelism between object and image plane; hence a single scale number for the whole satellite image was used.

Taking into account these parameters, the estimated accuracy of the Z object coordinates is 1.36 m and 0.31 m for the Pléiades and WorldView-3 images, respectively. These results suggest a minimum object height of 1 m as a reference height that guarantees a reasonable analysis. Since the smallest investigated cars have around 1.5 m height and buildings and single trees usually have more than 2 m height, the estimated elevation accuracy does not significantly influence our investigations.

In a final step, the reference object heights are computed by using the measured Z-coordinates and the elevations extracted from the LiDAR DTM (with 1 m resolution and σ_Z of 0.12 m) at each point's position. Assuming that the measured and extracted Z values have random and uncorrelated errors, according to the law of error propagation, the uncertainty associated with the manual measurements for reference object height computation is determined by the quadrature of input elevation uncertainties.

The geometric parameters, i.e., vehicle length/width, tree crown diameter and building length/width, were directly measured on the rectified images in a geographic information system (GIS) environment.

Three different classes of objects were separately investigated:

(a) Vehicles are classified into four groups depending on their length: (1) passenger and family car type, smaller than 5 m; (2) vans having lengths between 5 and 7.5 m; (3) trucks with lengths between 7.5 and 10 m; and (4) articulated lorries, trailers and multi-trailers usually having more than 10 m. The lengths, widths and the corresponding elevations at car's roofs of 50 vehicles are investigated. The computed mean reference heights are 1.5, 2.5, 3.7, and 4 m for family cars, vans, trucks, and articulated lorries, respectively. Related to average object height, the associated uncertainty of the manual height measurement varies between 8% for lorries and 22% for family cars for WorldView-3 and between 34% and 68% for Pléiades.

(b) Trees are typically classified into two categories: coniferous and deciduous. Coniferous trees are cone-shaped trees, represented mainly by spruce in our case. The second category is the broad-leaved trees with leaves that fall off on a seasonal basis, mostly represented by beech and oak. We needed to perform this classification due to the different acquisition times: one in June (leaf-on conditions) and the other one in April (leaf-off conditions). The diameter and elevations were measured for 100 trees (50 trees for each category: deciduous and coniferous). Depending on crown diameters, they were grouped into seven categories, beginning with trees with a crown smaller than 2.5 m, and ending at large trees with 10 m diameter. The computed mean reference heights for the seven coniferous tree classes are: 5.5, 7.8, 11.0, 14.6, 17.2, 23.4, and 28.7 m, with uncertainties between 1% and 6% from object height for WorldView-3 and between 5% and 24% for Pléiades. The mean reference heights for the deciduous tree classes are: 3.1, 5.4, 8.0, 12.6, 15.4, 16.2, and 18.5m, with uncertainties between 2% and 10% from object height for WorldView-3 and between 7% and 44% for Pléiades.

(c) For buildings, two geometrical parameters are taken into account: length and width. According to their size, built-up structures are grouped into several classes starting with very small (5 m in length and width) to large (50 m length and 25 m width). Therefore, lengths, widths, and roof ridge elevations were measured for 100 buildings in both Pléiades and WorldView-3 images. The mean reference height values varied from 2 m (small built-up structures) to 10–12 m (large industrial buildings), with associated uncertainties from 2% to 16% from object height for WorldView-3 and between 11% and 68% for Pléiades.

While identical trees and buildings were investigated in both Pléiades and WorldView-3 images, this was not possible for vehicles, since they are moving objects. Therefore, (parked) vehicles were randomly selected within the entire scene also using the non-overlapping parts.

3.3. Satellite Image Matching and 3D Reconstruction

Image matching algorithms were used to identify homologous objects or pixels within the oriented images. These algorithms can be divided into area-based (e.g., [33]), feature-based (e.g., [34]), and pixel-based matching (e.g., [35]). Alternatively, image-matching approaches are often classified into sparse and dense matching or into local and global matching methods. The automatic image matching and DSM reconstruction processes were individually performed for each tri-stereo-scene (Pléiades and WorldView-3) by using the specialized module of the Inpho software, called Match-T DSM. The DSM derivation was based on three matching strategies [30]: (a) least squares matching (LSM); (b) feature-based matching (FBM) [36]; and (c) cost-based matching (CBM). Like in most image matching procedures, where image pyramids were used to reduce computation time [37], in our case, the iterative processing chain contains ten pyramid levels. On each pyramid level, three processes were performed: the matching of homologous image points, 3D intersection in object space and DEM modelling. For the first seven pyramids, FBM was used, and the last three image pyramid levels were processed with CBM. CBM is a pixel-by-pixel matching technique similar to the semi-global matching algorithm [35,38]. The CBM strategy within the Match-T DSM module uses a search-path in a so-called 3D-cost-cube for finding the corresponding pixels in images. The cost functions (e.g., correlation coefficient) are used to find the minimum cost path, and each direction represents an X–Y movement in the image to match. Finding the pixel with the lowest cost generates a lowest-cost 3D model–surface model [30]. For each image pixel, the 3D object point coordinates were calculated by applying forward intersections, finally resulting in dense photogrammetric point clouds for the entire study area.

In the last step, high resolution DSMs were generated by using the robust moving planes interpolation method. For each grid node all points within a circular neighborhood are used to robustly estimate a best fitting tilted plane (minimizing the vertical distances); points classified as outliers are not considered in the plane fitting procedure. This step was performed with the scientific software OPALS (Orientation and Processing of Airborne Laser Scanning data) [39].

4. Results and Discussion

4.1. Block Triangulation

During the satellite triangulation of the tri-stereo scene, TPs were automatically extracted in each Pléiades image. They were homogenously distributed within the forward, nadir and backward images, with standard deviations of the adjusted coordinates for elevations ranging from 1.49 m (2.1 GSD) to 3.49 m (5.0 GSD) (Table 2). The standard deviations of the TP elevation obtained for the WorldView-3 scenes are clearly better relative to the GSD, with a maximum value of 0.74 m (2.5 GSD). The standard deviation of the satellite image block-triangulation was 0.54 pixels for Pléiades images and 0.46 pixels for WorldView-3.

Table 2. Standard deviation of adjusted tie points (TPs) coordinates.

Sensor Type	No. of TPs/Image	Standard Deviation (m/pixels)					
		Latitude (Along Track)		Longitude (Across Track)		Elevation	
		min	max	min	max	min	max
Pléiades	561/552/582	0.16/0.23	0.33/0.47	0.14/0.20	0.27/0.38	1.49/2.12	3.49/4.98
WorldView-3	556/585/554	0.09/0.30	0.19/0.63	0.08/0.26	0.20/0.66	0.38/1.26	0.74/2.46

We evaluate the accuracy of the estimated orientation with respect to the measured GCPs and CPs. The root mean square error values of their coordinates in units of meters and pixels are shown in Table 3. For both sensors the RMSE values of the GCPs for latitude, longitude and elevation are at sub-pixel level, showing that the block adjustment provides precise results for subsequent processing. The small planimetric discrepancies suggest that the GCP identification in images is better than 1/3rd of a pixel. The highest discrepancy values for the CP height are 0.85 m (1.21 GSD) for Pléiades and 0.28 m (0.93 GSD) for WorldView-3. The Pléiades elevation accuracy (0.85 m) is not as good as the one in [5], which resulted to 0.49 m, but is more favorable than the results in [7], where 1 m RMSE in elevation were reported. For WorldView-3, the vertical accuracy of 0.28 m fits to the results in [12], which reported an elevation bias of less than 0.5 m for 6001 ground LiDAR CPs. The residual errors for Pléiades images at CPs are comparable with the results obtained by [40] reporting RMSEs of 0.44 m, 0.60 m and 0.50 m at 40 GCPs for latitude, longitude and elevation, respectively.

Table 3. Root mean square error (RMSE) values of ground control point (GCP) and check point (CP) discrepancies.

Sensor Type	No. of GCPs/CPs	RMSE Values (m/pixels)		
		Latitude (Along Track)	Longitude (Across Track)	Elevation
Pléiades	43 GCPs	0.20/0.29	0.19/0.27	0.27/0.39
	50 CPs	0.44/0.63	0.53/0.76	0.85/1.21
WorldView 3	22 GCPs	0.06/0.21	0.08/0.29	0.11/0.37
	50 CPs	0.13/0.43	0.12/0.40	0.28/0.93

4.2. Dense Image Matching

The dense image matching algorithm was applied individually to the two sets of image triplets available as 4-band pansharpened products. Using four cores of a 3.50 GHz machine with 32 GB RAM, the 3D point clouds were generated in 10 and 33 h for Pléiades and WorldView-3 datasets, respectively. The direct output are dense 3D photogrammetric point clouds in the LAS file format with one point for each image pixel. In the overlapping area, the Pléiades-based point cloud consists of 169 million points, whereas the WorldView-3-based equivalent has 476 million points. Hence, it's almost three times denser. Overall, the reconstructed point clouds have a regular distribution on ground plane domain, with densities of 4 points/m^2 and of 12 points/m^2 for Pléiades and WorldView-3, respectively. A few small regions were not reconstructed due to occlusions (e.g., large elevation difference between buildings/trees and surrounding ground). Nevertheless, these holes were filled using interpolation algorithms in the following step.

Based on a neighborhood search radius of 1 m and a maximum number of 20 nearest neighbors, we interpolated a regular grid structure using robust moving planes interpolation. For each dataset, a digital surface model was generated using the same parameters. Since we wanted to use the direct output of the 3D reconstruction without losing any information, the point clouds were not edited or filtered before the interpolation step.

Usually, interpolation strategies tend to smooth the initial elevation values of the 3D point cloud. In order to minimize this effect, we selected a small size of the grid cell (0.5 m × 0.5 m) and a relatively small neighborhood definition for the interpolation. To determine the direct degree of smoothing, a raster containing the maximum elevations in each 0.5 m × 0.5 m cell was computed. Using this as a reference, the 2.5D Pléiades interpolated model showed an RMSE value of 0.079 m, while the WorldView-3 DSM had 0.098 m. The latter was slightly higher, but these values (at sub-decimetre level) still showed that the smoothing effect of the interpolation will not significantly influence further analysis.

The vertical quality of the photogrammetrically derived DSMs was evaluated against the available elevations of the GCPs and CPs (Table 4). The computed mean values for the CPs—0.07 m (Pléiades) and 0.01 m (WorldView-3) and the standard deviations of 0.30 m (Pléiades) and 0.13 m (WorldView-3)—are comparable with the results obtained by [13], who reported mean values ranging between 0.08 and 0.29 m and standard deviations between 0.22 and 0.53 m.

For Pléiades, we obtained an RMSE in the Z-direction of 0.29 m for GCPs and of 0.31 m for CPs. The WorldView-3 DSM showed a higher vertical accuracy, with RMSEs of 0.12 m and 0.13 m for GCPs and CPs, respectively. This is because the vertical accuracy of the DSM is directly influenced by the triplet acquisition geometry, especially by the intersection angle on the ground. As described in [41], the vertical accuracy of the DSMs from VHR satellite imagery is influenced by the acquisition geometry, where a wider angle of convergence (>15°) enhances the height accuracy. In our case, the Pléiades scenes with a narrow convergence angle (12°) show a lower vertical performance than the WorldView-3 scenes, with a larger convergence angle on the ground (23°).

Table 4. Vertical accuracy assessment of Pléiades and WorldView-3 tri-stereo digital surface models (DSMs).

Sensor Type	No. of GCPs/CPs	μ	σ	RMSE
Pléiades	43 GCPs	0.07	0.28	0.29
	50 CPs	0.07	0.30	0.31
WorldView-3	22 GCPs	0.03	0.18	0.12
	50 CPs	0.01	0.13	0.13

μ and σ are the mean and standard deviations. All values are given in meters.

The two high resolution DSMs derived from Pléiades and WorldView-3 tri-stereo scenes within the WGS84-UTM33 reference system were used as source data for the following single object investigations.

4.3. Object Height Differences

For a clear height estimation, in our work we considered only single objects, located in free, open and smooth areas, without any other entities in their close surroundings. Assuming that in the free, open areas, the DSM coincide with DTM elevations, we inspected the vertical quality of the photogrammetric derived surface models by computing the vertical offsets from the reference LiDAR DTM (1 m resolution and $\sigma_Z = 0.12$ m). The results showed a good correspondence of the DSMs with the reference DTM, featuring a RMSE of 0.32 m for Pléiades and of 0.20 m for WorldView-3. When it comes to individual objects, we sometimes observed a constant offset, but it was always below 0.30 m. Therefore, it did not significantly impact our height investigations.

The reconstructed individual object height was extracted from the interpolated surface models. Heights of buildings, vehicles and trees were computed by subtracting the DTM elevations from the elevations of the highest object points (located on roof ridges, on car roofs, and on treetops) at the same 2D location.

We consider the reference height (H) as being the real object height, which was computed by subtracting the DTM elevation from the manual measurements (Section 3.2) at the same position.

$$H = Z_M - Z_{DTM} \ (m) \tag{2}$$

$$h = Z_{DSM} - Z_{DTM} \ (m), \tag{3}$$

with H the reference object height, h the estimated object height, Z_M the manual elevation measurements, Z_{DTM} the elevation of the ground surface and Z_{DSM} the elevation of the reconstructed DSM.

In Equations (2) and (3) the Z_{DTM} elevation values are identical, since for computing the reference and the estimated heights, the same ground elevation at the same 2D position was considered. Based on the defined equations, we obtained a reference height (H) and two estimated heights for each individual object (from Pléiades and WorldView-3 DSMs). These values will be employed in further analysis.

4.4. Height Estimation of Single Objects

Within this investigation, we wanted to estimate the Pléiades and WorldView-3 DSMs heights at single objects as a function of object type and size. For this purpose, three object types were analyzed: non-moving vehicles, trees, and buildings. The main parameters taken into account were vehicle length/width, tree crown diameter, and building length/width.

For each single object, the estimated heights were compared with the monoscopic reference measurements based on a ratio measure. The percentage value derived by the following equation describes how much of the reference height was reconstructed:

$$p \ (\%) = h/H \times 100, \tag{4}$$

with H the reference height, h the estimated height, and p the reconstruction percentage.

4.4.1. Vehicles

By a visual inspection of the reconstructed Pléiades-based 3D point cloud, it was observable that points corresponding to small vehicles (vehicles with lengths smaller than 5 m) were not elevated against their surroundings. Larger vehicles were reconstructed in height, but not entirely (Figure 4).

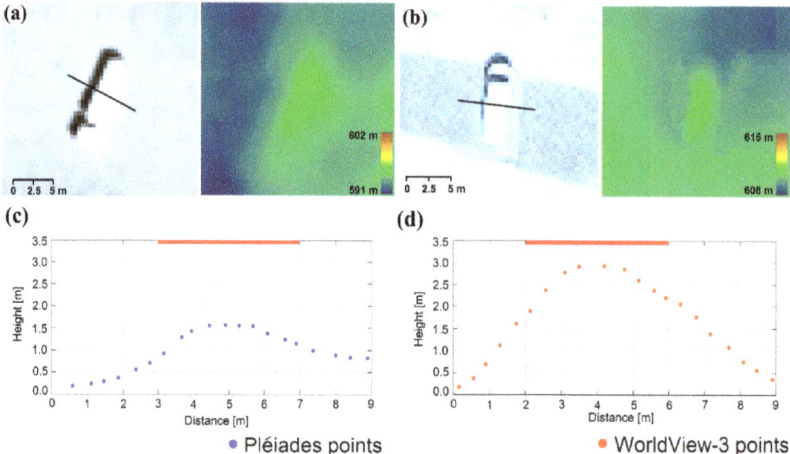

Figure 4. Height estimation for a truck. (**a**) Pléiades satellite image detail with profile location (left) and reconstructed DSM Z color-coded (right); (**b**) WorldView-3 satellite image detail with profile location (left) and reconstructed DSM Z color-coded (right); (**c**) Pléiades-; and (**d**) WV3-based vehicle profiles of 0.5 m width each from the matched point cloud; the red horizontal line is the corresponding measured reference height.

Specifically, Figure 4 represents a truck with a length of ~10 m and a real height of 3.5 m that is visible in both Pléiades (a) and WorldView-3 images (b). On the left-hand side are the magnified satellite image details of the vehicle and on the right-hand side are the corresponding reconstructed height color-coded DSM. From the DSMs and the cross profiles (c), it is clearly visible that in contrast to Pléiades, where there is only ~1 m elevation change, the WV3-based profile reconstructs the vehicle height up to 2.9 m, which represents ~83% of the real height.

Using Equation (4), the height percentages of 50 measured vehicles were computed for both cases: Pléiades and WorldView-3 sensors. As mentioned in Section 3.2, vehicles were classified into four different groups depending on their lengths, starting with small family cars, of approximately 5 m, followed by vans, trucks, and finally by large articulated lorries with more than 10 m length. For each group, containing between 10 and 13 vehicles, the mean value of the reconstruction height percentage and the standard deviation were determined. They are shown in Figure 5a for both Pléiades (blue) and WorldView-3 (yellow) data. By increasing length, the height percentage also increases, reaching up to 60% (Pléiades) and 92% (WorldView-3). In the case of the Pléiades imagery, small vehicles (<7 pixels length in image) do not show any height information, while for family cars, vans, and trucks the reconstruction height percentage is less than 30%. It exceeds 50% only for large vehicles (lengths >10 m, 14 pixels in the image). On the other hand, in WorldView-3 DSMs, vehicles have a good height reconstruction (reaching over 90% for lengths >10 m, 33 pixels in image). An exception are family cars (~16 pixels length), which have a percentage of less than 40%. The standard deviations of the estimated heights are notably smaller than the overall trend and they become smaller with increasing vehicle length.

A similar analysis was conducted by using vehicle width as the main parameter for vehicle grouping (Figure 5b). Because small family cars and vans have similar widths (~2 m), they form a single class, followed by trucks (2 to 2.5 m widths), and articulated lorries (2.5 to 3 m widths). Based on the WorldView-3 DSMs, vehicles with up to 2 m widths have ~41% of their real height reconstructed. This value is higher when compared with the percentage of the first length-class (37%) because it contains both family cars and vans, leading to an increased mean height percentage and standard deviation. The mean height percentages reach 59% (Pléiades) and 92% (WorldView-3) for the very large vehicles. The increased values of standard deviations suggest a higher variation of the reconstruction height percentages within the width-classes considered. Therefore, the length parameter is more suitable for the description of the vehicle's height estimation.

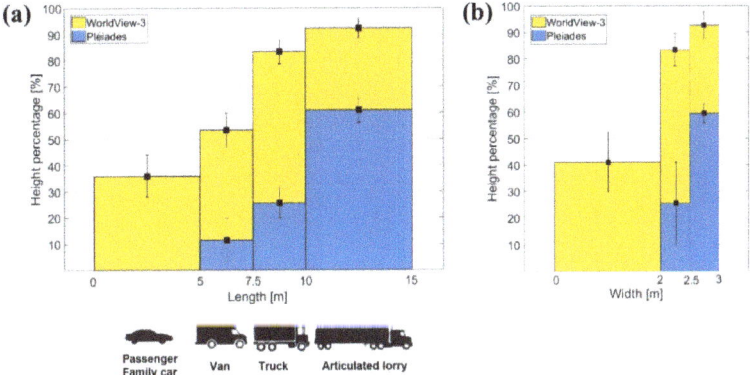

Figure 5. Interval-based percentage and standard deviation for height estimation of vehicles depending on (**a**) lengths and (**b**) widths.

4.4.2. Trees

Trees were investigated in two separate classes: deciduous trees, located especially on roadsides, parks and cities, and coniferous trees, found in forested areas and forest patches. Only single, isolated trees, not close to buildings or any other objects, were taken into consideration. In this case, the visible crown diameter was used as the geometric parameter. One hundred trees were investigated, and the same formula for height percentage computation, Equation (4), was applied. Based on diameter, they were grouped into seven different categories, each containing approximately 12 trees. The mean reconstruction percentages gradually increase with crown diameter and the standard deviations decrease (Figure 6). For WorldView-3, the heights of coniferous trees with crown diameters <2.5 m (8 pixels in image space) are reconstructed less than 50%, whereas trees with crown diameters larger than 5 m (16 pixels in image space) are over 95%. Generally, heights of coniferous trees are better estimated from WorldView-3 than from the Pléiades DSMs, where trees with diameters >7.5 m (11 pixels) barely reach 75% height reconstruction. For deciduous trees (Figure 6b) it can be observed that the situation is reversed: they get better height information from Pléiades DSM, than from WorldView-3 DSM. This is mainly because of the different acquisition times, June for Pléiades and April for WorldView-3, indicating leaf-on and leaf-off conditions which clearly influence the appearance and texture of the trees in the images. For the deciduous trees, only the stems and branches are visible in the WorldView-3 images, resulting in a poor dense image matching quality. For trees with a crown diameter larger than 9 m, heights of just over 50% of the tree values are reconstructed.

When analyzing the two tree types only from the Pléiades images (Figure 6), we could see that there is a slightly better height reconstruction for the very small and large deciduous trees, compared with the coniferous. The small deciduous trees in the first group (with crown diameters smaller than 2.5 m) have 26% of their real height reconstructed, whereas the small coniferous trees in the first group have only 21%. The same is true for the last group, where large deciduous trees (with crown diameters bigger than 8.75 m) have percentages close to 80%, whereas for coniferous they reach only 75%. On the other hand, for the coniferous trees of medium sizes (with crown diameters between 2.5 and 7.5 m) we obtained better height reconstruction percentages than for the deciduous.

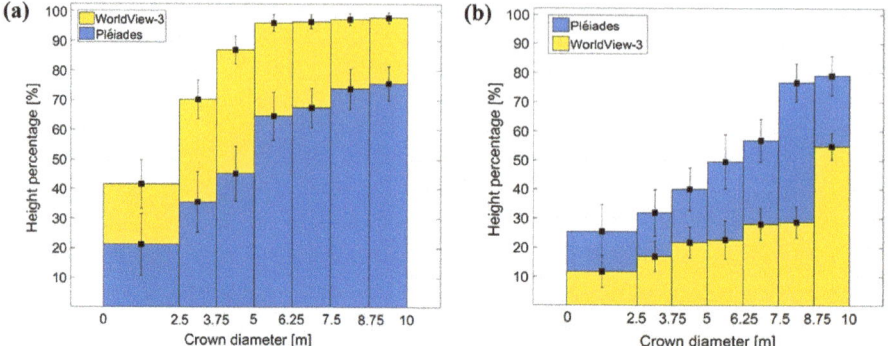

Figure 6. Interval-based percentage and standard deviation for height estimation of single trees (**a**) coniferous trees and (**b**) deciduous trees.

Two examples of height estimation of a coniferous (Figure 7) and a deciduous tree (Figure 8) are shown. They were comparatively analyzed for Pléiades and WorldView-3 satellite images based on both DSM and point cloud. For single coniferous trees with a 7 m crown diameter (Figure 7) we achieved different results based on the input data. From the true height of 12.2 m, the estimation resulted in approximately 12 m for WorldView-3 and only 8 m for Pléiades, which represents 98% and 65% of the real height.

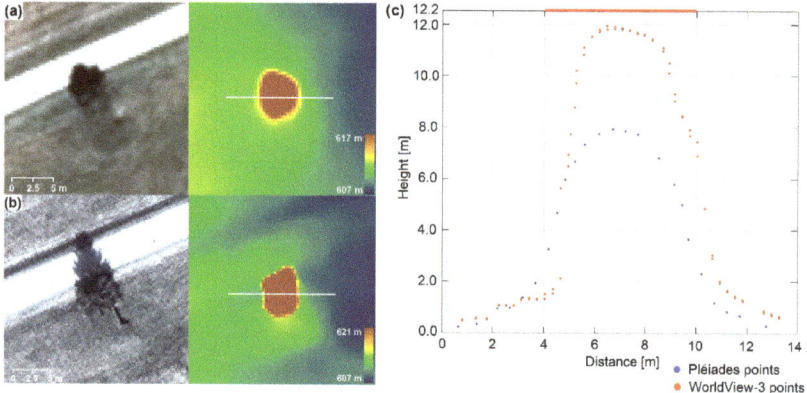

Figure 7. Height estimation of a coniferous tree with 7 m crown diameter: (**a**) Pléiades satellite image detail (left), corresponding height-colored DSM and profile location (right); (**b**) WorldView-3 satellite image detail (left), height-colored DSM with profile location (right); and (**c**) tree profile of 0.5 m width from the matched point clouds; the red horizontal line is the measured reference height.

A small area of the DSMs and the 3D point clouds generated from the Pléiades and WorldView-3 satellite imagery are shown in Figure 8. The largest deciduous tree with leaves (Figure 8a first row) has a reasonable appearance in the DSM and an estimated height of 14 m. Meanwhile, the same leafless tree in the WorldView-3 image (Figure 8a second row) has a height of only approximately 4 m. A visual analysis of height differences for this deciduous tree and a small building (next to the tree) is shown in two profiles (c) and (d) in Figure 8.

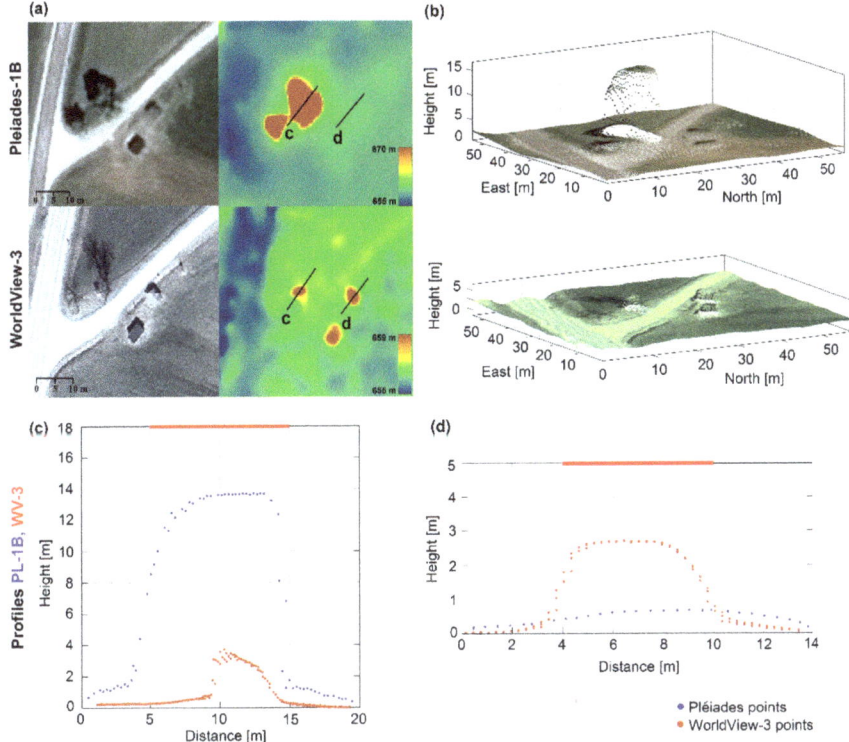

Figure 8. Height estimation of a deciduous tree and a small building: (**a**) Pléiades (first row) and WorldView-3 (second row) views with corresponding height–color coded DSMs; (**b**) RGB colored 3D point clouds, 1 m width height profiles from the matched point clouds; (**c**) for a deciduous tree; (**d**) for a small building; the red horizontal lines are the corresponding measured reference heights.

4.4.3. Buildings

The third category of objects investigated were single buildings, usually in suburban and rural areas. They usually had regular geometric shapes, such as rectangles or squares, which facilitated the visual identification of the corresponding dimensions in the rectified images. The building length/width were measured in the rectified images, based on their roof top view. For buildings with a T- or L-shaped plan, we considered the length as the longest dimension between extreme roof edges, whereas the width as the distance across the roof from one side to the other. One hundred buildings were chosen to cover a wide range of dimensions from 1 to 50 m length (large industrial buildings) and widths from 1 to 25 m. In the example shown in Figure 8b, the two small buildings are elevated against their surroundings in the WorldView-3 data, which is not the case for the Pléiades data. In addition, the profile (d) with two overlaid point clouds corresponding to a 3.6 by 2.5 m building clearly reveals the higher potential of WorldView-3 sensor for height reconstruction, as opposed to Pléiades. Even if the building edges appear very smooth, the corresponding main height is still reconstructed.

Again, Equation (4) was applied for investigating the potential of Pléiades and WorldView-3 DSMs for building's height estimation with respect to their lengths and widths (Figure 9). Based on their length, we defined ten different categories, each containing approximately ten buildings (Figure 9a). Like in the previous analyses for vehicles and trees, the computed mean height percentages for buildings show a similar trend, values gradually increasing with building length. In case of the Pléiades imagery, buildings with lengths smaller than 10 m (14 pixels in image space) have a height reconstruction percentage of less than 30%, whereas buildings with lengths greater than 5 m have percentages beyond 50%. For both Pléiades and WorldView-3, buildings with lengths >43 pixels have over 90% height reconstruction.

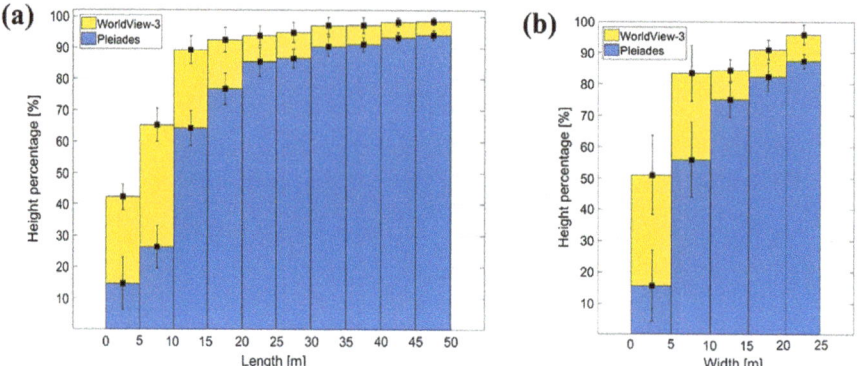

Figure 9. Interval-based percentage and standard deviation for height estimation of buildings depending on (**a**) lengths; (**b**) widths.

Due to the lower variation in building-width, we considered only five intervals, starting from 0–5 m to 20–25 m (Figure 9b). Compared with height-percentage analysis based on building length, for the width parameter the situation was different. When buildings were grouped based on their corresponding widths, we had an increased number of items within each interval (only 5 width-intervals). This led to a higher variation of height estimation percentages, and hence to higher values for the standard deviations within each interval. When analyzing the reconstruction height percentages based on the WorldView-3 images for both 0–5 m and 5–10 m length/width intervals, we could see higher mean values for the reconstruction percentages computed based on building widths. Nevertheless, the corresponding standard deviations are larger in contrast to those computed on the length basis. The standard deviations for the length-intervals (Figure 9a) become smaller with increasing building length, showing a clearer trend; therefore, the length parameter gives a better estimation when analyzing the height reconstruction of buildings.

The two scatter plots (Figure 10) show the relationships between building length and width and the reconstruction height percentage. As expected, height accuracy is better for WorldView-3 images than for the Pléiades data. The delineation lines separate the buildings with higher and lower reconstruction height percentage (p) than 50%. If for Pléiades data a 2D minimum building size of 7.5 m length by 6 m width is needed in order to get $p > 50\%$, for WorldView-3 data only 4 m × 2.5 m is needed.

The mean value for the residual heights corresponding to reconstructed buildings with $p > 50\%$ (of 2.02 m for Pléiades with 0.7 m GSD) is comparable with the residuals of 1.94 m obtained by [18] when using a DSM from a GeoEye stereo pair (0.5 m GSD).

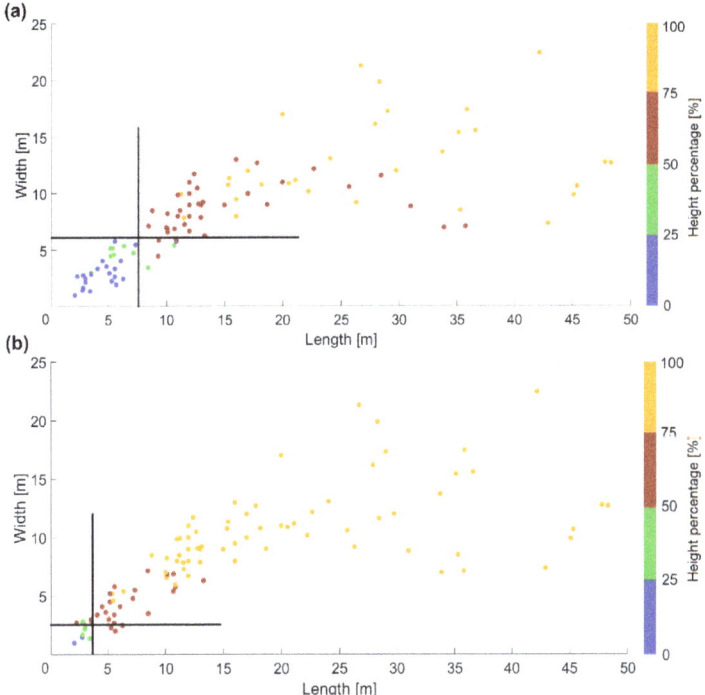

Figure 10. Height estimation of buildings: (**a**) for Pléiades; (**b**) for WorldView-3.

5. Conclusions

This study addresses the potential and limitations of both Pléiades and WorldView-3 tri-stereo reconstructed DSMs for height estimation of single objects. Satellite image orientation and dense image matching were performed. The tri-stereo images of each sensor were acquired in less than one minute, resulting in similar radiometric characteristics, a fact that is beneficial for any image matching algorithm. For obtaining higher accuracies (at sub-pixel level in image space and sub-meter level in object space) the direct sensor orientation available as RPCs is improved by GCPs through a bias-compensated RPC block adjustment. Evaluated at 50 CPs in open areas, elevation accuracies of 0.85 m and 0.28 m (RMSE) were obtained for Pléiades and WorldView-3, respectively.

Next, photogrammetric DSMs were generated from the Pléiades and WorldView-3 tri-stereo imagery using a dense image matching algorithm, forward intersections, and interpolation. The estimation of a single object's height depends on the accuracy of satellite image orientation, but also on the elevation accuracy of the derived DSMs and on the LiDAR reference DTM. We found that the vertical accuracy of the reference DTM w.r.t the RTK GCPs is 0.12 m, and the accuracies of the reconstructed photogrammetric models with respect to the DTM are 0.32 m for Pléiades and 0.20 m for WorldView-3 in open, smooth areas. The vertical accuracy of the DSMs from VHR satellite imagery is influenced by the acquisition geometry. The Pléiades scenes having a narrow convergence angle of 12° show a lower vertical performance compared to WorldView-3 scenes, which have a 23° convergence angle on the ground. In addition, the smoothness effect of the robust moving planes interpolation for deriving the DSMs is at sub-decimetre level. Under these conditions, the resulting height accuracy is reasonable and does not significantly influence our investigations, since the single objects measured are usually higher than 1.5 m. Based on the different GSDs and B/H ratios of the two satellite systems, it can be said that σ_0 of block adjustment, RMSE of check point elevation error and height reconstruction success for individual objects are similar, with deviations of up to 20%.

In order to investigate the potential of Pléiades and WorldView-3 tri-stereo DSMs for single object height estimation, 50 vehicles, 100 trees, and 100 buildings were analyzed. Vehicle length/width, tree crown diameter and building length/width were identified as the main geometric parameters for classifying single objects into different groups. The results clearly show a gradual increase of the estimated heights with object dimensions. Starting from very small objects (whose heights are not reconstructed) the values for the height reconstruction percentage reached almost 100% for very large single objects (e.g., buildings with lengths >35 m in the case of WorldView-3 DSM). Hence, individual objects' reconstructed heights strongly depend on object type, size, and the sensor being used. As an example, if we consider the $p = 50\%$ threshold for an appropriate height reconstruction, then we obtain the following minimum size values for the single objects: 10 m (Pléiades) and 5 m (WorldView-3) for vehicle lengths; 5 m (Pléiades) and 2.5 m (WorldView-3) for tree crown diameters and 7.5 m × 6 m (Pléiades) and 3.5 m × 2.5 m (WorldView-3) for building length/width. The ratio of these dimensions fits well to the GSD ratio of 70 cm to 31 cm.

Generally, for all single objects except deciduous trees, the results achieved with WorldView-3 data were better than those achieved by using the Pléiades data. Hence, the ability of WorldView-3-derived DSM to perform height estimation of single objects is higher compared with Pléiades DSM. This is mainly because of the very high resolution (0.31 m) that leads (through photogrammetric processing) to very dense 3D point clouds, describing the real ground surface and the objects on it more accurately. The acquisition times also play a significant role in the photogrammetric procedure. For stable objects such as buildings, the different periods of acquisition do not affect the 3D reconstruction, but for deciduous trees, the automatic reconstructed heights are highly underestimated in the leaf-off season. Their leafless appearance in the images brings difficulties to the image matching process when finding correct correspondences for tree branches. From this observation, we can conclude that the image appearance and texture of trees (and other single objects) are also very important for the 3D reconstruction and height estimation. For all objects investigated, the resulting histograms (Figures 5, 6 and 9) show interesting trends, opening new investigation options for refinement such as adding the influence of object orientation, color, and texture in the feature reconstruction ability of heights, but these are outside the scope of the present research. The poor performance on small individual objects is mainly caused by the continuity constraint of dense matching algorithms. A topic for future research would be to investigate if a combination of different matching strategies can deliver better results on single object heights.

In our investigation, the current findings are valid for Pléiades and WorldView-3 DSMs derived from the tri-stereo images with their specific acquisition geometry. Nevertheless, the acquisition geometry could be changed, but this is only to a very small extend under the control of the user. A higher B/H ratio and implicit convergence angle to the ground will increase the geometric intersection quality, leading to a higher vertical accuracy of the photogrammetric DSM. This will have a positive effect on the single object height signature in the DSM.

From the analyses and investigations performed, the results suggest that the acquisition geometry clearly has an effect on accuracy, but the object height estimation based on automatically derived DSMs primarily depends on object size. Objects of a few pixel in size are hardly mapped in the DSM, whereas, with some dependency on object type, objects of 12 pixels in size are typically mapped with approximately 50% of their real height. To reach 90% of the real height, a minimum object size between 15 and 30 pixel is necessary.

Author Contributions: Conceptualization, A.-M.L., J.O.-S. and N.P.; Formal analysis, A.L.; Investigation, A.-M.L.; Methodology, A.-M.L., J.O.-S. and N.P.; software, J.O.-S.; Project administration, J.O.-S.; Software, J.O.-S.; Supervision, N.P. and J.O.-S.; Validation, A.-M.L., J.O.-S. and N.P.; Visualization, A.-M.L.; Resources, J.O.-S. and N.P.; Data curation, A.-M.L., J.O.-S. and N.P.; Writing—original draft preparation, A.-M.L.; Writing—review and editing, J.O.-S. and N.P.; funding acquisition, N.P. All authors have read and agreed to the published version of the manuscript.

Funding: The work was partly funded by the Austrian Research Promotion Agency (FFG), Vienna, Austria within the research project "3D Reconstruction and Classification from Very High Resolution Satellite Imagery (ReKlaSat 3D)" (grant agreement No. 859792). Open Access Funding by TU Wien.

Acknowledgments: The authors acknowledge Open Access Funding by TU Wien. We would like to thank the Institut für Militärisches Geowesen (Maximilian Göttlich and Clemens Strauß) for support in satellite image acquisition and the company Vermessung Schmid ZT GmbH (Karl Zeitlberger) for providing the RTK ground control point coordinates used in this study.

Conflicts of Interest: The authors declare no conflict of interest.

References

1. Toutin, T.; Cheng, P. Comparison of automated Digital Elevation Model extraction results using along-track aster and across-track Spot stereo images. *Opt. Eng.* **2002**, *41*, 2102–2106.
2. Sadeghian, S.; Zoej, M.J.V.; Delavar, M.R.; Abootalebi, A. Precision rectification of high resolution satellite imagery without ephemeris data. *Int. J. Appl. Earth Obs. Geoinf.* **2001**, *3*, 366–371. [CrossRef]
3. Astrium. *Pléiades Imagery—User Guide v 2.0*; Astrium: Toulouse, France, 2012.
4. DigitalGlobe. Digitalglobe Core Imagery Products Guide. Available online: http://www.digitalglobe.com/downloads/DigitalGlobeCoreImageryProductsGuide.pdf (accessed on 15 November 2019).
5. Bernard, M.; Decluseau, D.; Gabet, L.; Nonin, P. 3D capabilities of Pleiades satellite. *Xxii ISPRS Congr. Tech. Comm. Iii* **2012**, *39-B3*, 553–557. [CrossRef]
6. Poli, D.; Remondino, F.; Angiuli, E.; Agugiaro, G. Evaluation of Pleiades-1A triplet on Trento testfield. *ISPRS Hannover Workshop* **2013**, *40-1*, 287–292. [CrossRef]
7. Perko, R.; Raggam, H.; Gutjahr, K.; Schardt, M. Assessment of the mapping potential of Pléiades stereo and triplet data. *ISPRS Ann. Photogramm. Remote Sens. Spat. Inf. Sci.* **2014**, *2*, 103. [CrossRef]
8. Alabi, T.; Haertel, M.; Chiejile, S. Investigating the use of high resolution multi-spectral satellite imagery for crop mapping in Nigeria crop and landuse classification using WorldView-3 high resolution multispectral imagery and Landsat8 data. In Proceedings of the 2nd International Conference on Geographical Information Systems Theory, Applications and Management (GISTAM), Rome, Italy, 26–27 April 2016; pp. 109–120.
9. Asadzadeh, S.; de Souza, C.R. Investigating the capability of WorldView-3 superspectral data for direct hydrocarbon detection. *Remote Sens. Environ.* **2016**, *173*, 162–173. [CrossRef]
10. Bostater, C.R.; Oney, T.S.; Rotkiske, T.; Aziz, S.; Morrisette, C.; Callahan, K.; Mcallister, D. Hyperspectral signatures and WorldView-3 imagery of indian river Lagoon and Banana river estuarine water and bottom types. *Remote Sens. Ocean Sea Ice Coast. Waters Large Water Reg.* **2017**, *2017*, 10422.
11. Koenig, J.; Gueguen, L. A comparison of land use land cover classification using superspectral WorldView-3 vs hyperspectral imagery. In Proceedings of the 2016 8th Workshop on Hyperspectral Image and Signal Processing: Evolution in Remote Sensing (WHISPERS), Los Angeles, CA, USA, 21–24 August 2016.
12. Hu, F.; Gao, X.M.; Li, G.Y.; Li, M. DEM extraction from WorldView-3 stereo-images and accuracy evaluation. *Xxiii ISPRS Congr. Comm. I* **2016**, *41*, 327–332.
13. Rupnik, E.; Pierrot-Deseilligny, M.; Delorme, A. 3D reconstruction from multi-view VHR-satellite images in Micmac. *ISPRS J. Photogramm.* **2018**, *139*, 201–211. [CrossRef]
14. Abduelmula, A.; Bastos, M.L.M.; Gonçalves, J.A. High-accuracy satellite image analysis and rapid DSM extraction for urban environment evaluations (Tripoli-Libya). *World Acad. Sci. Eng. Technol. Int. J. Comput. Electr. Autom. Control Inf. Eng.* **2015**, *9*, 661–666.
15. Flamanc, D.; Maillet, G. *Evaluation of 3D City Model Production from Pleiades hr Satellite Images and 2D Ground Maps*; URBAN: Tempe, AZ, USA, 2005.
16. Poli, D.; Remondino, F.; Angiuli, E.; Agugiaro, G. Radiometric and geometric evaluation of GeoEye-1, WorldView-2 and Pléiades-1A stereo images for 3D information extraction. *ISPRS J. Photogramm.* **2015**, *100*, 35–47. [CrossRef]
17. Sirmacek, B.; Taubenbock, H.; Reinartz, P.; Ehlers, M. Performance evaluation for 3-D city model generation of six different DSMs from air-and spaceborne sensors. *IEEE J.-STARS* **2012**, *5*, 59–70. [CrossRef]
18. Macay Moreia, J.; Nex, F.; Agugiaro, G.; Remondino, F.; Lim, N. From DSM to 3D building models: A quantitative evaluation. *ISPRS-Int. Arch. Photogramm. Remote Sens. Spat. Inf. Sci.* **2013**, *1*, 213–219. [CrossRef]

19. Panagiotakis, E.; Chrysoulakis, N.; Charalampopoulou, V.; Poursanidis, D. Validation of Pleiades tri-stereo DSM in urban areas. *ISPRS Int. J. Geo-Inf* **2018**, *7*, 118. [CrossRef]
20. Piermattei, L.; Marty, M.; Karel, W.; Ressl, C.; Hollaus, M.; Ginzler, C.; Pfeifer, N. Impact of the acquisition geometry of Very High-Resolution Pléiades imagery on the accuracy of canopy height models over forested alpine regions. *Remote Sens.-Basel* **2018**, *10*, 1542. [CrossRef]
21. Stentoumis, C.; Karkalou, E.; Karras, G. A review and evaluation of penalty functions for semi-global matching. In Proceedings of the 2015 IEEE International Conference on Intelligent Computer Communication and Processing (ICCP), Cluj-Napoca, Romania, 3–5 September 2015; pp. 167–172.
22. Fraser, C.S.; Dial, G.; Grodecki, J. Sensor orientation via RPCs. *ISPRS J. Photogramm.* **2006**, *60*, 182–194. [CrossRef]
23. Perko, R.; Raggam, H.; Schardt, M.; Roth, P.M. Very high resolution mapping with the Pleiades satellite constellation. *Am. J. Remote Sens.* **2018**, *6*, 2019. [CrossRef]
24. Jacobsen, K.; Topan, H. DEM generation with short base length Pleiades triplet. *Int. Arch. Photogramm. Remote Sens. Spat. Inf. Sci. ISPRS Arch.* **2015**, *40*, 81–86. [CrossRef]
25. Panem, C.; Bignalet-Cazalet, F.; Baillarin, S. Pleiades-hr system products performance after in-orbit commissioning phase. *Int. Arch. Photogramm. Remote Sens. Spat. Inf. Sci.* **2012**, *39*, B1. [CrossRef]
26. Orfeo Toolbox. Available online: https://www.orfeo-toolbox.org/ (accessed on 8 August 2019).
27. Kuester, M. *Radiometric Use of Worldview-3 Imagery—Technical Note*; DigitalGlobe: Longmont, CO, USA, 2016.
28. Tong, X.; Liu, S.; Weng, Q. Bias-corrected Rational Polynomial Coefficients for high accuracy geo-positioning of Quickbird stereo imagery. *ISPRS J. Photogramm.* **2010**, *65*, 218–226. [CrossRef]
29. Dial, G.; Grodecki, J. Block adjustment with rational polynomial camera models. In Proceedings of the ASPRS 2002 Conference, Washington, DC, USA, 19–26 April 2002; pp. 22–26.
30. Trimble. *Match-T DSM Reference Manual*; Trimble Inc.: Sunnyvale, CA, USA, 2016.
31. Geospatial. Trimble Inpho Brochure. Available online: http://www.inpho.de (accessed on 31 August 2018).
32. Kraus, K. *Photogrammetrie: Photogrammetrie: Geometrische Informationen aus Photographien und Laserscanneraufnahmen*; Walter de Gruyter: Berlin, Germany; New York, NY, USA, 2004.
33. Förstner, W. On the geometric precision of digital correlation. In Proceedings of the Symposium of the ISPRS Commision III, Helsinki, France, 7–11 June 1982; pp. 176–189.
34. Förstner, W.; Gülch, E. A fast operator for detection and precise location of distinct points, corners and centres of circular features. In Proceedings of the ISPRS Intercommission Conference on Fast Processing of Photogrammetric Data, Interlaken, Switzerland, 2–4 June 1987; pp. 281–305.
35. Hirschmüller, H. Stereoprocessing by Semi Global Matching and mutual information. *IEEE Trans. Pattern Anal. Mach. Intell.* **2008**, *30*, 328–341. [CrossRef] [PubMed]
36. Förstner, W. A feature based correspondence algorithm for image matching. *Int. Arch. Photogramm.* **1986**, *26*, 150–166.
37. Heipke, C. A global approach for least-squares image matching and surface reconstruction in object space. *Photogramm. Eng. Remote Sens.* **1992**, *58*, 317.
38. Wenzel, K.; Rothermel, M.; Haala, N.; Fritsch, D. Sure–the ifp software for dense image matching. In Proceedings of the Photogrammetric Week, Stuttgart, Germany, 9–13 September 2013; pp. 59–70.
39. Pfeifer, N.; Mandlburger, G.; Otepka, J.; Karel, W. Opals–a framework for airborne laser scanning data analysis. *Comput. Environ. Urban Syst.* **2014**, *45*, 125–136. [CrossRef]
40. Stumpf, A.; Malet, J.P.; Allemand, P.; Ulrich, P. Surface reconstruction and landslide displacement measurements with Pleiades satellite images. *ISPRS J. Photogramm.* **2014**, *95*, 1–12. [CrossRef]
41. Poli, D.; Caravaggi, I. 3D modeling of large urban areas with stereo VHR satellite imagery: Lessons learned. *Nat. Hazards* **2013**, *68*, 53–78. [CrossRef]

© 2020 by the authors. Licensee MDPI, Basel, Switzerland. This article is an open access article distributed under the terms and conditions of the Creative Commons Attribution (CC BY) license (http://creativecommons.org/licenses/by/4.0/).

Article

Soil Moisture Content Retrieval from Remote Sensing Data by Artificial Neural Network Based on Sample Optimization

Qixin Liu [1,2], Xingfa Gu [1,2,3], Xinran Chen [1,2], Faisal Mumtaz [1,2], Yan Liu [1], Chunmei Wang [1], Tao Yu [1], Yin Zhang [4], Dakang Wang [5] and Yulin Zhan [1,*]

- [1] Aerospace Information Research Institute, Chinese Academy of Sciences, Beijing 100094, China; liuqx@radi.ac.cn (Q.L.); guxf@radi.ac.cn (X.G.); chenxr@aircas.ac.cn (X.C.); faisal@aircas.ac.cn (F.M.); liuyan@aircas.ac.cn (Y.L.); wangcm@aircas.ac.cn (C.W.); yutao@radi.ac.cn (T.Y.)
- [2] University of Chinese Academy of Sciences, Beijing 100049, China
- [3] School of Remote Sensing and Information Engineering, North China Institute of Aerospace Engineering, Langfang 065000, China
- [4] Beijing Institute of Space Long March Vehicle, Beijing 100076, China; zhangyin@radi.ac.cn
- [5] School of Environmental Science and Engineering, Southern University of Science and Technology, Shenzhen 518055, China; wangdk@aircas.ac.cn
- * Correspondence: zhanyl@radi.ac.cn

Citation: Liu, Q.; Gu, X.; Chen, X.; Mumtaz, F.; Liu, Y.; Wang, C.; Yu, T.; Zhang, Y.; Wang, D.; Zhan, Y. Soil Moisture Content Retrieval from Remote Sensing Data by Artificial Neural Network Based on Sample Optimization. *Sensors* **2022**, *22*, 1611. https://doi.org/10.3390/s22041611

Academic Editors: Francesco Pirotti and Francesco Mancini

Received: 24 January 2022
Accepted: 14 February 2022
Published: 18 February 2022

Publisher's Note: MDPI stays neutral with regard to jurisdictional claims in published maps and institutional affiliations.

Copyright: © 2022 by the authors. Licensee MDPI, Basel, Switzerland. This article is an open access article distributed under the terms and conditions of the Creative Commons Attribution (CC BY) license (https://creativecommons.org/licenses/by/4.0/).

Abstract: Soil moisture content (SMC) plays an essential role in geoscience research. The SMC can be retrieved using an artificial neural network (ANN) based on remote sensing data. The quantity and quality of samples for ANN training and testing are two critical factors that affect the SMC retrieving results. This study focused on sample optimization in both quantity and quality. On the one hand, a sparse sample exploitation (SSE) method was developed to solve the problem of sample scarcity, resultant from cloud obstruction in optical images and the malfunction of in situ SMC-measuring instruments. With this method, data typically excluded in conventional approaches can be adequately employed. On the other hand, apart from the basic input parameters commonly discussed in previous studies, a couple of new parameters were optimized to improve the feature description. The Sentinel-1 SAR and Landsat-8 images were adopted to retrieve SMC in the study area in eastern Austria. By the SSE method, the number of available samples increased from 264 to 635 for ANN training and testing, and the retrieval accuracy could be markedly improved. Furthermore, the optimized parameters also improve the inversion effect, and the elevation was the most influential input parameter.

Keywords: soil moisture content; artificial neural network; sample optimization; synthetic aperture radar; optical remote sensing image

1. Introduction

The soil moisture content (SMC) refers to the volume of water present in the gaps between surface soil granules. The SMC is a critical parameter for investigating and predicting the factors associated with climate change. It also plays a key role in various fields of science such as ecology, hydrology, and agriculture [1–3]. However, the measurement and acquisition processes of SMC are pretty challenging. Although conventional measurement methods, such as time-domain reflectometry and gravimetric technique, may yield relatively precise SMC values at monitoring sites, they can hardly provide soil moisture information in the case of large areas, making it difficult to describe the spatial heterogeneity pattern of soils. In addition, such field measurements require a considerable workforce and lead to the deterioration of the local soil environment [4]. Remote sensing (RS) techniques have been rapidly developed in recent decades, featuring fast data acquisition and low effort consumption in their application to land surface investigation. Among other RS techniques, synthetic aperture radar (SAR) has been proven to be promising. Apart from such optical sensors, the SAR can collect ground surface information even at night and under cloudy weather conditions. The competitive penetrating power and the direct relationship between SMC and the SAR observations also make the estimation of SMC much more reliable [5]. Researchers have fully exploited this advantage; therefore, the SAR has been extensively employed for SMC retrieval [6–9].

Regarding microwave data, theoretical and semi-empirical models have been established for SMC estimation, such as the integral equation model (IEM) [10], advanced integral equation model (AIEM) [11], Oh model [12], Dubois model [13], Michigan microwave canopy scattering model (MIMICS) [14], water-cloud model (WCM) [15], and tau–omega model [16]. In applying these microwave models, considering the impact of land surface vegetation on microwave RS data [17], the effect of vegetation should be accurately quantified for a more precise SMC estimation. Because optical RS data are more sensitive to land surface vegetation, the combination of optical and microwave detection has emerged as an intuitive approach. Instead of deploying a single model for SMC retrieval, researchers have attempted to modify the original models by integrating them with optical information, hence carrying out tasks such as synergistic SMC inversion using both optical and SAR images [18–21]. Zhang et al. [22] built a radar backscattering coefficient database based on advanced integral equation model (AIEM) simulation, eliminated the vegetation effect using the WCM, and acquired the SMC by minimizing the difference between the observed bare soil backscattering coefficient and the simulated one. Han et al. [23] put forward a model-coupling method using GF-3 and GF-1 data by incorporating a series of models and achieved high-precision soil moisture mapping. Khabazan et al. [24] compared the capabilities of the IEM, Oh model, and Dubois model for surface soil moisture retrieval with C-band and L-band data to analyze the different conditions of vegetation land cover systematically. Overall, the application of theoretical and semi-empirical models can help represent the physical transmission processes more accurately. However, there are evident drawbacks. Most of the models stated above have complex structures and variables. Determining the values of some of these variables, such as the surface roughness and vegetation water content, requires laborious field experiments, and precise outcomes can hardly be ensured [23,25].

As a non-linear empirical model, the artificial neural network (ANN) can build an implicit relationship between input data and output targets, and it has been proven effective for SMC retrieval [26]. Studies on SMC estimation using ANNs with microwave and optical RS data have been conducted. For example, Baghdadi et al. [27] combined Radarsat-2 and Landsat data and inputted them to an ANN for simultaneous SMC and leaf area index estimation; the merits and demerits of radar data in dual- and full-polarization modes were also highlighted. El Hajj et al. [28] mainly focused on agricultural areas and depicted high-resolution SMC maps of bare and vegetation-covered farmlands using the backscattering coefficient and normalized difference vegetation index (NDVI) as input parameters. El Hajj et al. [29] combined "vegetation descriptors" derived from optical images and backscattering coefficients as ANN training and testing samples, and three different inversion configurations were compared in terms of their performances.

As we know, samples are the key elements of ANN. To obtain ideal retrieval results, both the quantity and quality of the samples for ANN training and testing should be guaranteed. That is to say, not only should the sample pool be large enough, but also the input parameters of the samples should be inclusive of the features that are helpful to accurate SMC retrieval. As for the quantitative optimization, sufficient samples are conducive to improving the training accuracy and representing various geographical situations [30–32]. In many previous studies, when choosing samples, it was often required that the data of each monitoring site in the entire research area be "complete" at one specific time, entailing remote sensing images and in situ measurements of sound quality [28,33–35]. However, such conditions are hard to meet.

For one thing, the use of optical images is associated with contamination from clouds, thick fogs, and mists [36], which may lead to a shortage of optical RS data. For another, there are temporal discrepancies between in situ measurements in a study area because instrument malfunctions make it impossible to acquire data of some parts of the monitoring sites in specific periods, which may also lead to the shortage of in situ data [37]. Therefore, gathering enough samples for ANN training and testing is difficult. As for qualitative optimization, it is of importance to determine the input parameters of ANN. In previous studies, some common variables, including the radar incidence angle, VH/VV backscattering coefficients, and NDVI, were investigated about the effectiveness of being used as inputs of the ANN for SMC retrieval [27,33]. In fact, in addition to these commonly considered ones, variables about other factors, such as local land use, topography, and phenology, can also be influential in local soil moisture and deserve to be given close attention.

To address the problem of quantitative optimization, a novel sparse sample exploitation (SSE) method was proposed, whereby a part of the samples that were otherwise excluded could be sufficiently utilized and incorporated into the SMC retrieval procedure. To address the problem of qualitative optimization, we extended the array of input parameters of ANN for SMC retrieval. Apart from the radar incidence angle, VH/VV backscattering coefficients and NDVI, which were included in this paper as the basic input parameters, parameters such as LST, land cover type, elevation, slope, and data acquisition time, are likewise considered as the inputs of ANN in this paper. The sensitivity of SMC retrieval to these parameters was discussed.

The rest of this paper is organized as follows. In Section 2, the study area and raw data involved in this study are introduced in detail. In Section 3, the methodology of the SSE is described, the array of input parameters is specified, and the entire ANN-based SMC retrieval process is demonstrated. In Section 4, the results are discussed regarding the retrieval accuracy improvement brought by the SSE and the respective influences of the ANN input parameters and their combinations on SMC retrieval. Finally, Section 5 presents the conclusions drawn from the study results.

2. Study Area and Dataset

2.1. Study Area and Ground Truth Data

The study area is located in the eastern part of Austria (Figure 1). Compared with the Eastern Alps region in the middle and west of the country, the topography in the study area is flatter, but hilly terrain still exists. The winter is often cold, but temperatures can be relatively high in summer, and the continental climate features dominate, thus, the precipitation tends to be low [38]. The ground surface is prevalently covered by vegetation, and land use types mainly comprise cropland, forest, and grassland. The croplands are rainfed, and the staple crops are wheat and corn. Closed forests feature in the study area, with the fractional vegetation cover (FVC) > 0.4. The principal tree species contain oak, hornbeam, and beech. As for the hydrological conditions, surface water in this region is closely related to the groundwater [39].

Ground truth data come from The International Soil Moisture Network (ISMN), which was implemented in 2009 aiming exclusively to validate and calibrate SMC retrieval with RS techniques. The data are qualitatively controlled after collecting them from the networks and then distributed on the website portal (https://ismn.geo.tuwien.ac.at/, accessed on 20 January 2022) [40,41]. This study selected monitoring sites from WEGENERNET and GROW, two soil moisture networks in Austria. The WEGENERNET network is situated in Styria State with nine monitoring sites, and the GROW network is located in Lower Austria State with 13 monitoring sites. WEGENERNET is a durable network with relatively continuous SMC data acquisition dating from 2007. We adopted data from January 2016 to May 2020 for our research. In contrast, for GROW, the data were available only between May 2017 and June 2019 in an intermittent manner. The SMC data at a depth of 0–5 cm was chosen considering the detecting ability of remote sensing techniques used in this study. Table 1 shows the coordinates (latitude and longitude), the network, and each site's land cover type. As the ground truth data, the SMC observations were recorded with acquisition times in accordance with the corresponding acquisition times of the SAR images (described below).

Table 1. Information of monitoring sites in the study area.

#	Lat. and Long.	Network	Landcover	#	Lat. and Long.	Network	Landcover
1	46.91691° N 15.78112° E	WEGENERNET	farmland	12	48.15202° N 15.15303° E	GROW	farmland
2	46.97232° N 15.81499° E	WEGENERNET	farmland	13	48.15257° N 15.15104° E	GROW	farmland
3	46.99726° N 15.85507° E	WEGENERNET	farmland	14	48.15356° N 15.14857° E	GROW	farmland
4	46.98299° N 15.87115° E	WEGENERNET	farmland	15	48.15403° N 15.15299° E	GROW	farmland
5	46.93296° N 15.90710° E	WEGENERNET	farmland	16	48.15474° N 15.14844° E	GROW	farmland
6	46.93291° N 15.92462° E	WEGENERNET	grassland	17	48.15562° N 15.14804° E	GROW	farmland
7	46.97970° N 15.94122° E	WEGENERNET	grassland	18	48.15645° N 15.14799° E	GROW	farmland
8	46.92135° N 16.03337° E	WEGENERNET	farmland	19	48.15709° N 15.13658° E	GROW	farmland
9	46.93427° N 16.04056° E	WEGENERNET	farmland	20	48.15725° N 15.15149° E	GROW	farmland
10	48.15117° N 15.15417° E	GROW	farmland	21	48.15804° N 15.14731° E	GROW	farmland
11	48.15179° N 15.15424° E	GROW	farmland	22	48.18776° N 15.98071° E	GROW	grassland

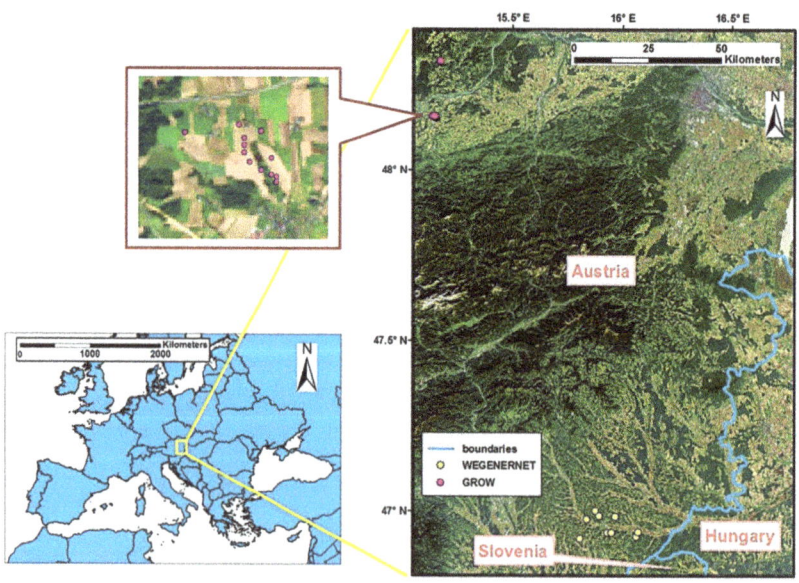

Figure 1. Location of the study area and monitoring sites.

2.2. Remote Sensing Data

The optical RS data employed in this study was obtained by the Landsat-8 satellite. Onboard the Landsat-8 satellite were two sensors, namely operational land imager (OLI) and thermal infrared sensor (TIRS), which help obtain multi-band data in the form of visible and infrared spectra with a fine resolution. We chose Landsat-8 images considering the synchronization of optical and land surface temperature data. The Landsat-8 images were utilized to extract optical data and calculate land surface temperature by the thermal infrared band. In this study, OLI-TIRS Level-1 images, downloaded from the United States Geological Survey (USGS) data archive (https://earthexplorer.usgs.gov/, accessed on 20 January 2022), were selected. The span period was from January 2016 to April 2020, and the spatial resolution of the images was 30 m. Based on the method described in Section 3.1, images were selected as long as they covered at least one monitoring site that was clear and without cloud obstruction on the date of image acquisition. The optical RS data were then subject to preprocessing procedures, including radiometric correction, FLAASH atmospheric correction, and band calculation. Finally, the NDVI and LST values in the monitoring sites were derived and recorded.

The microwave RS data employed in this study came from the Sentinel-1 satellite. Sentinel-1 provides VH and VV polarization modes C-band images with relatively high spatiotemporal resolution and radiometric accuracy. The imaging data played a crucial part in dynamic hydrological processes and SMC monitoring [42–46]. Here, the interferometric wave (IW) mode images were utilized with a spatial resolution of 10 m and a revisit period of 6 days. The images were downloaded from https://search.asf.alaska.edu/ (accessed on 20 January 2022) by courtesy of the Alaska Satellite Facility (ASF). We chose Sentinel-1 images of the study area based on their acquisition times to ensure that the radar and optical data were approximately synchronous in pairs. The temporally nearest microwave image was selected for each optical image collected already. It was confirmed that the acquisition times of the microwave images were less than five days away from the acquisition times of their optical counterparts. Furthermore, we checked the intervals between the acquisition times of each microwave image and their corresponding optical image to ensure that no precipitation event had occurred during the gaps. Subsequently, the microwave images underwent preprocessing as well. The preprocessing procedures included multi-looking, filtering, topographical correction, geocoding, and radiometric calibration. Finally, the backscattering coefficients in the VH and VV polarization modes of each monitoring site were derived, and the radar incidence angles were recorded.

Table 2 shows the acquisition times of the RS data used in this study. The dates of the radar and optical images were given in pairs.

Table 2. Acquisition times of RS images used in the study.

#	Dates of Radar Images	Dates of Optical Images	#	Dates of Radar Images	Dates of Optical Images	#	Dates of Radar Images	Dates of Optical Images
1	18 January 2016	18 January 2016	24	24 June 2017	22 June 2017	47	3 February 2019	4 February 2019
2	26 January 2016	27 January 2016	25	31 July 2017	31 July 2017	48	27 February 2019	27 February 2019
3	30 March 2016	31 March 2016	26	11 August 2017	9 August 2017	49	23 March 2019	24 March 2019
4	18 April 2016	16 April 2016	27	4 November 2017	4 November 2017	50	30 March 2019	31 March 2019
5	23 April 2016	23 April 2016	28	20 November 2017	20 November 2017	51	16 April 2019	16 April 2019
6	4 July 2016	5 July 2016	29	5 December 2017	6 December 2017	52	27 April 2019	25 April 2019
7	12 July 2016	12 July 2016	30	24 February 2018	24 February 2018	53	2 May 2019	2 May 2019
8	23 July 2016	21 July 2016	31	21 April 2018	22 April 2018	54	18 May 2019	18 May 2019
9	29 August 2016	29 August 2016	32	28 April 2018	29 April 2018	55	3 June 2019	3 June 2019
10	22 September 2016	23 September 2016	33	31 May 2018	31 May 2018	56	14 June 2019	12 June 2019
11	29 September 2016	30 September 2016	34	2 July 2018	2 July 2018	57	19 June 2019	19 June 2019
12	16 October 2016	16 October 2016	35	18 July 2018	18 July 2018	58	27 June 2019	28 June 2019
13	1 November 2016	1 November 2016	36	26 July 2018	27 July 2018	59	4 July 2019	5 July 2019
14	9 November 2016	10 November 2016	37	2 August 2018	3 August 2018	60	14 August 2019	15 August 2019
15	3 December 2016	3 December 2016	38	19 August 2018	19 August 2018	61	2 September 2019	31 August 2019
16	14 December 2016	12 December 2016	39	30 August 2018	28 August 2018	62	8 October 2019	9 October 2019
17	20 January 2017	20 January 2017	40	19 September 2018	20 September 2018	63	20 October 2019	18 October 2019
18	5 February 2017	5 February 2017	41	28 September 2018	29 September 2018	64	1 November 2019	25 October 2019
19	9 March 2017	9 March 2017	42	6 October 2018	6 October 2018	65	5 January 2020	6 January 2020
20	2 April 2017	3 April 2017	43	22 October 2018	22 October 2018	66	9 March 2020	10 March 2020
21	9 April 2017	10 April 2017	44	30 October 2018	31 October 2018	67	2 April 2020	2 April 2020
22	27 May 2017	28 May 2017	45	11 November 2018	7 November 2018	68	10 April 2020	11 April 2020
23	13 June 2017	13 June 2017	46	15 November 2018	16 November 2018	69	26 April 2020	27 April 2020

2.3. Auxiliary Data

The auxiliary data contained a digital elevation model (DEM) and land cover product. This study used DEM from Shuttle Radar Topography Mission (SRTM) downloaded from the USGS website (http://gdex.cr.usgs.gov/gdex/, accessed on 20 January 2022). The slope data were then derived from DEM using the "Slope" tool integrated into the ArcMap 10.5 software. Both the elevation and slope of each monitoring site were extracted and recorded. We obtained land cover data referring to "Global Land Cover with Fine Classification System at 30 m" (GLC_FCS30) downloaded from http://data.casearth.cn/ (accessed on 20 January 2022). The land cover types of the monitoring sites were collected. Because of the evident attenuation effect of dense vegetation canopies on C-band radar backscattering [47,48], we eliminated the monitoring sites located in the forests. Considering the subsequent operations of ANN training and testing, the land cover types were transformed into numerical data. "Cropland" and "Grassland" were substituted with "1" and "2," respectively.

2.4. Sample Pool

After the processing procedures, the data were used to form a collection of samples. If one monitoring site had "complete" data on one particular date, with optical RS data, microwave RS data, auxiliary data, and in situ SMC measurement all accessible, then the corresponding sample will be assembled. Each sample can be considered a 10-dimensional vector, comprising 9 parameters derived from RS and auxiliary data and one ground-truth SMC observation (specified below in Section 3.2). The samples were placed in the sample pool (635 in aggregate) and ready to be designated as training/validation/testing datasets in the subsequent ANN training and testing processes.

3. Methodology

3.1. Sample Quantity Optimization: Sparse Sample Exploitation

In this section, the SSE method is put forward in detail. In essence, the SSE is a sort of data expansion technique over the time scale. By taking full advantage of the available images and observations, it manages to gather more samples derived over a wider time frame, thereby transferring more valuable information into the sample pool, and helping to accomplish SMC retrieval with higher precision. Briefly, the SSE involves 2 steps:

1. For dates when the sky above the study area is clear and no in situ observation is absent, all the samples are recorded in the sample pool.
2. For dates when the study area is partially blocked by clouds or in situ observations are absent, the "sparse samples" with available optical, microwave data, and ground truth observations are recorded similarly in the sample pool.

To illustrate this method straightforwardly, we take Figure 2 as an example. In Figure 2, the images are the optical RS images covering the region of interest on six different dates, namely $d_1, d_2, \ldots, d_6$. Points A, B, C, and D denote the locations of the monitoring sites, of which the RS data and in situ SMC observations are expected. The points in pink indicate that in situ data are available, whereas the points in yellow indicate that in situ data are missing.

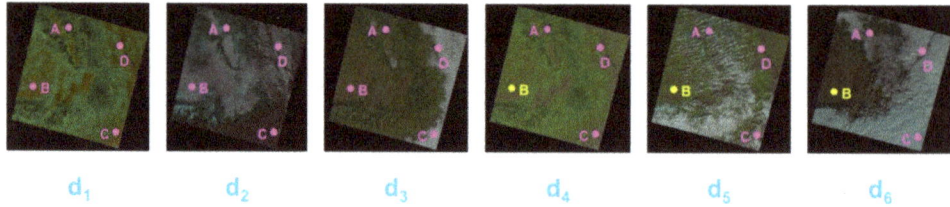

Figure 2. Schematic of sample collection process of SSE method.

As for d_1, the image is cloud-free, and every monitoring site has its SMC observations; hence, the samples derived from the four sites are all valid for the sample pool. For d_2 and d_3, the clouds start to interfere. For the traditional sample-picking method mentioned in previous studies, data in these dates should be dismissed because optical RS data corresponding to specific sites are unavailable, and not all sites have complete data. However, it can be found that points such as C in the image of d_2 and A and B in the image of d_3 are clear in optical RS images and can still yield complete data. The samples corresponding to these points are designated as "sparse samples." For the SSE method, these samples are considered to be included in the sample pool.

Similarly, for d_4, d_5, and d_6, when point B has no available in situ SMC observation due to, hypothetically, instrument power failure, the data from point B are consequently eliminated. For the traditional sample-picking method, the whole data in these dates will again be abandoned due to the data's incompleteness. For the SSE method, however, because samples can still be formed from the complete data of points A, C, D on date d_4 and D on date d_5, these samples are thus collected in the sample pool. On date d_6, no sample can be collected.

Table 3 lists the comparison of sample selection via the traditional and SSE methods. It is evident that for the traditional method, the quantity of samples is severely limited, owing to the requirement of data completeness in the entire study area when collecting samples. Therefore, the samples from the four points in d_1 will be the only valid ones. In contrast, the SSE method manages to enlarge the sample pool by making full use of the sparse samples. In this study, a total of 635 samples can be collected by the SSE method, but only 264 out of the 635 samples can be collected if the traditional method is implemented.

Table 3. Comparison of samples selection via traditional method and SSE method based on Figure 2.

Date	d_1	d_2	d_3	d_4	d_5	d_6
traditional method	ABCD	-	-	-	-	-
SSE method	ABCD	C	AB	ACD	D	-

3.2. Sample Quality Optimization: Input Parameter Selection

For more accurate SMC retrieval results, the combination of inputs of the ANN is supposed to contain enough variables to represent the main features [32]. In addition to these commonly discussed parameters, including radar backscattering coefficient, radar incidence angle, and NDVI, some other SMC-related variables, such as data acquisition time, land surface temperature, elevation, slope, and the land cover type, are worth considering as well.

1. Data acquisition time: the data acquisition time was strongly correlated to the surface soil hydraulic conductivities [49]. Meanwhile, the phenological traits of vegetation follow a circulation of alteration on an annual basis [50,51], which plays an essential role in vegetation effect elimination during the process of SMC retrieval in vegetation-covered areas.
2. Land surface temperature (LST): previous studies have proven the correlation of variation between the SMC and temperature vegetation dryness index (TVDI) [52,53]. The synergy of LST and vegetation indices (such as NDVI) on SMC retrieval has also been stressed [54–56].
3. Elevation and slope: soil moisture was closely related to the local topographical heterogeneity. The landscape shapes physically controlled the hydrological processes and SMC time stability [57,58], with upland water moving to the groundwater and lowland water coming from the groundwater, and water content increasing from the top to the bottom of a slope in a non-linear pattern [59,60].
4. Land cover type: the land use was analyzed as a factor influencing soil hydraulic attributes and SMC distribution. For example, human activities such as grazing,

plowing, and urban development impact the macropores and the continuity of the macropore network of soil, thus altering the mode of local soil water supply and SMC distribution [49,61].

By referring to these existing studies, we here selected 9 parameters derived from the RS and auxiliary data, namely the data acquisition time (month), radar incidence angle (θ), VH backscattering coefficient (σ_{VH}), VV backscattering coefficient (σ_{VV}), NDVI, LST, elevation, slope, and land cover type, as inputs of the ANN. The acquisition of the parameters was explained in the previous sections, and we introduced the ordinal number of the data acquisition month to present the data acquisition time for the ANN calculations.

Furthermore, to investigate the effects of the input parameters and their combinations on SMC retrieval, a total of 7 scenarios were considered, as shown in Table 4. In Scenario 0, all the 9 parameters were taken into account; in Scenario 1, the 4 commonly discussed parameters, i.e., θ, σ_{VH}, σ_{VV}, and NDVI, were included as the basic inputs; in Scenario 2–6, the other 5 parameters were added individually into the basic input parameters. By comparing the SMC retrieving results of these scenarios, the sensitivity of SMC to specific input parameters was assessed and analyzed.

Table 4. Scenarios of input parameter combinations for ANN SMC retrieval.

Scenario	Input Parameters
0	θ, σ_{VH}, σ_{VV}, NDVI, month, LST, elevation, slope, land cover
1	θ, σ_{VH}, σ_{VV}, NDVI
2	θ, σ_{VH}, σ_{VV}, NDVI, month
3	θ, σ_{VH}, σ_{VV}, NDVI, LST
4	θ, σ_{VH}, σ_{VV}, NDVI, elevation
5	θ, σ_{VH}, σ_{VV}, NDVI, slope
6	θ, σ_{VH}, σ_{VV}, NDVI, land cover

3.3. ANN and SMC Retrieval

After selecting data using the SSE method and determining input parameters, a group of samples was obtained. The ANN approach was then adopted to retrieve the SMC. The ANN is the abstraction of the neural network of human brains from the perspective of data processing. With the nodes of neurons connected sequentially, the ANN is organized into a layered structure. As the data are inputted to the ANN, neurons perform weighted computations and pass on the results to other neurons until reaching the output layer, which yields the final result [31,62]. In terms of SMC estimation, the ANN approach provides a better solution than conventional theoretical and semi-empirical models owing to its capacity for describing non-linear relationships [63].

Moreover, the independence of the ANN from a priori knowledge and radiative transfer information relieves the estimation process of explicit physical mechanism and complicated parameters, and the parameters or combinations involved can be more flexible [64,65]. Figure 3 shows the flowchart of the SMC retrieval process developed in this study, with all 9 input parameters mentioned above utilized. Here, a feed-forward perceptron model was employed, and the ANN has a 3-layer structure comprising input, hidden, and output layers.

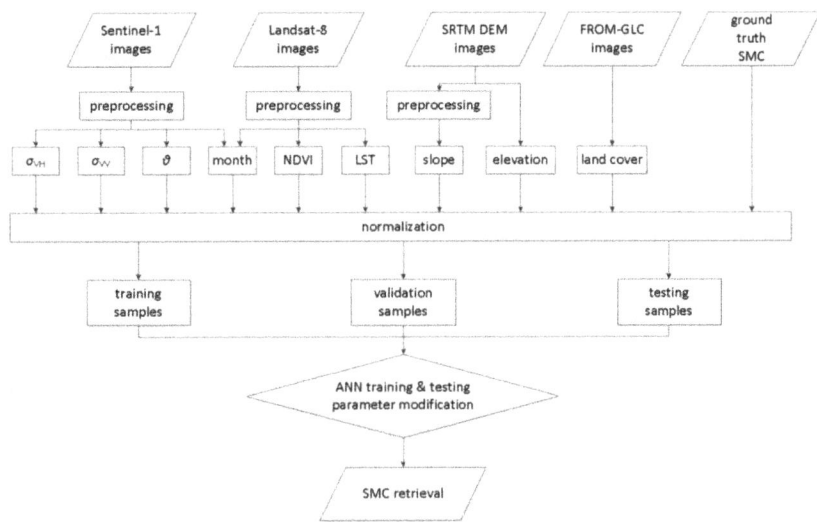

Figure 3. Flowchart of the SMC retrieval by ANN.

The number of neurons in the hidden layer is another essential characteristic. Too few or too many neurons may lead to underfitting or overfitting, thus affecting the accuracy of SMC retrieval [66]. In this study, 10 neurons were contained in the hidden layer, determined through a trial-and-error method. The SMC ground-truth observations were set as outputs. Next, both the inputs and outputs were normalized to 0 to 1 based on their respective range of values. The normalization procedure can improve the training speed and help prevent the outcomes from getting stuck in local minimums to a certain extent [66].

The samples in the sample pool were randomly partitioned into training, validation, and testing datasets in proportions of 80%, 10%, and 10%, respectively, for the following ANN training and testing. The purpose of ANN training was to iteratively modify the weights of correlation between the inputs and outputs thus that the differences can be minimized. The training process was accomplished with training samples as well as validation samples. The validation samples here were aimed at ensuring the generalization capacity of the ANN and avoiding overfitting during the training process [64]. The Levenberg–Marquardt method was chosen as the training algorithm. This method provides an optimal solution for a certain minimizing problem [67]. Numerous iterations were conducted in search of an optimal solution during the training process, and the maximum number of iterations was set as 1000. The training process was stopped either when the generalization capacity of the ANN began to level off, which indicates that more training processes cannot improve the accuracy, or when the maximum number of iterations was reached. The testing process was performed with the testing samples by comparing the ground truth SMC with the estimated SMC derived from the corresponding inputs using the trained ANN. The training and testing processes were conducted in MATLAB, and the well-trained ANN was deployed for SMC mapping in the entire study area.

3.4. Statistical Metrics

The retrieval accuracy was evaluated using 2 statistical metrics: the root-mean-square error (RMSE) and the correlation coefficient (*r*), which can be expressed as follows:

$$\text{RMSE} = \sqrt{\frac{1}{n}\sum_{i=1}^{n}\left(SMC_i - S\hat{M}C_i\right)^2} \quad (1)$$

$$r = \frac{\sum_{i=1}^{n}(SMC_i - \overline{SMC})(S\hat{M}C_i - \overline{S\hat{M}C})}{\sqrt{\sum_{i=1}^{n}(SMC_i - \overline{SMC})^2}\sqrt{\sum_{i=1}^{n}(S\hat{M}C_i - \overline{S\hat{M}C})^2}} \quad (2)$$

where SMC_i and $\overline{SMC}$ represent the ith sample's ground-truth and mean ground-truth SMC values of the relevant samples; $S\hat{M}C_i$ and $\overline{S\hat{M}C}$ represent the ith sample's estimated SMC value and the mean estimated SMC values of all the relevant samples, respectively. The RMSE and r were calculated based on the training, validation, and testing results.

4. Results and Discussion

4.1. Evaluation of Overall Accuracy

First, the overall accuracy of the ANN was evaluated. All the 635 samples were used and divided into training, validation, and testing datasets. The training/testing process was conducted once, with all nine parameters being involved as the inputs of the ANN (Scenario 0 in Section 3.2). Figure 4 shows the scatter plots of the SMC estimation results for the training, validation, testing datasets, and the entire samples. The correlation coefficient (r) values were also given above each plot. The results were promising, with the testing dataset r and overall r reaching 0.85. Table 5 shows the corresponding RMSE values, which seem favorable, with RMSE values of 0.048 m^3m^{-3}, 0.054 m^3m^{-3}, and 0.052 m^3m^{-3} on the training, validation, and testing datasets, respectively. The ground-truth SMC values range from 0.024 to 0.477 m^3m^{-3} with an average value of 0.336 m^3m^{-3}, whereas the estimated SMC values ranged from 0.039 to 0.470 m^3m^{-3} with the average value of 0.335 m^3m^{-3}. In comparison with the work conducted by Alexakis et al. [33], our study quantitatively expanded the sample pool and qualitatively improved the accuracy of SMC retrieval, with the testing dataset r rising from 0.508 to 0.848 and the overall r rising from 0.803 to 0.850.

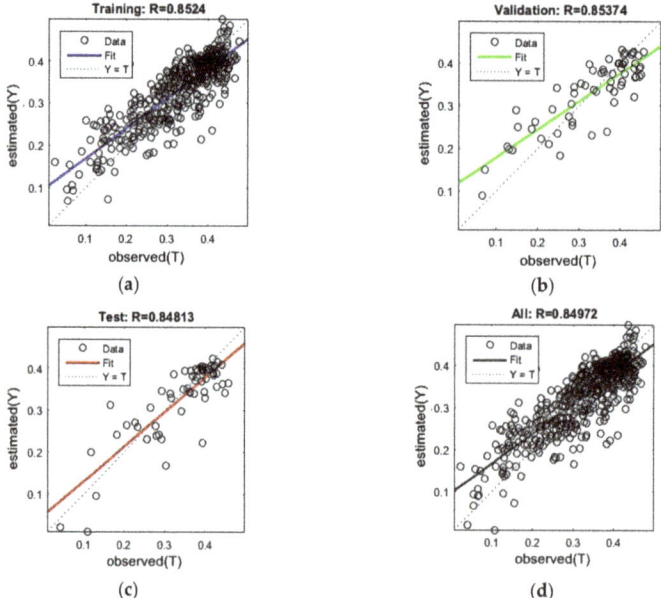

Figure 4. Scatter plots of SMC estimations for training (**a**), validation (**b**), testing dataset (**c**), and the entire samples (**d**). Corresponding correlation coefficients are placed above each plot.

Table 5. RMSE values on training, validation, and testing datasets.

Dataset	Training	Validation	Testing
RMSE (m^3m^{-3})	0.048	0.054	0.052

4.2. Evaluation of SSE Method

The effectiveness of the proposed SSE method was evaluated by comparing the SMC retrieval results with and without sparse sample exploitation. Concerning the study by Holtgrave et al. [68], considering that different schemes of the random division of training/validation/testing datasets might give rise to different SMC retrieval outcomes, we repeated the ANN training/testing process 50 times to assess the average SMC retrieval performance. Since the accuracy of the testing datasets was more worthwhile in terms of the ANN performance, only the statistical metrics on the testing datasets were discussed in the remainder of the paper. Similarly, all nine parameters were set as the inputs of the ANN (Scenario 0 in Section 3.2). Table 6 lists the average RMSE and r values on the testing datasets for SMC retrieval with and without the SSE method.

Table 6. Statistical metrics on testing dataset for SMC retrieval with and without the SSE method.

	Without SSE	With SSE
RMSE (m^3m^{-3})	0.090	0.068
r	0.635	0.736

The results indicate a striking increase in the SMC retrieval accuracy when introducing the SSE method, with the RMSE decreasing from 0.090 m^3m^{-3} to 0.068 m^3m^{-3} and r increasing from 0.635 to 0.736. The main reason could be the efficient utilization of the dismissed samples in the images where optical or in situ data were partially missing, and the sample pool could consequently be expanded. For the empirical SMC retrieval methods, such as ANN, large samples were required during the training process. Thus, the precise relationship between the inputs and outputs could be established [69]. Although the SSE did not conventionally ensure that all the monitoring sites had identical time series of data acquisition, it nonetheless enlarged the sample capacity by gathering more samples derived over a broader period. Meanwhile, the information provided by these samples could be made full use of, and the features of the training dataset were enriched, consequently enhancing the representativeness of the samples as well as the stability of the ANN. Therefore, the training precision of the ANN was improved.

4.3. Sensitivity Analysis of Input Parameters

As mentioned in Section 3.2, several scenarios were considered for the sensitivity analysis of different input parameters. Table 7 shows scenarios 1–6 of input parameter combinations and their statistical metrics for SMC retrieval. During the sensitivity analysis, all the 635 samples were employed. Similarly, the ANN training/testing process was repeated 50 times after random divisions of each scenario's training/validation/testing datasets. The mean statistical metrics on the testing dataset representing the average performances were evaluated. As listed in Table 7, the first scenario was the combination of basic input parameters, including the VH/VV backscattering coefficients, NDVI, and radar incidence angle. For the rest of the scenarios, the data acquisition time, LST, elevation, slope, and land cover were added individually to the basic combination.

Table 7. Scenarios of different input parameter combinations and corresponding performances of SMC retrieval. Ticks indicate that the parameters are chosen as input scenarios systems.

Scenarios	Input Parameters									Statistical Metrics	
	Month	σ_{VH}	σ_{VV}	NDVI	LST	Elevation	Slope	Land Cover	θ	RMSE ($m^3 m^{-3}$)	r
1		✓	✓	✓					✓	0.089	0.588
2	✓	✓	✓	✓					✓	0.078	0.637
3		✓	✓	✓	✓				✓	0.084	0.616
4		✓	✓	✓		✓			✓	0.070	0.689
5		✓	✓	✓			✓		✓	0.083	0.639
6		✓	✓	✓				✓	✓	0.091	0.599

4.3.1. Data Acquisition Time

Comparing scenarios 1 and 2: after the data acquisition time (i.e., the data acquisition month in this study) was added as the input parameter, r increased from 0.588 to 0.637, and RMSE declined from 0.089 $m^3 m^{-3}$ to 0.078 $m^3 m^{-3}$.

The influences of adding data acquisition time on each sample were investigated for further analysis. For each sample participating in SMC retrieval, "accuracy improvement." I_i was proposed with the expressions below:

$$I_i = \varepsilon_{i_{basic}} - \varepsilon_{i_{new}} \tag{3}$$

$$\varepsilon_{i\_basic} = \frac{|SMC_i - SMC_{i\_basic}|}{SMC_i} \times 100\% \tag{4}$$

$$\varepsilon_{i\_new} = \frac{|SMC_i - SMC_{i\_new}|}{SMC_i} \times 100\% \tag{5}$$

where SMC_i denotes the ground-truth SMC of the ith sample, SMC_{i_basic} denotes the estimated SMC of the ith sample with only basic input parameters as the ANN inputs, and SMC_{i_new} denotes the estimated SMC of the ith sample with a new parameter incorporated into the basic ones as the ANN inputs. ε_{i_basic} and ε_{i_new} denote the corresponding relative errors of the ith sample. I_i, the value of accuracy improvement is the difference between the two errors. When I_i is positive, it means that the error of SMC retrieval with the new input parameter is lower than that of SMC retrieval by basic input parameters, indicating a real accuracy improvement of SMC retrieval; conversely, when I_i is negative, it means adding the new input parameter in the ANN brings about a worse result. In addition, $\bar{I}$ was used to denote the average accuracy improvement of corresponding i samples:

$$\bar{I} = \sum_i I_i \tag{6}$$

Table 8 lists the results of accuracy improvement of SMC retrieval by adding data acquisition time as the input parameter. Because of the distinctive phenological pattern of cropland, we paid extra attention to the cropland samples. For all 635 samples, the number of samples with $I_i > 0$ reached 372. Among these samples, 287 were cropland samples, accounting for 77.2%. As for $\bar{I}$, the average accuracy improvement of all samples was 6.64%, whereas for cropland samples, the $\bar{I}$ was 5.66%, accounting for 85.2% of the total gain. These results indicate that the addition of data acquisition time as the input parameter improves the SMC retrieval performance of cropland samples, thus driving up the retrieving accuracy of the entire samples.

Table 8. Accuracy improvement by adding data acquisition time over total samples and cropland samples.

	Of All Samples	Of Cropland Samples	Percentage of Cropland Samples
number of samples with $I_i > 0$	372	287	77.2%
$\bar{I}$	6.64%	5.66%	85.2%

In fact, the season or data acquisition time was strongly correlated to the plant growth condition and the corresponding SMC ground-truth data in the vegetation-covered regions. Here, we chose three monitoring sites of which the SMC observations were continuous and long-lasting, and the SMC time series are displayed in Figure 5. These SMC time series generally present a periodic pattern of annual variation, respectively. For site #5, SMC observations are high in winter and spring, begin to fluctuate in summer and keep relatively low in August and September. For site #7, the fluctuations in summer were more drastic, and sharp declines occurred around May in three consecutive years (2016, 2017, and 2018). For site #9, the SMC variation is not so regular; however, some annual patterns, such as the plateaus in February and March, the significant dips after summer, and the rises in November, are still observable.

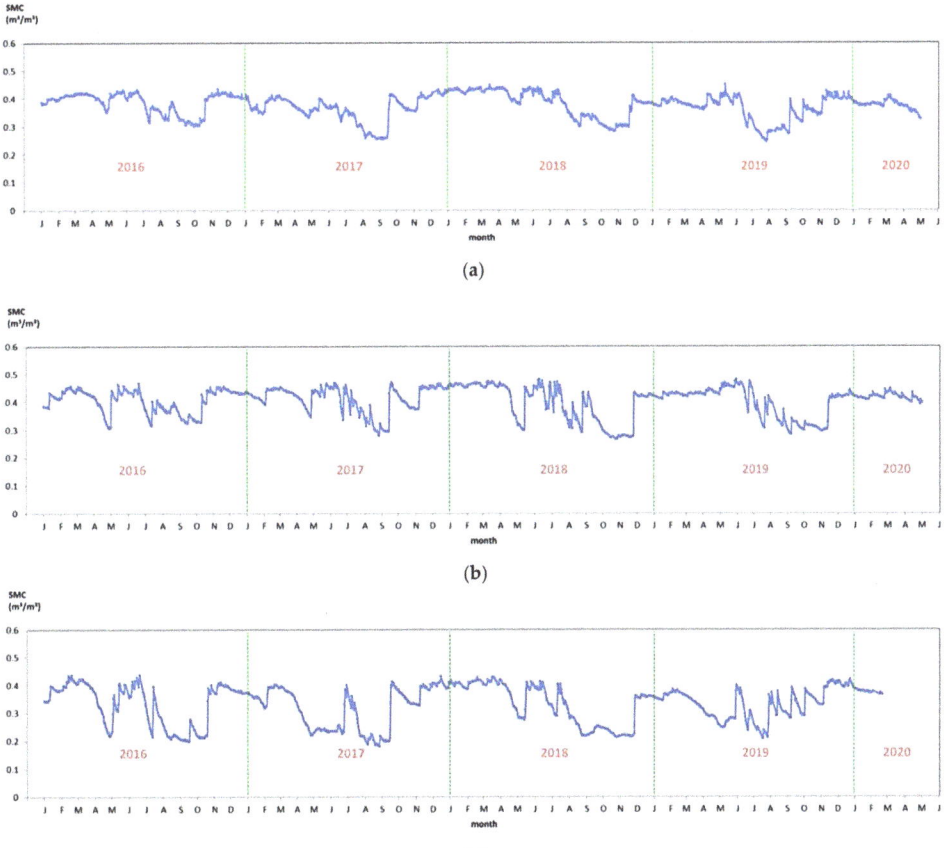

Figure 5. Ground-truth SMC time series of some monitoring sites: (**a**) Site #5 (cropland), (**b**) Site #7 (grassland), (**c**) Site #9 (cropland).

Hence, as an input parameter, the data acquisition time contributes to a more delicate description of the vegetation phenological features, and better SMC retrieval outcomes can thus be obtained.

4.3.2. LST

Comparing scenarios 1 and 3: The addition of LST helped increase r to 0.616 and decrease the RMSE to 0.084 m^{3}m^{-3}.

Sandholt et al. [53] defined the temperature vegetation dryness index (TVDI) as:

$$TVDI = \frac{LST - LST_{min}}{LST_{max} - LST_{min}} \quad (7)$$

where LST_{min} and LST_{max} are the minimum and the maximum land surface temperatures, respectively, corresponding to a specific NDVI value in the LST-NDVI space. The correlation of TVDI and SMC suggests the rationality of SMC retrieval with the synergy of NDVI and LST.

After the addition of LST as the input parameter, for those samples with positive I_i, the scatter plot of the relationship between TVDI and ground-truth SMC is shown in Figure 6, and the negative correlation is evident.

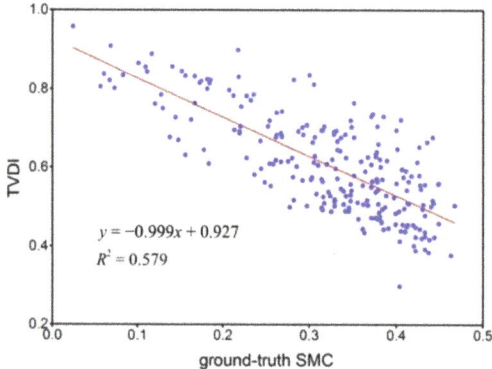

Figure 6. Scatter plot of the relationship between TVDI and ground-truth SMC for those samples with positive I_i after the addition of LST as the input parameter.

Furthermore, Schmugge [70] claimed that the soil's surface temperature was the function of both internal and external factors. The thermal conductivity and heat capacity, which belonged to the internal factors, both increased with the rise of SMC. As a factor reflecting the intensity of evapotranspiration, the spatial distribution of the LST varied significantly with the land surface water.

In this study, the ANN managed to retrieve the SMC with higher accuracy with the aid of the LST. This result was further proof for the conclusions made in the studies mentioned above in Section 1.

4.3.3. Elevation and Slope

Comparing scenarios 1, 4, and 5: the retrieval accuracy improved remarkably after incorporating the elevation into the input parameter pool. The r-value increased to 0.689, and the RMSE decreased to 0.070 m^{3}m^{-3}. The slope promoted accuracy, with r up to 0.639 and RMSE falling to 0.083 m^{3}m^{-3}.

For further explanation, the accuracy improvement of those samples improving SMC retrieval accuracy ($I_i > 0$) by virtue of incorporating topographic factors is illustrated in Figures 7 and 8. Figure 7 displays the accuracy improvement by adding elevation as the input parameter for different samples of elevation and slope values. In contrast, Figure 8 indicates the accuracy improvement by adding slope as the input parameter in relation to samples of different elevation and slope values. In each figure, samples are arranged in descending order of their corresponding I_i

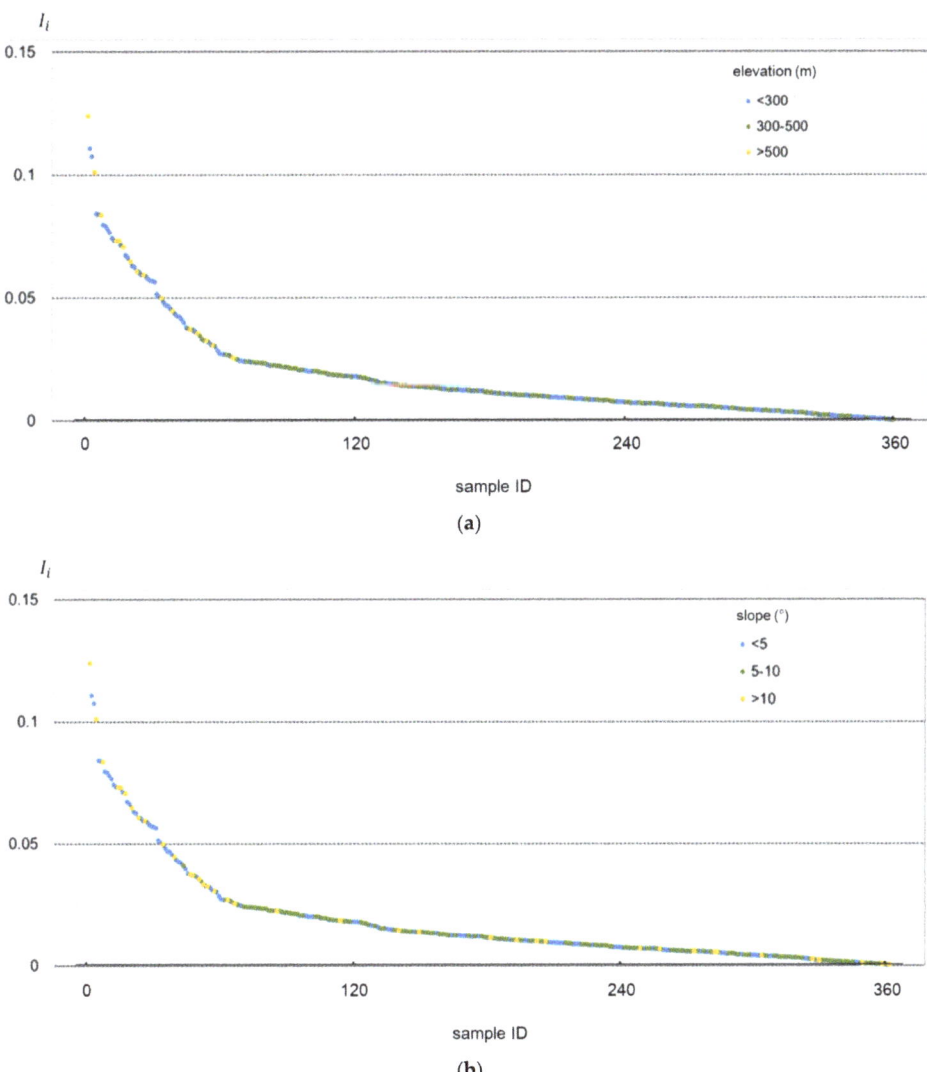

Figure 7. The accuracy improvement of samples by adding elevation as the input parameter. In (**a**), the samples are categorized into three groups by elevation and (**b**) by the slope.

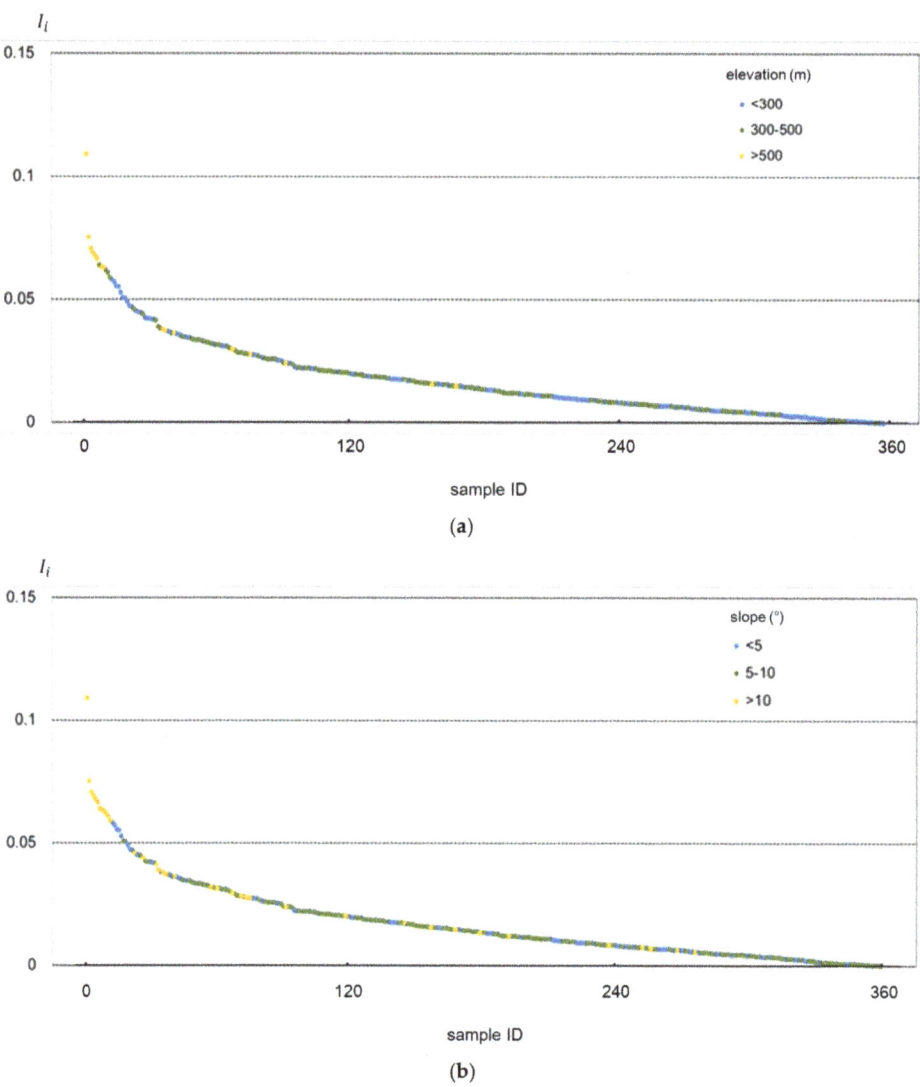

Figure 8. The accuracy improvement of samples by adding slope as the input parameter. In (**a**), the samples are categorized into three groups by elevation and (**b**) by the slope.

It can be observed from Figures 7 and 8 that, no matter for adding elevation or adding slope as the input parameter, samples with relatively higher elevation (>500 m) and steeper slope (>10°) tended to yield better accuracy improvement results, with most of the corresponding samples gathering in the left of the figures.

Previous studies claimed that local topographical heterogeneity reinforced the variation in the soil moisture distribution. Due to gravity and overland flow, locations with a high elevation and steep slope were more prone to SMC change. In contrast, low and flat locations were more inclined to SMC invariability [57,58,60,71]. Analogous to these studies, the difference in the topography of our study area was noticeable enough to impact the soil moisture distribution as well. Therefore, taking the elevation and slope into consideration during SMC retrieval was rational.

4.3.4. Land Cover Type

Comparing scenarios 1 and 6: when land cover type was considered an input parameter, the accuracy of SMC retrieval failed to improve as expected. Despite the existing studies emphasizing the influences of land use on SMC distribution [72,73], the outcome of ANN-based SMC retrieval with the assistance of land cover type did not show any improvement. The r slightly increased to 0.599, whereas the RMSE rose marginally to 0.091 m^{3}m^{-3}. This was probably attributed to the poor land cover categorization of the samples. In this study, after eliminating forest, the samples merely fell into two land cover types; in practice, the ground-truth geographical conditions of the study area could be quite intricate. The land cover categorization could not adequately improve the accuracy of SMC retrieval, and a refined land cover map was required.

4.4. SMC Mapping

Figure 9 shows the map of the SMC retrieval outcome at a depth of 5 cm for the study area on 6 October 2018. Considering the representativeness of the training samples, regions of forests and high elevation (>500 m) were masked. In addition, the water bodies and residential areas where no soil existed were masked as well. The gray patches indicate masked regions. The soil moisture distribution was visually plausible based on the map, with shades of blue and green (high SMC) mainly representing cropland and grassland, while red or yellow ones (low SMC) representing relatively bare land.

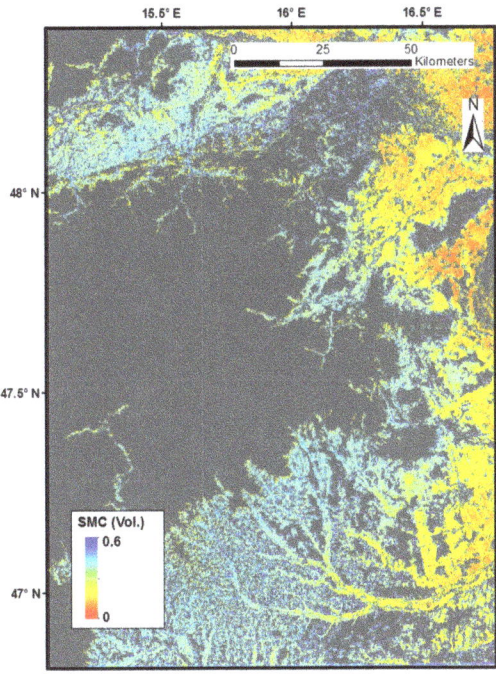

Figure 9. Volumetric SMC mapping of the study area.

5. Conclusions

An ANN approach for SMC retrieval using microwave RS data (Sentinel-1 SAR images) and optical RS data (Landsat-8 images) was demonstrated, and a novel SSE methodology was proposed. With the SSE, the problem of data deficiency due to cloud contamination in optical images and in situ instrument malfunction was resolved. Complete data were fully utilized in the ANN training/testing procedure, and the enlarged sample pool was beneficial to SMC retrieval with high precision.

The sample volume could be increased from 264 to 635 by the SSE, and the SMC retrieval accuracy was significantly enhanced. Regarding the average statistical metrics corresponding to 50 ANN training/testing iterations, r increased from 0.635 to 0.736, and the RMSE decreased from 0.090 m^3m^{-3} to 0.068 m^3m^{-3}.

A couple of variables were newly considered about the inputs of ANN for SMC retrieval. As for the sensitivity analysis of the ANN inputs, the parameters, such as the elevation, slope, data acquisition time, LST, and the land cover type, influenced the SMC retrieval accuracy to varying degrees. Among these parameters, the elevation had the most significant impact on the results, as evidenced by the increase in the r-value from 0.588 to 0.689 and the decrease in the RMSE from 0.089 m^3m^{-3} to 0.070 m^3m^{-3}. Other parameters were also advantageous to SMC retrieval, except for the land cover type, which barely promoted the accuracy due to the lack of refined land cover categorization. Notably, overall, the SMC retrieval statistical metrics of Scenario 0, where all nine relevant input parameters were considered (the situation "with SSE" discussed in Section 4.2), proved to be much more favorable than those of the scenarios analyzed in Section 4.3. This signifies that, to some degree, more relevant input parameters tend to improve retrieval accuracy.

The study results show that SSE is a promising method for ANN-based SMC retrieval. However, certain limitations need to be addressed. Because study areas overseas are beyond our reach and field surveys on topography and land cover are challenging to implement, the inconsistency between ground truth data and RS data cannot be excluded. The consequent biases in the SMC retrieval are inevitable. Moreover, the SMC mapping lacks additional in situ data for further validation.

We plan to focus on sample exploitation over the spatial dimension in the future. In other words, for study areas without sufficient samples for ANN training, synchronous data from another site of geographical resemblance with sufficient samples will be considered for SMC retrieval. The accuracy and conditions for the application of this method will be investigated. Additionally, we intend to utilize state-of-the-art RS data from Chinese satellites, such as GF-3 and GF-1, and explore their applicability in SMC retrieval problems.

Author Contributions: Conceptualization, Q.L.; Y.Z. (Yulin Zhan) and X.C.; data curation, Q.L.; formal analysis, Y.L., C.W., Y.Z. (Yin Zhang) and D.W.; funding acquisition, X.G. and T.Y.; methodology, Q.L., Y.Z. (Yulin Zhan) and X.C.; writing—original draft, Q.L.; writing—review and editing, Y.Z. (Yulin Zhan), F.M., X.C., Y.L., C.W., X.G. and T.Y. All authors have read and agreed to the published version of the manuscript.

Funding: This research was funded by National Key R&D Program of China (Grant No. 2020YFE0200700, 2019YFE0127300), Major Special Project-the China High-Resolution Earth Observation System (Grant No. 30-Y30F06-9003-20/22), and the Guangxi Science and Technology Development Project of Major Projects (Guike AA18118048-2).

Institutional Review Board Statement: Not applicable.

Informed Consent Statement: Not applicable.

Data Availability Statement: The data that support the findings of this study are available from the corresponding author upon reasonable request.

Acknowledgments: We thank Fengli Zhang and Xiaofei Mi for the special efforts they have made. We also sincerely thank the editors and reviewers.

Conflicts of Interest: The authors declare no conflict of interest.

References

1. Petropoulos, G.P.; Ireland, G.; Srivastava, P.K.; Ioannou-Katidis, P. An appraisal of the accuracy of operational soil moisture estimates from SMOS MIRAS using validated in situ observations acquired in a Mediterranean environment. *Int. J. Remote Sens.* **2014**, *35*, 5239–5250. [CrossRef]

2. Hirschi, M.; Seneviratne, S.I.; Schär, C. Seasonal variations in terrestrial water storage for major midlatitude river basins. *J. Hydrometeorol.* **2006**, *7*, 39–60. [CrossRef]
3. Robinson, D.A.; Campbell, C.S.; Hopmans, J.W.; Hornbuckle, B.K.; Jones, S.B.; Knight, R.; Ogden, F.; Selker, J.; Wendroth, O. Soil moisture measurement for ecological and hydrological watershed-scale observatories: A review. *Vadose Zone J.* **2008**, *7*, 358–389. [CrossRef]
4. Seneviratne, S.I.; Corti, T.; Davin, E.L.; Hirschi, M.; Jaeger, E.B.; Lehner, I.; Orlowsky, B.; Teuling, A.J. Investigating soil moisture–climate interactions in a changing climate: A review. *Earth-Sci. Rev.* **2010**, *99*, 125–161. [CrossRef]
5. Wang, J. The dielectric properties of soil-water mixtures at microwave frequencies. *Radio Sci.* **1980**, *15*, 977–985. [CrossRef]
6. Jackson, T.J.; Bindlish, R.; Cosh, M.H.; Zhao, T.; Starks, P.J.; Bosch, D.D.; Seyfried, M.; Moran, M.S.; Goodrich, D.C.; Kerr, Y.H.; et al. Validation of soil moisture and ocean salinity (SMOS) soil moisture over watershed networks in the US. *IEEE Trans. Geosci. Remote Sens.* **2012**, *50*, 1530–1543. [CrossRef]
7. Sahebi, M.R.; Angles, J. An inversion method based on multi-angular approaches for estimating bare soil surface parameters from RADARSAT-1. *Hydrol. Earth Syst. Sci.* **2010**, *14*, 2355–2366. [CrossRef]
8. Zhang, X.; Chen, B.; Fan, H.; Huang, J.; Zhao, H. The potential use of multi-band SAR data for soil moisture retrieval over bare agricultural areas: Hebei, China. *Remote Sens.* **2016**, *8*, 7. [CrossRef]
9. Gao, Q.; Zribi, M.; Escorihuela, M.J.; Baghdadi, N. Synergetic use of Sentinel-1 and Sentinel-2 data for soil moisture mapping at 100 m resolution. *Sensors* **2017**, *17*, 1966. [CrossRef]
10. Fung, A.K.; Li, Z.; Chen, K.S. Backscattering from a randomly rough dielectric surface. *IEEE Trans. Geosci. Rremote Sens.* **1992**, *30*, 356–369. [CrossRef]
11. Wu, T.D.; Chen, K.S.; Shi, J.; Fung, A.K. A transition model for the reflection coefficient in surface scattering. *IEEE Trans. Geosci. Remote Sens.* **2001**, *39*, 2040–2050. [CrossRef]
12. Oh, Y.; Sarabandi, K.; Ulaby, F.T. An empirical model and an inversion technique for radar scattering from bare soil surfaces. *IEEE Trans. Geosci. Remote Sens.* **1992**, *30*, 370–381. [CrossRef]
13. Dubois, P.C.; Van Zyl, J.; Engman, T. Measuring soil moisture with imaging radars. *IEEE Trans. Geosci. Remote Sens.* **1995**, *33*, 915–926. [CrossRef]
14. Ulaby, F.T.; Sarabandi, K.; McDonald, K.Y.; Whitt, M.; Dobson, M.C. Michigan microwave canopy scattering model. *Int. J. Remote Sens.* **1990**, *11*, 1223–1253. [CrossRef]
15. Attema, E.P.W.; Ulaby, F.T. Vegetation modeled as a water cloud. *Radio Sci.* **1978**, *13*, 357–364. [CrossRef]
16. Mo, T.; Choudhury, B.J.; Schmugge, T.J.; Wang, J.R.; Jackson, T.J. A model for microwave emission from vegetation-covered fields. *J. Geophys. Rese. Oceans* **1982**, *87*, 11229–11237. [CrossRef]
17. Shi, J.; Chen, K.S.; Li, Q.; Jackson, T.J.; O'Neill, P.E.; Tsang, L. A parameterized surface reflectivity model and estimation of bare-surface soil moisture with L-band radiometer. *IEEE Trans. Geosci. Remote Sens.* **2002**, *40*, 2674–2686. [CrossRef]
18. Dabrowska-Zielinska, K.; Gruszczynska, M.; Kowalik, W.; Stankiewicz, K. Application of multisensor data for evaluation of soil moisture. *Adv. Space Res.* **2002**, *29*, 45–50. [CrossRef]
19. Notarnicola, C.; Angiulli, M.; Posa, F. Use of radar and optical remotely sensed data for soil moisture retrieval over vegetated areas. *IEEE Trans. Geosci. Remote Sens.* **2006**, *44*, 925–935. [CrossRef]
20. Kong, J.; Yang, J.; Zhen, P.; Li, J.; Yang, L. A coupling model for soil moisture retrieval in sparse vegetation covered areas based on microwave and optical remote sensing data. *IEEE Trans. Geosci. Remote Sens.* **2018**, *56*, 7162–7173. [CrossRef]
21. He, B.; Xing, M.; Bai, X. A synergistic methodology for soil moisture estimation in an alpine prairie using radar and optical satellite data. *Remote Sens.* **2014**, *6*, 10966–10985. [CrossRef]
22. Zhang, L.; Meng, Q.; Yao, S.; Wang, Q.; Zeng, J.; Zhao, S.; Ma, J. Soil moisture retrieval from the Chinese GF-3 satellite and optical data over agricultural fields. *Sensors* **2018**, *18*, 2675. [CrossRef] [PubMed]
23. Han, L.; Wang, C.; Yu, T.; Gu, X.; Liu, Q. High-precision soil moisture mapping based on multi-model coupling and background knowledge, over vegetated areas using chinese Gf-3 and GF-1 satellite data. *Remote Sens.* **2020**, *12*, 2123. [CrossRef]
24. Khabazan, S.; Motagh, M.; Hosseini, M. Evaluation of radar backscattering models IEM, OH, and dubois using L and C-Bands SAR Data over different vegetation canopy covers and soil depths. In Proceedings of the International Archives of the Photogrammetry, Remote Sensing and Spatial Information Sciences, Volume XL-1/W3 2013, SMPPR 2013, Tehran, Iran, 5–8 October 2013; pp. 225–230.
25. Zeng, J.; Li, Z.; Chen, Q.; Bi, H. Method for soil moisture and surface temperature estimation in the Tibetan Plateau using spaceborne radiometer observations. *IEEE Geosci. Remote Sens. Lett.* **2015**, *12*, 97–101. [CrossRef]
26. Yao, P.; Shi, J.; Zhao, T.; Lu, H.; Al-Yaari, A. Rebuilding long time series global soil moisture products using the neural network adopting the microwave vegetation index. *Remote Sens.* **2017**, *9*, 35. [CrossRef]
27. Baghdadi, N.N.; El Hajj, M.; Zribi, M.; Fayad, I. Coupling SAR C-band and optical data for soil moisture and leaf area index retrieval over irrigated grasslands. *IEEE J. Sel. Top. Appl. Earth Obs. Remote Sens.* **2016**, *9*, 1229–1243. [CrossRef]
28. El Hajj, M.; Baghdadi, N.; Zribi, M.; Bazzi, H. Synergic use of Sentinel-1 and Sentinel-2 images for operational soil moisture mapping at high spatial resolution over agricultural areas. *Remote Sens.* **2017**, *9*, 1292. [CrossRef]
29. El Hajj, M.; Baghdadi, N.; Zribi, M.; Belaud, G.; Cheviron, B.; Courault, D.; Charron, F. Soil moisture retrieval over irrigated grassland using X-band SAR data. *Remote Sens. Environ.* **2016**, *176*, 202–218. [CrossRef]

30. Del Frate, F.; Ferrazzoli, P.; Schiavon, G. Retrieving soil moisture and agricultural variables by microwave radiometry using neural networks. *Remote Sens. Environ.* **2003**, *84*, 174–183. [CrossRef]
31. Kolassa, J.; Gentine, P.; Prigent, C.; Aires, F. Soil moisture retrieval from AMSR-E and ASCAT microwave observation synergy. Part 1: Satellite data analysis. *Remote Sens. Environ.* **2016**, *173*, 1–14. [CrossRef]
32. Santi, E.; Paloscia, S.; Pettinato, S.; Notarnicola, C.; Pasolli, L.; Pistocchi, A. Comparison between SAR Soil Moisture Estimates and Hydrological Model Simulations over the Scrivia Test Site. *Remote Sens.* **2013**, *5*, 4961–4976. [CrossRef]
33. Alexakis, D.D.; Mexis, F.D.; Vozinaki, A.E.; Daliakopoulos, I.N.; Tsanis, I.K. Soil moisture content estimation based on Sentinel-1 and auxiliary earth observation products. A hydrological approach. *Sensors* **2017**, *17*, 1455. [CrossRef] [PubMed]
34. Rodríguez-Fernández, N.J.; Aires, F.; Richaume, P.; Kerr, Y.H.; Prigent, C.; Kolassa, J.; Cabot, F.; Jimenez, C.; Mahmoodi, A.; Drusch, M. Soil Moisture Retrieval Using Neural Networks: Application to SMOS. *IEEE Trans. Geosci. Remote Sens.* **2015**, *53*, 5991–6007. [CrossRef]
35. Cui, Y.; Long, D.; Hong, Y.; Zeng, C.; Zhou, J.; Han, Z.; Liu, R.; Wan, W. Validation and reconstruction of FY-3B/MWRI soil moisture using an artificial neural network based on reconstructed MODIS optical products over the Tibetan Plateau. *J. Hydrol.* **2016**, *543*, 242–254. [CrossRef]
36. Xing, C.; Chen, N.; Zhang, X.; Gong, J. A machine learning based reconstruction method for satellite remote sensing of soil moisture images with in situ observations. *Remote Sens.* **2017**, *9*, 484. [CrossRef]
37. Liu, Y.; Dorigo, W.; Parinussa, R.; de Jeu, R.; Wagner, W.; McCabe, M.; Evans, J.; van Dijk, A. Trend-preserving blending of passive and active microwave soil moisture retrievals. *Remote Sens. Environ.* **2012**, *123*, 280–297. [CrossRef]
38. British Broadcasting Corporation. Average Conditions, Vienna, Austria. 2006. Available online: https://web.archive.org/web/20101202042009/http://www.bbc.co.uk/weather/world/city_guides/results.shtml?tt=TT000033 (accessed on 20 January 2022).
39. Händel, F.; Liu, G.; Fank, J.; Friedl, F.; Liedl, R.; Dietrich, P. Assessment of small-diameter shallow wells for managed aquifer recharge at a site in southern Styria, Austria. *Hydrogeol. J.* **2016**, *24*, 2079–2091. [CrossRef]
40. Al-Yaari, A.; Wigneron, J.P.; Dorigo, W.; Colliander, A.; Pellarin, T.; Hahn, S.; Mialon, A.; Richaume, P.; Fernandez-Moran, R.; Fan, L.; et al. Assessment and inter-comparison of recently developed/reprocessed microwave satellite soil moisture products using ISMN ground-based measurements. *Remote Sens. Environ.* **2019**, *224*, 289–303. [CrossRef]
41. Dorigo, W.A.; Xaver, A.; Vreugdenhil, M.; Gruber, A.; Hegyiova, A.; Sanchis-Dufau, A.D.; Zamojski, D.; Cordes, C.; Wagner, W.; Drusch, M. Global automated quality control of in situ soil moisture data from the International Soil Moisture Network. *Vadose Zone J.* **2013**, *12*, vzj2012.0097. [CrossRef]
42. Gruber, A.; Wagner, W.; Hegyiova, A.; Greifeneder, F.; Schlaffer, S. Potential of Sentinel-1 for high-resolution soil moisture monitoring. In Proceedings of the IEEE International Geoscience and Remote Sensing Symposium-IGARSS, Melbourne, VIC, Australia, 21–26 July 2013; IEEE: Piscataway, NJ, USA, 2013; pp. 4030–4033.
43. Paloscia, S.; Pettinato, S.; Santi, E.; Notarnicola, C.; Pasolli, C.; Reppucci, A. Soil moisture mapping using Sentinel-1 images: Algorithm and preliminary validation. *Remote Sens. Environ.* **2013**, *134*, 234–248. [CrossRef]
44. Hornacek, M.; Wagner, W.; Sabel, D.; Truong, H.L.; Snoeij, P.; Hahmann, T.; Diedrich, E.; Doubkova, M. Potential for high resolution systematic global surface soil moisture retrieval via change detection using Sentinel-1. *IEEE J. Sel. Top. Appl. Earth Obs. Remote Sens.* **2012**, *5*, 1303–1311. [CrossRef]
45. Attarzadeh, R.; Amini, J.; Notarnicola, C.; Greifeneder, F. Synergetic Use of Sentinel-1 and Sentinel-2 data for soil moisture mapping at plot scale. *Remote Sens.* **2018**, *10*, 1285. [CrossRef]
46. El Hajj, M.; Baghdadi, N.; Zribi, M.; Angelliaume, S. Analysis of Sentinel-1 radiometric stability and quality for land surface applications. *Remote Sens.* **2016**, *8*, 406. [CrossRef]
47. Millard, K.; Richardson, M. Quantifying the relative contributions of vegetation and soil moisture conditions to polarimetric C-Band SAR response in a temperate peatland. *Remote Sens. Environ.* **2018**, *206*, 123–138. [CrossRef]
48. Taconet, O.; Vidal-Madjar, D.; Emblanch, C.; Normand, M. Taking into account vegetation effects to estimate soil moisture from C-band radar measurements. *Remote Sens. Environ.* **1996**, *56*, 52–56. [CrossRef]
49. Zhou, X.; Lin, H.S.; White, E.A. Surface soil hydraulic properties in four soil series under different land uses and their temporal changes. *Catena* **2008**, *73*, 180–188. [CrossRef]
50. Chen, J.; Jönsson, P.; Tamura, M.; Gu, Z.; Matsushita, B.; Eklundh, L. A simple method for reconstructing a high-quality NDVI time-series data set based on the Savitzky–Golay filter. *Remote Sens. Environ.* **2004**, *91*, 332–344. [CrossRef]
51. Arvor, D.; Jonathan, M.; Meirelles, M.S.; Dubreuil, V.; Durieux, L. Classification of MODIS EVI time series for crop mapping in the state of Mato Grosso, Brazil. *Int. J. Remote Sens.* **2011**, *32*, 7847–7871. [CrossRef]
52. Srivastava, P.K.; Han, D.; Ramirez, M.R.; Islam, T. Machine learning techniques for downscaling SMOS satellite soil moisture using MODIS land surface temperature for hydrological application. *Water Resour. Manag.* **2013**, *27*, 3127–3144. [CrossRef]
53. Sandholt, I.; Rasmussen, K.; Andersen, J. A simple interpretation of the surface temperature/vegetation index space for assessment of surface moisture status. *Remote Sens. Environ.* **2002**, *79*, 213–224. [CrossRef]
54. Goetz, S.J. Multi-sensor analysis of NDVI, surface temperature and biophysical variables at a mixed grassland site. *Int. J. Remote Sens.* **1997**, *18*, 71–94. [CrossRef]
55. Holzman, M.E.; Rivas, R.; Bayala, M. Subsurface soil moisture estimation by VI–LST method. *IEEE Geosci. Remote Sens. Lett.* **2014**, *11*, 1951–1955. [CrossRef]

56. Nemani, R.; Pierce, L.; Running, S.; Goward, S. Developing satellite-derived estimates of surface moisture status. *J. Appl. Meteor.* **1993**, *32*, 548–557. [CrossRef]
57. Joshi, C.; Mohanty, B.P.; Jacobs, J.M.; Ines, A.V. Spatiotemporal analyses of soil moisture from point to footprint scale in two different hydroclimatic regions. *Water Resour. Res.* **2011**, *47*, W01508. [CrossRef]
58. Mohanty, B.P.; Skaggs, T.H. Spatio-temporal evolution and time-stable characteristics of soil moisture within remote sensing footprints with varying soil, slope, and vegetation. *Adv. Water Resour.* **2001**, *24*, 1051–1067. [CrossRef]
59. Lin, H.; Bouma, J.; Wilding, L.; Richardson, J.; Kutilek, M.; Nielsen, D. Advances in hydropedology. *Adv. Agron.* **2005**, *85*, 1–89. [CrossRef]
60. Hawley, M.E.; Jackson, T.J.; McCuen, R.H. Surface soil moisture variation on small agricultural watersheds. *J. Hydrol.* **1983**, *62*, 179–200. [CrossRef]
61. Buczko, U.; Bens, O.; Huttl, R. Tillage effects on hydraulic properties and macroporosity in silty and sandy soils. *Soil Sci. Soc. Am. J.* **2006**, *70*, 1998–2007. [CrossRef]
62. Santi, E.; Pettinato, S.; Paloscia, S.; Pampaloni, P.; Macelloni, G.; Brogioni, M. An algorithm for generating soil moisture and snow depth maps from microwave spaceborne radiometers: HydroAlgo. *Hydrol. Earth Syst. Sci.* **2012**, *16*, 3659–3676. [CrossRef]
63. Tapoglou, E.; Karatzas, G.P.; Trichakis, I.C.; Varouchakis, E.A. A spatio-temporal hybrid neural network-Kriging model for groundwater level simulation. *J. Hydrol.* **2014**, *519*, 3193–3203. [CrossRef]
64. Said, S.; Kothyari, U.C.; Arora, M.K. ANN-based soil moisture retrieval over bare and vegetated areas using ERS-2 SAR data. *J. Hydrol. Eng.* **2008**, *13*, 461–475. [CrossRef]
65. Santi, E.; Paloscia, S.; Pettinato, S.; Fontanelli, G. Application of artificial neural networks for the soil moisture retrieval from active and passive microwave spaceborne sensors. *Int. J. Appl. Earth Obs. Geoinf.* **2016**, *48*, 61–73. [CrossRef]
66. Chai, S.S.; Walker, J.P.; Makarynskyy, O.; Kuhn, M.; Veenendaal, B.; West, G. Use of soil moisture variability in artificial neural network retrieval of soil moisture. *Remote Sens.* **2010**, *2*, 166–190. [CrossRef]
67. Gavin, H. The Levenberg-Marquardt Method for Nonlinear Least Squares Curve-Fitting Problems. Department of Civil and Environmental Engineering, Duke University. 9 October 2013; pp. 1–17. Available online: https://www.academia.edu/9985415/The_Levenberg_Marquardt_method_for_nonlinear_least_squares_curve_fitting_problems (accessed on 20 January 2022).
68. Holtgrave, A.K.; Förster, M.; Greifeneder, F.; Notarnicola, C.; Kleinschmit, B. Estimation of soil moisture in vegetation-covered floodplains with sentinel-1 SAR data using support vector regression. *PFG–J. Photogram. Remote Sens. Geoinf. Sci.* **2018**, *86*, 85–101. [CrossRef]
69. Baghdadi, N.; Holah, N.; Zribi, M. Soil moisture estimation using multi-incidence and multi-polarization ASAR data. *Int. J. Remote Sens.* **2006**, *27*, 1907–1920. [CrossRef]
70. Schmugge, T. Remote Sensing of Surface Soil Moisture. *J. Appl. Meteor.* **1978**, *17*, 1549–1557. [CrossRef]
71. Charpentier, M.; Groffman, P. Soil moisture variability within remote sensing pixels. *J. Geophys. Res.* **1992**, *97*, 18987–18995. [CrossRef]
72. Grossman, R.; Harms, D.; Seybold, C.; Herrick, J. Coupling use-dependent and use-invariant data for soil quality evaluation in the United States. *J. Soil Water Conserv.* **2001**, *56*, 63–68. [CrossRef]
73. Wagner, W.; Reimer, C.; Bauer-Marschallinger, B.; Enenkel, M.; Hahn, S.; Melzer, T.; Naeimi, V.; Paulik, C.; Dorigo, W. Long-term soil moisture time series analyses based on active microwave backscatter measurements. In Proceedings of the International Archives of Photogrammetry, Remote Sensing and Spatial Information Sciences, Volume XL-7/W3, 2015 36th International Symposium on Remote Sensing of Environment, Berlin, Germany, 11–15 May 2015; pp. 545–550.

Article

High-Precision Automatic Calibration Modeling of Point Light Source Tracking Systems

Ruijin Li [1,2], Liming Zhang [1,*], Xianhua Wang [1], Weiwei Xu [1], Xin Li [1], Jiawei Li [1,2] and Chunhui Hu [1,2]

[1] Key Laboratory of Optical Calibration and Characterization, Anhui Institute of Optics and Fine Mechanics, Hefei Institutes of Physical Science, Chinese Academy of Sciences, Hefei 230031, China; ruijinli@mail.ustc.edu.cn (R.L.); xhwang@aiofm.ac.cn (X.W.); weilxu@aiofm.ac.cn (W.X.); lixin110@aiofm.ac.cn (X.L.); jiawei19@mail.ustc.edu.cn (J.L.); sariell@mail.ustc.edu.cn (C.H.)

[2] Science Island Branch, Graduate School, University of Science and Technology of China, Hefei 230026, China

* Correspondence: lmzhang@aiofm.ac.cn

Abstract: To realize high-precision and high-frequency unattended site calibration and detection of satellites, automatic direction adjustment must be implemented in mirror arrays. This paper proposes a high-precision automatic calibration model based on a novel point light source tracking system for mirror arrays. A camera automatically observes the solar vector, and an observation equation coupling the image space and local coordinate systems is established. High-precision calibration of the system is realized through geometric error calculation of multipoint observation data. Moreover, model error analysis and solar tracking verification experiments are conducted. The standard deviations of the pitch angle and azimuth angle errors are 0.0176° and 0.0305°, respectively. The root mean square errors of the image centroid contrast are 2.0995 and 0.8689 pixels along the x- and y-axes, respectively. The corresponding pixel angular resolution errors are 0.0377° and 0.0144°, and the comprehensive angle resolution error is 0.0403°. The calculated model values are consistent with the measured data, validating the model. The proposed point light source tracking system can satisfy the requirements of high-resolution, high-precision, high-frequency on-orbit satellite radiometric calibration and modulation transfer function detection.

Keywords: radiometric calibration; modeling; geometric error; high-precision calibration

1. Introduction

With the rapid development of remote-sensing technology, China's satellite remote-sensing technology can realize global and multisatellite network observations, thereby enabling comprehensive global observation with three-dimensional and high-, medium-, and low-resolution imaging, which has gradually penetrated all aspects of the national economy, social life, and national security [1]. Radiometric calibration is the process of establishing the functional response relationship between the absolute value of the radiance at the entrance pupil of the remote sensor and the digital number of the output image of the remote sensor and determining the radiometric calibration coefficient of the remote sensor data [2,3]. With the development of global remote-sensing quantitative applications, it has become increasingly urgent to improve the level of quantitation in remote-sensing applications of satellite data. On-orbit radiometric calibration and modulation transfer function (MTF) detection by satellite remote sensors are the basis of satellite remote-sensing quantitative applications. Therefore, higher requirements are put forward for the accuracy of remote sensor radiometric calibration and MTF detection [4–7]. Vicarious calibration, which is not affected by the space environment or satellite state, can account for atmospheric transmission and environmental impacts. This approach, which can help facilitate authenticity and model accuracy tests of on-orbit remote sensors, has been developed rapidly [8]. As a kind of high-spatial resolution satellite site for vicarious calibration equipment, point light sources are light-weight and small and exhibit excellent

optical characteristics. Their layout is flexible, and they can be moved easily. The aperture of the convex mirror depends on the pointing accuracy of the system. To ensure reliability, it is desirable to increase the pointing accuracy, reduce the aperture size, and reduce the volume and weight of the point light source. Furthermore, it is desirable to change the number of mirrors to realize on-orbit radiometric calibration and MTF detection of point light sources with different energy levels [9,10]. Point light source radiometric calibration mainly uses the point light source equipment to reflect sunlight into the entrance pupil of the satellite. Upon calculating the equivalent entrance pupil radiance of the satellite combined with the target response value of the remote-sensing image, the calibration coefficient is calculated according to the remote sensor calibration equation. Because this procedure simplifies the radiative transfer process, it has been widely used [11–15].

According to literature research, so far, few countries have carried out on-orbit radiation calibration and MTF detection of point light sources. The United States was the first to carry out this work, followed by France and China. France has adopted active point light source equipment, mainly using high-energy spotlight for on-orbit MTF detection of SPOT5 [16]. The United States and China mainly use reflective point light source equipment to carry out the corresponding experiments [17–23]. The key to high-resolution satellite on-orbit radiation calibration based on point light sources is to control the direction of the central optical axis of the point light source reflector. When the central optical axis of the reflector points to the sun, the sunlight enters the convex mirror perpendicularly, the reflected light spot is in a divergent state, and the direction points toward the sun. When the central optical axis of the reflector points toward the position of the bisector of the angle between the satellite and the sun, the reflected light spot is reflected toward the satellite direction in a divergent state. If the pointing position of the optical axis at the edge is reflected toward the direction of the satellite due to low pointing accuracy, the satellite may not observe the point light source or may observe only part of the reflected light spot, which may cause the radiation calibration to fail. Therefore, the success or failure of the point light source on-orbit experiment depends on the pointing accuracy, and the pointing accuracy depends on the tracking accuracy of the system. To improve the pointing accuracy of the system, it is necessary to improve the tracking accuracy of the system. The pointing accuracy of the reflector equipment used by American researchers Schiller et al. [24] to implement the SPARC method (specular array radiometric calibration) of radiation calibration is better than $\pm 0.5°$. In particular, a large convex mirror is used to compensate for the lack of pointing accuracy to ensure that the reflection spot enters the pupil of the satellite. However, the processing accuracy of large convex mirrors is difficult to ensure, and this approach is not convenient for engineering practice and application promotion. In China, the Anhui Institute of Optics and Fine Mechanics, Chinese Academy of Sciences, successively conducted on-orbit radiometric calibration experiments and MTF detection based on point light sources [7,12,13,22]. Initially, a large plane mirror was used as the reflection point light source to perform experiments involving medium- and high-orbit satellites on orbit [22,23]. At present, we mainly carry out on-orbit experiments of point light sources based on convex mirrors. Compared with existing foreign point light source systems, the difference is that we use a smaller convex mirror to overcome the disadvantages associated with larger convex mirrors. The advantage of this approach is that it is easy to change the number of mirrors to produce different energy levels of reflected light, which is suitable for different resolutions in satellite radiometric calibration and MTF detection [13]. However, the disadvantage is that the reflection spot decreases due to the reduction of the aperture of the convex mirror, which increases the difficulty of the satellite reliably receiving the reflected spot. Therefore, to ensure that the reflected light spot is reliably incident on the entrance pupil of the satellite, the key technological improvement that needs to be addressed when using a smaller convex mirror is improving the pointing accuracy. Therefore, to improve the pointing accuracy of the system, a high-precision calibration modeling method for a point light turntable based on a solar vector was established [9]. Compared with previous-generation equipment [22], the integrated pointing accuracy of the system could be enhanced; however, a camera with an automatic observation ability was not introduced in the modeling process. Consequently, the system cannot realize automatic calibration, and it is difficult to realize the high-precision calibration of large-scale automatic cooperative work. To realize automatic calibration, the literature [10] proposed a mirror normal calibration method based on the centroid of the solar image; however, in the initial stage of the model, the influencing factors such as equipment placement errors and camera distortion corrections are not considered. Consequently, the calibration accuracy is affected by single-point calibration and the solar image, and the calibration accuracy needs to be further increased.

The abovementioned calibration techniques based on convex mirrors can achieve satisfactory results in radiometric calibration and MTF detection; however, such approaches cannot meet the requirements of high precision, high frequency and use of existing high-resolution satellites. Nevertheless, unattended multipoint automatic and high-precision pointing adjustment technology can satisfy these requirements. Therefore, in this study, based on the development of a point light source turntable tracking system, an automatic calibration modeling method is developed. Moreover, a high-precision automatic geometric calibration model is established. The system can realize network-based remote control, achieve high-precision pointing of the point light source array tracking system, and realize high frequency and high-efficiency orbit radiation calibration and MTF detection of high-spatial resolution satellites.

The tracking accuracy described in this paper is the basic guaranteed accuracy required to achieve a comprehensive system design accuracy better than $0.1°$; therefore, the design accuracy of our system needs to be better than $0.1°$. To realize automatic calibration of the point light source array and achieve the purpose of high-precision tracking of the point light source system, this paper focuses on the establishment of a high-precision calibration model of the point light source system. Starting from the composition of the point light source system, the establishment of a coordinate system and the principle of geometric calibration modeling, this paper studies the establishment of a simplified calibration model of the point light source system. On the basis of the simplified calibration model, considering the geometric error parameters and camera lens distortion parameters that affect the tracking accuracy of the system, the automatic high-precision geometric calibration model is further established. Based on the theoretical verification and solution of the model, the inverse solution algorithm of the calibration model is proposed for experimental verification of the calibrated model. Finally, the experimental verification and system tracking accuracy analysis are carried out.

2. Principle of Geometric Calibration Modeling

2.1. Equipment System Composition and Coordinate System Establishment

The turntable system of the point light source is composed mainly of a posture control module, mirror assembly, camera and electric control system. The posture control module includes a pitching component and an azimuth component. The pitching component adopts a "U"-shaped forked arm structure. The pitch motor drives a pitching turbine through a two-stage reduction device to drive a mirror to rotate around the pitch axis. The azimuth component is driven by an azimuth motor through the two-stage reduction mechanism to cause the rotary table to rotate around the azimuth axis. The reduction ratio of the second reduction device is 1:360. The pitch and azimuth terminals of the equipment are equipped with an encoder detection device to feed back the rotation angle of the rotary table terminal. The detection accuracy of the encoder is $0.02°$. The mirror assembly is arranged between the "U"-shaped forked arms to form a pitching rotation axis. The camera is fixed to the top of the mirror assembly to maintain the camera plane parallel to the mirror plane. The field of view is $23° \times 17°$. The image resolution is 1280×1024 pixels. The resolutions of the azimuth and pitch pixel angles are $0.018°$ and $0.0166°$, respectively. The electric control system is arranged at the base and two fork arms. The abovementioned components compose a point light turntable system, as shown in Figure 1a.

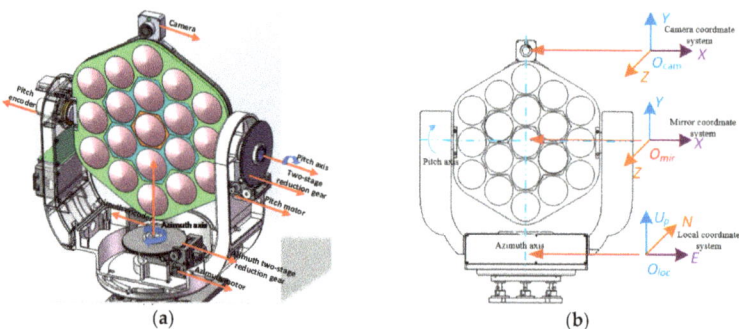

Figure 1. (a) Composition of the point light source system; (b) coordinate system establishment.

To conveniently describe the coordinate position of the sun and a satellite observed from a certain point on Earth's surface, a coordinate system is established based on the position of the point light source on Earth's surface. This system is named the northeast upper coordinate system, which is expressed as loc and described as $[\ E\ \ N\ \ U_p\ ]$. E points due east in the positive direction. N points due north in the positive direction. U_p points in the vertical upward direction against the geocenter in the positive direction. The mirror coordinate system is fixed on the turntable. The right-hand rectangular coordinate system is composed of the z-axis of the central light axis of the mirror, which is described as $[\ x_{mir}\ \ y_{mir}\ \ z_{mir}\ ]$. In addition, x_{mir} is based on the pitch axis of the turntable and points to the east, and y_{mir} takes the azimuth axis of the turntable as the baseline, which is consistent with the U_p direction, with z_{mir} pointing to the north. The camera coordinate system is established in accordance with the mirror coordinate system, which is described as $[\ x_{cam}\ \ y_{cam}\ \ z_{cam}\ ]$. The establishment of the coordinate system is shown in Figure 1b.

2.2. Principle of Geometric calibration Modeling

Based on the principle of central projection and perspective transformation [25,26], in the same coordinate system, a collinear condition equation is established using the collinear condition, and a geometric calibration model is established based on this equation. A rotation transformation relationship between the image plane of the image space coordinate system and object plane of the local coordinate system is established by using the camera to observe the solar vector. Moreover, considering the angle readings of the pitch and azimuth encoders, centroid coordinates of the solar image and solar position parameters at different positions at different times, a multipoint observation equation is established, and the least squares method is used to solve the model. Geometric calibration of the equipment is conducted to determine the initial positions of the azimuth and pitch encoders. The mirror normal vector diagram is shown in Figure 2.

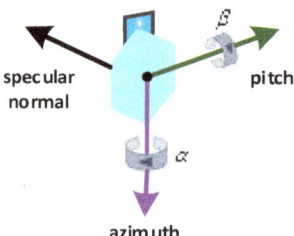

Figure 2. Mirror normal vector diagram.

Assuming that the point light source is placed horizontally in the initial position, the pitch axis is orthogonal to the azimuth axis, and the central light axis of the reflector points to the north. This configuration is expressed as $\begin{bmatrix} 0 & 1 & 0 \end{bmatrix}^T_{loc}$ and $\begin{bmatrix} 0 & 0 & 1 \end{bmatrix}^T_{mir}$ in the northeast upper coordinate system and reflector coordinate system, respectively. At a certain moment, if the azimuth and altitude angles of the incident sunlight are $a_{azimuth}$ and $a_{altitude}$, respectively, the turntable rotates anticlockwise and clockwise around the pitch X-axis and azimuth axis, respectively. At this time, the central optical axis vector of the reflector coincides with the solar vector in the northeast upper coordinate system. In this case, in the local coordinate system, the transformation process from the optical axis vector of the mirror center to the coordinate rotation consistent with the solar vector can be expressed as

$$\begin{bmatrix} X \\ Y \\ Z \end{bmatrix}_{loc} = \begin{bmatrix} \cos(\alpha - \alpha_0) & \sin(\alpha - \alpha_0) & 0 \\ -\sin(\alpha - \alpha_0) & \cos(\alpha - \alpha_0) & 0 \\ 0 & 0 & 1 \end{bmatrix} \begin{bmatrix} 1 & 0 & 0 \\ 0 & \cos(\beta - \beta_0) & -\sin(\beta - \beta_0) \\ 0 & \sin(\beta - \beta_0) & \cos(\beta - \beta_0) \end{bmatrix} \begin{bmatrix} 0 \\ 1 \\ 0 \end{bmatrix}_{loc} \quad (1)$$

where α and β are the readings of the azimuth and elevation encoders at a certain time, respectively; α_0 and β_0 are the initial position readings.

According to the definition of the coordinate system, if the mirror coordinate system is rotated anticlockwise by 90° around the axis, the local coordinate system coincides with the mirror coordinate system. According to the rotation matrix relationship of the coordinate transformation, the coordinate transformation relationship can be established at any point as follows:

$$\begin{bmatrix} X \\ Y \\ Z \end{bmatrix}_{loc} = R_X^{-1}(\frac{\pi}{2}) \begin{bmatrix} x \\ y \\ z \end{bmatrix}_{mir}. \quad (2)$$

Combining the coordinate rotation relation expressed in Equation (1) with the coordinate transformation and rotation relation expressed in Equation (2) yields

$$\begin{bmatrix} X \\ Y \\ Z \end{bmatrix}_{loc} = \begin{bmatrix} \cos(\alpha - \alpha_0) & \sin(\alpha - \alpha_0) & 0 \\ -\sin(\alpha - \alpha_0) & \cos(\alpha - \alpha_0) & 0 \\ 0 & 0 & 1 \end{bmatrix} \begin{bmatrix} 1 & 0 & 0 \\ 0 & \cos(\beta - \beta_0) & -\sin(\beta - \beta_0) \\ 0 & \sin(\beta - \beta_0) & \cos(\beta - \beta_0) \end{bmatrix} R_X^{-1}(\frac{\pi}{2}) \begin{bmatrix} x \\ y \\ z \end{bmatrix}_{mir}. \quad (3)$$

In particular, when the optical axis vector of the mirror center is consistent with the solar vector, the coordinates of the solar vector in the mirror coordinate system are $\begin{bmatrix} 0 & 0 & 1 \end{bmatrix}^T_{mir}$, and the unit vector coordinates in the local coordinate system are $\begin{bmatrix} X & Y & Z \end{bmatrix}^T_{loc}$. According to Equation (3), the solar vector under the reflector can be transformed to the vector in the local coordinate system. Based on this aspect, the coordinate transformation relationship between the mirror and local coordinate systems is established based on the solar vector.

3. Geometric Calibration Modeling of the Turntable

3.1. Basic Calibration Model of the Turntable

In terms of the initial position of the point light source in the basic calibration model of the turntable, the X- and Z-axes in the mirror coordinate system coincide with the E- and N-axes in the local coordinate system, respectively. The central optical axis of the reflector points true north. The camera is affixed to the mirror assembly bracket, and the definition of its coordinate system is consistent with the mirror coordinate system. Therefore, the central optical axis vector of the reflector is replaced by the camera center optical axis vector. When the camera coordinate system is transformed to the local coordinate system, the relationship between the two coordinate systems must be established by multiplying the left side by the rotation matrix $R_X^{-1}(\frac{\pi}{2})$, as follows:

$$\begin{bmatrix} X \\ Y \\ Z \end{bmatrix}_{loc} = R_X^{-1}(\frac{\pi}{2}) \begin{bmatrix} x - x_0 \\ y - y_0 \\ f \end{bmatrix}_{cam}. \quad (4)$$

By combining Equations (3) and (4), the relationship between the camera and local coordinate systems can be established as

$$\begin{bmatrix} X_i \\ Y_i \\ Z_i \end{bmatrix}_{loc} = R_Z(\alpha_i - \alpha_0)R_X(\beta_i - \beta_0)R_X^{-1}(\frac{\pi}{2})\lambda \begin{bmatrix} x_i - x_0 \\ y_i - y_0 \\ f \end{bmatrix} \quad (5)$$

where

$$R_Z(\alpha_i - \alpha_0) = \begin{bmatrix} \cos(\alpha_i - \alpha_0) & \sin(\alpha_i - \alpha_0) & 0 \\ -\sin(\alpha_i - \alpha_0) & \cos(\alpha_i - \alpha_0) & 0 \\ 0 & 0 & 1 \end{bmatrix}, R_X(\beta_i - \beta_0) = \begin{bmatrix} 1 & 0 & 0 \\ 0 & \cos(\beta_i - \beta_0) & -\sin(\beta_i - \beta_0) \\ 0 & \sin(\beta_i - \beta_0) & \cos(\beta_i - \beta_0) \end{bmatrix}.$$

α_i and β_i are the azimuth and pitch encoder values corresponding to the encoder at a certain moment, respectively; x_i and y_i are the coordinates of the centroid of the solar image in the pixel coordinate system at a certain moment; and λ is the imaging scale factor. Moreover, x_0 and y_0 are the camera main point coordinates, and i represents the camera imaging time serial number or the solar position serial number at different times, with $i = 1 \cdots n$.

We define $R_{cam}^{loc} = R_Z(\alpha_i - \alpha_0)R_X(\beta_i - \beta_0)R_X^{-1}(\frac{\pi}{2})$. Consequently, Equation (5) can be rewritten as

$$(R_{cam}^{loc})^{-1} \begin{bmatrix} X_i \\ Y_i \\ Z_i \end{bmatrix} = \lambda \begin{bmatrix} x_i - x_0 \\ y_i - y_0 \\ f \end{bmatrix} \quad (6)$$

where $\begin{bmatrix} X_i \\ Y_i \\ Z_i \end{bmatrix} = \begin{bmatrix} \sin a_{azimuth} \cos a_{altitude} \\ \cos a_{azimuth} \cos a_{altitude} \\ \sin a_{altitude} \end{bmatrix}$, X_i represents the east (E) component of the sun in the local coordinate system, Y_i represents the component of the sun due north (N) in the local coordinate system, and Z_i represents the upward (U_p) component of the sun perpendicular to the earth plane in the local coordinate system.

Equation (6) represents the rotation transformation relationship between the image plane in the image space coordinate system and object plane in the local coordinate system. By dividing the first and second expressions of Equation (6) by the third expression, $x_i - x_0 = a(n_{xi} - n_{x0})$ and $y_i - y_0 = a(n_{yi} - n_{y0})$, where a is the pixel size and n is the number of pixels. Upon substituting this content into Equation (6), the basic calibration model of the turntable can be expressed as

$$\begin{cases} \frac{a}{f}(n_{xi} - n_{x0}) = \frac{\cos(\alpha - \alpha_0)X_i - \sin(\alpha - \alpha_0)Y_i}{\sin(\alpha - \alpha_0)\cos(\beta - \beta_0)X_i + \cos(\alpha - \alpha_0)\cos(\beta - \beta_0)Y_i + \sin(\beta - \beta_0)Z_i} \\ \frac{a}{f}(n_{yi} - n_{y0}) = \frac{\sin(\alpha - \alpha_0)\sin(\beta - \beta_0)X_i + \cos(\alpha - \alpha_0)\sin(\beta - \beta_0)Y_i - \cos(\beta - \beta_0)Z_i}{\sin(\alpha - \alpha_0)\cos(\beta - \beta_0)X_i + \cos(\alpha - \alpha_0)\cos(\beta - \beta_0)Y_i + \sin(\beta - \beta_0)Z_i} \end{cases} \quad (7)$$

The right and left sides of the equation represent the calculation formula of the solar vector and optical axis vector of the turntable mirror center, respectively. When $x = x_0$ and $y = y_0$, the optical axis vector of the reflector points toward the sun. In this case, $a_{altitude} = \beta - \beta_0$ and $a_{azimuth} = \alpha - \alpha_0$. When $x \neq x_0$ and $y \neq y_0$, the optical axis vector of the reflector points toward a certain angle in space. In this case, $\theta_{altitude} = \beta - \beta_0$ and $\varphi_{azimuth} = \alpha - \alpha_0$.

In this manner, the relationship between the camera coordinate system and local coordinate system can be established by using the camera to observe the solar vector. Thus, any vector in the image space coordinate system can be transformed to the local coordinate system through the coordinate rotation transformation relationship. The solar vector observed by the camera represents the optical axis vector of the reflector. The control turntable uses the camera to realize data acquisition and automatic calibration in the local coordinate system.

3.2. High-Precision Geometric Calibration Model of the Turntable

The basic calibration model of the turntable is based on the assumption that the turntable is placed horizontally, the pitch axis is orthogonal to the azimuth axis, and the camera is positioned vertically. However, regardless of whether the actual turntable is horizontal, the pitch axis is vertical to the azimuth axis, and the camera is vertical. The levelness error, perpendicularity error, and camera placement perpendicularity error must be considered in the high-precision control system. In particular, to realize high-precision automatic calibration control of the turntable, it is necessary to establish a high-precision calibration model of the turntable and examine the geometric error parameters of the turntable obtained considering the basic calibration model. We consider that the error matrix of the turntable placement levelness is R^L, the orthogonal error matrix of the pitch and azimuth axes is R^S, and the vertical error matrix of the camera placement is R^C. In this case, the high-precision calibration model can be expressed as

$$\begin{bmatrix} X_i \\ Y_i \\ Z_i \end{bmatrix}_{\text{loc}} = \lambda R^L R_Z(\alpha_i - \alpha_0) R^S R_X(\beta_i - \beta_0) R^C R_X^{-1}(\frac{\pi}{2}) \begin{bmatrix} x_i - x_0 \\ y_i - y_0 \\ f \end{bmatrix} \quad (8)$$

where $R^L = R_X^L R_Y^L R_Z^L$, $R^S = R_Z^S R_Y^S R_X^S$, and $R^C = R_X^C R_Y^C R_Z^C$.

According to the rotation matrix, the same kind of rotation can be combined in the same direction. Equation (8) can be simplified to obtain a high-precision calibration model of the turntable as

$$\begin{bmatrix} X_i \\ Y_i \\ Z_i \end{bmatrix}_{\text{loc}} = \lambda R_X^L R_Y^L R_Z(\alpha_i - \alpha_0) R_Y^S R_X(\beta_i - \beta_0) R_Y^C R_X^{-1}(\frac{\pi}{2}) \begin{bmatrix} x_i - x_0 \\ y_i - y_0 \\ f \end{bmatrix} \quad (9)$$

where R_X^L, R_Y^L, and R_Z^L represent the rotation matrix around the X, Y, and Z axes from the mirror coordinate system to the local coordinate system, respectively; R_Z^S, R_Y^S, and R_X^S represent the rotation matrix around the Z, X, and Y axes from the pitch axis coordinate system to the azimuth axis coordinate system, respectively; and R_X^C, R_Y^C, and R_Z^C represent the rotation matrix around the X, Y, and Z axes from the camera coordinate system to the mirror coordinate system, respectively. Consequently,

$$R_X^L R_Y^L = \begin{bmatrix} 1 & 0 & 0 \\ 0 & \cos\mu_0 & -\sin\mu_0 \\ 0 & \sin\mu_0 & \cos\mu_0 \end{bmatrix} \begin{bmatrix} \cos\nu_0 & 0 & \sin\nu_0 \\ 0 & 1 & 0 \\ -\sin\nu_0 & 0 & \cos\nu_0 \end{bmatrix} R_Y^S = \begin{bmatrix} \cos\omega_0 & 0 & \sin\omega_0 \\ 0 & 1 & 0 \\ -\sin\omega_0 & 0 & \cos\omega_0 \end{bmatrix}$$

$$R_Y^C = \begin{bmatrix} \cos\gamma_0 & 0 & \sin\gamma_0 \\ 0 & 1 & 0 \\ -\sin\gamma_0 & 0 & \cos\gamma_0 \end{bmatrix}$$

where μ_0 and ν_0 represent the level offset error of the turntable installation, ω_0 represents the geometric error of the verticality of the pitch axis and azimuth axis of the turntable, and γ_0 represents the verticality offset error of the camera placement.

We define $R_{\text{cam}}^{\text{loc}} = R_X^L R_Y^L R_Z(\alpha_i - \alpha_0) R_Y^S R_X(\beta_i - \beta_0) R_Y^C R_X^{-1}(\frac{\pi}{2}) = \begin{bmatrix} a_1 & b_1 & c_1 \\ a_2 & b_2 & c_2 \\ a_3 & b_3 & c_3 \end{bmatrix}$.

By inserting Equation (9), we obtain

$$\begin{cases} \frac{a}{f}(n_{xi} - n_{x0}) = \frac{X_i a_1 + Y_i a_2 + Z_i a_3}{X_i c_1 + Y_i c_2 + Z_i c_3} \\ \frac{a}{f}(n_{yi} - n_{y0}) = \frac{X_i b_1 + Y_i b_2 + Z_i b_3}{X_i c_1 + Y_i c_2 + Z_i c_3} \end{cases} \quad (10)$$

Thus, a high-precision calibration model considering the geometric error of the system is established. However, in the process of automatic system calibration, camera lens distortion may produce errors, which may limit the increase in the calibration accuracy.

Therefore, it is necessary to correct the lens distortion to further reduce the error sources. Considering the calibration model expressed in Equation (10), the chessboard calibration results are incorporated [27], and the lens distortion correction term is added. The first term approximation of the Taylor series expansion is adopted to correct the radial distortion error of the lens

$$\begin{cases} (x_i - x_0) + \Delta x = f_x \frac{\overline{X}}{\overline{Z}} \\ (y_i - y_0) + \Delta y = f_y \frac{\overline{Y}}{\overline{Z}} \end{cases} \quad (11)$$

where x_i and y_i are the coordinates of the image centroid in the pixel coordinate system; x_0 and y_0 are the camera main point coordinates; Δx and Δy are the radial distortion errors of the camera; f_x and f_y are the focal lengths of the camera in the x and y directions, respectively; and $\overline{X} = X_i a_1 + Y_i a_2 + Z_i a_3$, $\overline{Y} = X_i b_1 + Y_i b_2 + Z_i b_3$, and $\overline{Z} = X_i c_1 + Y_i c_2 + Z_i c_3$.

According to the camera physical calibration model [28,29], the radial distortion error of the camera can be defined as follows:

$$\Delta x = \overline{x} k_1 r^2, \quad \Delta y = \overline{y} k_1 r^2 \quad (12)$$

where $\overline{x} = (x_i - x_0)$, $\overline{y} = (y_i - y_0)$, and $r^2 = (x_i - x_0)^2 + (y_i - y_0)^2$. Here, k_1 is the radial distortion coefficient of the camera, and r is the radial distance of the actual image point.

Substituting Equation (12) into Equation (11) yields a high-precision geometric error calibration model with camera distortion correction, as follows:

$$\begin{cases} \frac{a}{f_x}(n_{x_i} - n_{x_0})\left\{1 + a^2 k_1 \left[(n_{x_i} - n_{x_0})^2 + (n_{y_i} - n_{y_0})^2\right]\right\} = \frac{\overline{X}}{\overline{Z}} \\ \frac{a}{f_y}(n_{y_i} - n_{y_0})\left\{1 + a^2 k_1 \left[(n_{x_i} - n_{x_0})^2 + (n_{y_i} - n_{y_0})^2\right]\right\} = \frac{\overline{Y}}{\overline{Z}} \end{cases} \quad (13)$$

Equation (13) represents the conversion of the solar vector in the local coordinate system to the representation in the image space coordinate system. Thus, the relationship between the solar vector observed by the camera in the image space coordinate system is established, and transformation from any vector in the image space system to the local coordinate system is realized. Finally, through actual camera observations, multipoint data are collected to establish multipoint observation equations to achieve high-precision calibration of the system installation geometric errors and verify the corresponding error parameters μ_0, ν_0, ω_0, and γ_0, encoder initial positions α_0 and β_0, and camera principal point and principal distance values x_0, y_0, f_x, and f_y, among other factors. In this manner, high-precision calibration of the turntable system in the local coordinate system can be realized, leading to increased pointing accuracy.

4. Model Verification and Solution

4.1. Verification of the Model Coordinate Rotation Transformation Relationship

When the central light axis of the reflector points toward the sun, the coordinates of the solar vector in the mirror coordinate system are $\begin{bmatrix} 0 & 0 & 1 \end{bmatrix}^T_{mir}$, and the unit vector coordinates in the local coordinate system are $\begin{bmatrix} X & Y & Z \end{bmatrix}^T_{loc}$. First, forward verification is conducted according to Equation (3). By substituting $\begin{bmatrix} 0 & 0 & 1 \end{bmatrix}^T_{mir}$ and multiplying the three terms on the right side, we can obtain the vector representation of the sun in the local coordinate system, as follows:

$$\begin{bmatrix} X \\ Y \\ Z \end{bmatrix}_{loc} = \begin{bmatrix} \cos(\alpha - \alpha_0) & \sin(\alpha - \alpha_0)\sin(\beta - \beta_0) & \sin(\alpha - \alpha_0)\cos(\beta - \beta_0) \\ -\sin(\alpha - \alpha_0) & \cos(\alpha - \alpha_0)\sin(\beta - \beta_0) & \cos(\alpha - \alpha_0)\cos(\beta - \beta_0) \\ 0 & -\cos(\beta - \beta_0) & \sin(\beta - \beta_0) \end{bmatrix} \begin{bmatrix} 0 \\ 0 \\ 1 \end{bmatrix}_{mir} = \begin{bmatrix} \sin(\alpha - \alpha_0)\cos(\beta - \beta_0) \\ \cos(\alpha - \alpha_0)\cos(\beta - \beta_0) \\ \sin(\beta - \beta_0) \end{bmatrix} \quad (14)$$

where $a_{azimuth} = \alpha - \alpha_0$, and $a_{altitude} = \beta - \beta_0$. The result is the same as that of the solar unit vector $\begin{bmatrix} X \\ Y \\ Z \end{bmatrix}_{loc} = \begin{bmatrix} \sin a_{azimuth} \cos a_{altitude} \\ \cos a_{azimuth} \cos a_{altitude} \\ \sin a_{altitude} \end{bmatrix}$ in the local coordinate system.

Thus, the accuracy of the rotation matrix is preliminarily verified. Second, the vector representation of the sun in the local coordinate system is substituted into Equation (3) to calculate the vector representation of the sun in the mirror coordinate system, as follows:

$$(R_{cam}^{loc})^{-1} \begin{bmatrix} X_{loc} \\ Y_{loc} \\ Z_{loc} \end{bmatrix} = \begin{bmatrix} \cos(\alpha - \alpha_0) & -\sin(\alpha - \alpha_0) & 0 \\ \sin(\alpha - \alpha_0)\sin(\beta - \beta_0) & \cos(\alpha - \alpha_0)\sin(\beta - \beta_0) & -\cos(\beta - \beta_0) \\ \sin(\alpha - \alpha_0)\cos(\beta - \beta_0) & \cos(\alpha - \alpha_0)\cos(\beta - \beta_0) & \sin(\beta - \beta_0) \end{bmatrix} \begin{bmatrix} \sin(\alpha - \alpha_0)\cos(\beta - \beta_0) \\ \cos(\alpha - \alpha_0)\cos(\beta - \beta_0) \\ \sin(\beta - \beta_0) \end{bmatrix} = \begin{bmatrix} 0 \\ 0 \\ 1 \end{bmatrix} \quad (15)$$

The calculation result for Equation (15) is the same as the vector representation $\begin{bmatrix} 0 & 0 & 1 \end{bmatrix}_{mir}^T$ of the sun in the mirror coordinate system when the optical axis of the reflector is aligned with the sun. Both the forward and reverse verification calculation results are the same as the predicted results, which demonstrates the accuracy of the coordinate rotation transformation matrix of the basic calibration model. The coordinate rotation transformation verification diagram for the calibration model is shown in Figure 3.

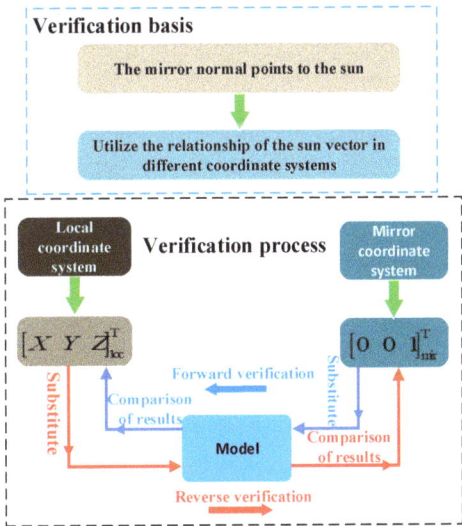

Figure 3. Coordinate rotation transformation verification of the calibration model.

4.2. Model Solution

According to Equation (13), the geometric error parameters of the system to be calibrated are (μ_0, ν_0, ω_0, and γ_0), the initial position parameters of the encoder are (α_0 and β_0), and the camera parameters are (x_0, y_0, f_x, f_y, and k_1). In total, 11 parameters exist. To solve the model, multipoint observations are needed. To this end, the multipoint observation equation is established, and the least squares method is used to solve the unknown parameters iteratively until the accuracy requirements are met. The solution process is as follows:

$$\begin{matrix} w_x = \frac{a}{f_x}(n_{x_i} - n_{x_0})\left\{1 + a^2 k_1 \left[(n_{x_i} - n_{x_0})^2 + (n_{y_i} - n_{y_0})^2\right]\right\} - \frac{\overline{X}}{\overline{Z}} \\ w_y = \frac{a}{f_y}(n_{y_i} - n_{y_0})\left\{1 + a^2 k_1 \left[(n_{x_i} - n_{x_0})^2 + (n_{y_i} - n_{y_0})^2\right]\right\} - \frac{\overline{Y}}{\overline{Z}} \end{matrix}. \quad (16)$$

The first-order Taylor linearization expansion of Equation (16) is carried out at the initial value $\begin{bmatrix} \mu_0 & v_0 & \omega_0 & \gamma_0 & x_0 & y_0 & f_x & f_y & k_1 & \alpha_0 & \beta_0 \end{bmatrix}_i^T$, and the error equation is established:

$$\begin{aligned} w'_x &= \frac{\partial w_x}{\partial \mu_0}\Delta\mu_0 + \frac{\partial w_x}{\partial v_0}\Delta v_0 + \frac{\partial w_x}{\partial \omega_0}\Delta\omega_0 + \frac{\partial w_x}{\partial \gamma_0}\Delta\gamma_0 + \frac{\partial w_x}{\partial x_0}\Delta x_0 + \frac{\partial w_x}{\partial y_0}\Delta y_0 \\ &+ \frac{\partial w_x}{\partial f_x}\Delta f_x + \frac{\partial w_x}{\partial f_y}\Delta f_y + \frac{\partial w_x}{\partial k_1}\Delta k_1 + \frac{\partial w_x}{\partial \alpha_0}\Delta\alpha_0 + \frac{\partial w_x}{\partial \beta_0}\Delta\beta_0 \\ w'_y &= \frac{\partial w_y}{\partial \mu_0}\Delta\mu_0 + \frac{\partial w_y}{\partial v_0}\Delta v_0 + \frac{\partial w_y}{\partial \omega_0}\Delta\omega_0 + \frac{\partial w_y}{\partial \gamma_0}\Delta\gamma_0 + \frac{\partial w_y}{\partial x_0}\Delta x_0 + \frac{\partial w_y}{\partial y_0}\Delta y_0 \\ &+ \frac{\partial w_y}{\partial f_x}\Delta f_x + \frac{\partial w_y}{\partial f_y}\Delta f_y + \frac{\partial w_y}{\partial k_1}\Delta k_1 + \frac{\partial w_y}{\partial \alpha_0}\Delta\alpha_0 + \frac{\partial w_y}{\partial \beta_0}\Delta\beta_0 \end{aligned} \quad (17)$$

This equation is expressed in matrix form as

$$\begin{bmatrix} w'_x \\ w'_y \end{bmatrix} = \begin{bmatrix} \frac{\partial w_x}{\partial \mu_0} & \cdots & \frac{\partial w_x}{\partial \beta_0} \\ \frac{\partial w_y}{\partial \mu_0} & \cdots & \frac{\partial w_y}{\partial \beta_0} \end{bmatrix} \begin{bmatrix} \Delta\mu_0 \\ \vdots \\ \Delta\beta_0 \end{bmatrix}. \quad (18)$$

By using the camera multipoint observation, the multipoint observation equation is established as follows:

$$L_1 = \begin{bmatrix} w'_{x,1} \\ w'_{y,1} \end{bmatrix}^0, L_n = \begin{bmatrix} w'_{x,n} \\ w'_{y,n} \end{bmatrix}^0, x_0 = \begin{bmatrix} \Delta\mu_0 \\ \vdots \\ \vdots \\ \Delta\beta_0 \end{bmatrix}^i, A_1 = \begin{bmatrix} \frac{\partial w_{x,1}}{\partial \mu_0} & \cdots & \frac{\partial w_{x,1}}{\partial \beta_0} \\ \frac{\partial w_{y,1}}{\partial \mu_0} & \cdots & \frac{\partial w_{y,1}}{\partial \beta_0} \end{bmatrix}^0, A_n = \begin{bmatrix} \frac{\partial w_{x,n}}{\partial \mu_0} & \cdots & \frac{\partial w_{x,n}}{\partial \beta_0} \\ \frac{\partial w_{y,n}}{\partial \mu_0} & \cdots & \frac{\partial w_{y,n}}{\partial \beta_0} \end{bmatrix}^0.$$

We define

$$L = \begin{bmatrix} L_1 \\ \vdots \\ L_n \end{bmatrix}, A = \begin{bmatrix} A_1 \\ \vdots \\ A_n \end{bmatrix},$$

where $\begin{bmatrix} \vdots \end{bmatrix}^0$ represents the value at $\begin{bmatrix} \mu_0 & v_0 & \omega_0 & \gamma_0 & x_0 & y_0 & f_x & f_y & k_1 & \alpha_0 & \beta_0 \end{bmatrix}_i^T$. L_1 and L_n denote the difference matrix between the solar vector observed by the camera and the calculated solar vector at the first and nth moment, respectively. In addition, $w'_{x,1}$ and $w'_{y,1}$ are the error components of the azimuth and pitch directions of the solar vector observed by the camera and the calculated solar vector at the first moment, respectively; $w'_{x,n}$ and $w'_{y,n}$ denote the error components of the azimuth and pitch directions of the solar vector observed by the camera and calculated solar vector at the nth moment, respectively; and x_0 is the matrix of the difference between the values of each variable and each corresponding expansion point. A_1 and A_n denote the error equation at the first and nth moments, respectively, which are used to calculate the partial derivative matrix of each variable.

In this case, $L = Ax^0$, and we perform double left multiplication of A^T. After the positive definite treatment and matrix inversion, we obtain $x^0 = (A^T A)^{-1} A^T L$. Subsequently, x^0 is substituted into the following expression to obtain the parameters to be solved:

$$\begin{bmatrix} \mu_0 & v_0 & \omega_0 & \gamma_0 & x_0 & y_0 & f_x & f_y & k_1 & \alpha_0 & \beta_0 \end{bmatrix}_{i+1}^T = \begin{bmatrix} \mu_0 & v_0 & \omega_0 & \gamma_0 & x_0 & y_0 & f_x & f_y & k_1 & \alpha_0 & \beta_0 \end{bmatrix}_i^T + x^0$$

where $\begin{bmatrix} \mu_0 & v_0 & \omega_0 & \gamma_0 & x_0 & y_0 & f_x & f_y & k_1 & \alpha_0 & \beta_0 \end{bmatrix}_i^T$ is the first-order Taylor expansion point value from the 0th to ith points ($i = 0 \cdots n$).

Next, the Taylor expansion point is moved to the latest solution point $\begin{bmatrix} \mu_0 & v_0 & \omega_0 & \gamma_0 & x_0 & y_0 & f_x & f_y & k_1 & \alpha_0 & \beta_0 \end{bmatrix}_{i+1}^{T}$ expansion, and x_{i+1}^0 is solved again. The solution is iteratively found until the accuracy requirements are met.

4.3. Inverse Calculation of the Calibration Model

After solving the model, it is necessary to verify the results. After applying the calibration model, the encoder position coordinates α and β are calculated as the target value when the mirror normal vector and solar vector point in the same direction. Next, the servo motor is driven and controlled to move to the target position, and the camera collects the data for further verification. The model inverse solution algorithm after calibration is as follows.

According to the high-precision geometric calibration model, since the main point of the camera coincides with the image centroid coordinates when the mirror normal vector points toward the sun, that is, $x_i = x_0$ and $y_i = y_0$, the left term of the model is equal to zero. The right side of the model has a denominator $X_i c_1 + Y_i c_2 + Z_i c_3 \neq 0$. Therefore, the following formula is established, and the inverse solution algorithm model can be expressed as

$$\begin{cases} X_i a_1 + Y_i a_2 + Z_i a_3 = 0 \\ X_i b_1 + Y_i b_2 + Z_i b_3 = 0 \end{cases} . \tag{19}$$

According to Equation (19), the azimuth and pitch α and β of the encoder, respectively, can be calculated by the least squares method when the normal of the reflector at different positions points toward the sun at different times. We define

$$\begin{cases} \overline{X}_{re} = X_i a_1 + Y_i a_2 + Z_i a_3 \\ \overline{Y}_{re} = X_i b_1 + Y_i b_2 + Z_i b_3 \end{cases} .$$

In this case, the α and β values satisfying the accuracy requirement can be determined using the following formula:

$$\min_{\alpha,\ \beta} \left(\overline{X}_{re}^2 + \overline{Y}_{re}^2 \right). \tag{20}$$

5. Experimental Results and Analysis

5.1. Reliability Analysis of Measured Data

Before obtaining the experimental data, the equipment is placed at the initial position, and the central light axis direction of the reflector is initially determined to be due north. To accelerate the calibration progress, reduce the calibration time, and test the encoder's large-scale and multiple-angle motion characteristics, solar images at different positions of the camera array are collected. These images are used to perform the calibration model calculation and provide basic data to ensure accurate calibration. Using three techniques, three groups of data are collected to analyze the universality of the model solution. For the first group, the system moves from the right end to the left and collects two relatively irregular sets of pixel coordinate point data spread over the image plane of the detector. For the second group, the system moves from the right end to the left and collects a group of pixel coordinate points evenly distributed in the image plane of the detector. For the third group, the system moves from the left end to the right and collects a group of pixel coordinate points that are evenly distributed in the image plane of the detector. Moreover, the corresponding pitch, azimuth encoder readings and solar position parameters are recorded. The data acquisition path is shown in Figure 4.

Before the model is solved, the reliability of the experimental data is analyzed. The geometric parameters μ_0, v_0, ω_0, and γ_0 to be calibrated are set as 0, the calculated solar vector value of the three groups of data is considered the ordinate, the actual observation value of the optical axis vector of the mirror center is considered the abscissa for fitting analysis, and the calculated value of the solar vector is compared with the actual observation value.

The comparison results are shown in Figures 5–7, where $x_i - x_0$ and $y_i - y_0$ represent the actual solar vector pitch and azimuth components observed by the camera, respectively.

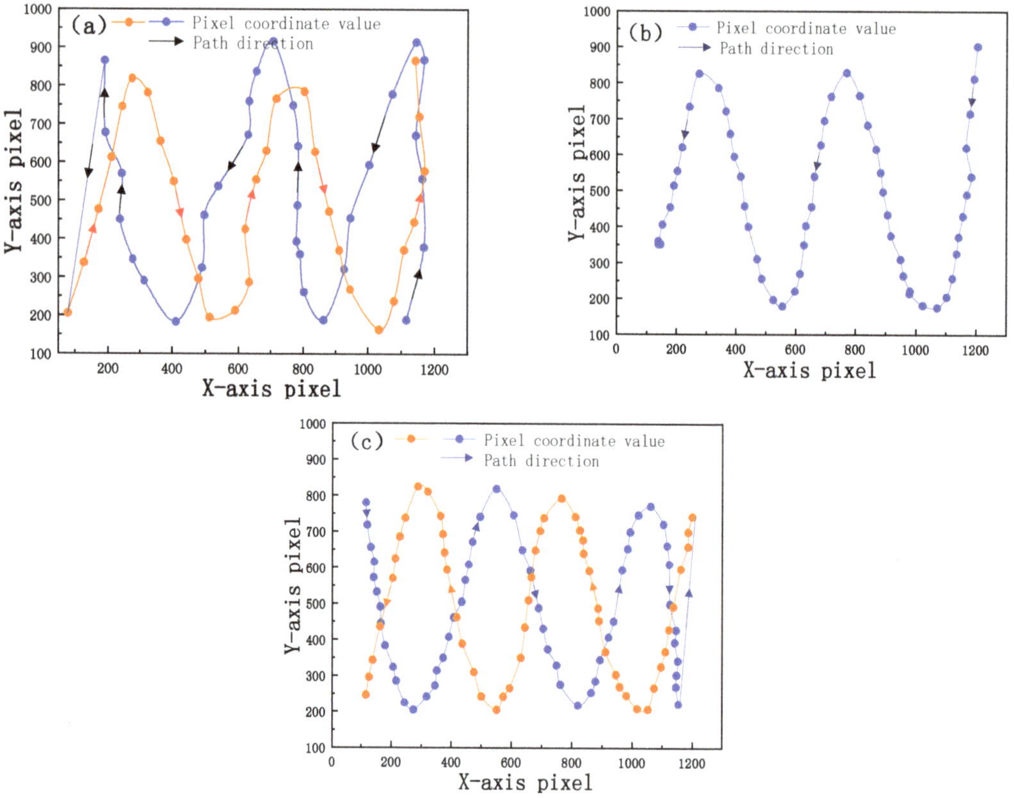

Figure 4. (**a**) First set of data; (**b**) second set of data; (**c**) third set of data.

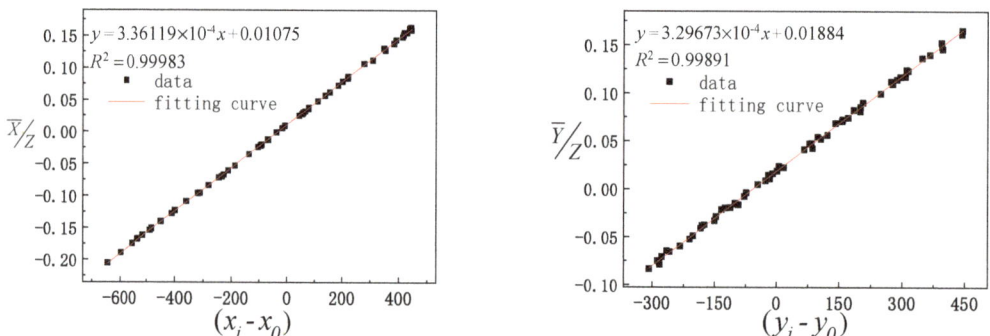

Figure 5. Fitting between the calculated solar vectors of the first group of data and actual observation values of the optical axis vector of the mirror center.

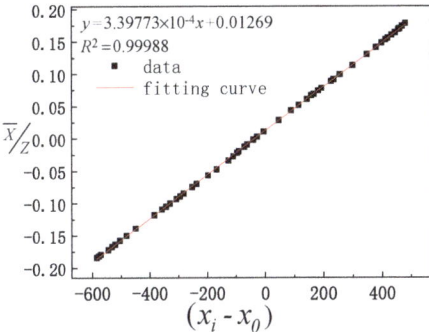 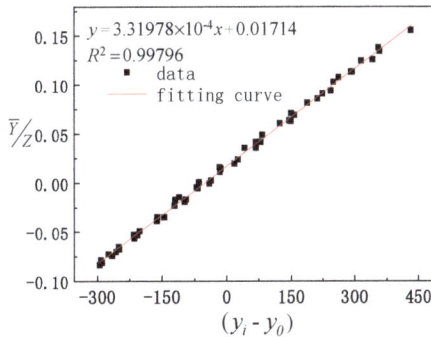

Figure 6. Fitting between the calculated values of the second group of data and actual observation values of the pointing mirror center optical axis.

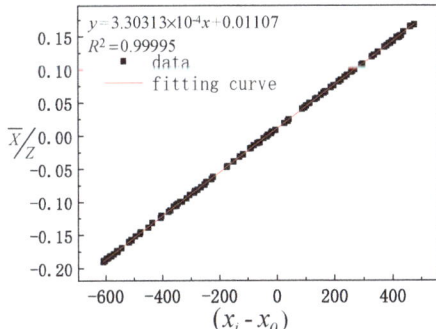 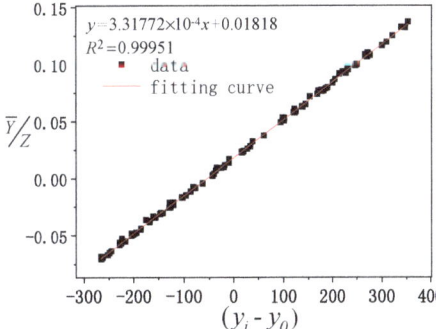

Figure 7. Fitting between the calculated values of the third group of data and actual observation values of the pointing mirror center optical axis.

It can be seen from Figures 5–7 that the data fitting results for the three groups of different paths indicate that the linear fitting correlation coefficient values between the calculated value of the solar vector and optical axis vector value of the mirror center observed by the camera are greater than 0.99. The linear fitting results are ideal, which further verifies the reliability of the experimental data and provides reliable basic data to solve the model.

5.2. Model Calculation and Theoretical Verification

The verified data solution model is used. The data of the model are shown in Table 1. Only 8 sets of data are listed in the table. The first row indicates the time of data collection. The second row indicates the corresponding pitch and azimuth encoder readings when the solar image is located at a certain position of the camera array. The third row indicates the altitude and azimuth of the sun in the local coordinate system corresponding to the data acquisition time.

In total, 105 sets of data are extracted from 221 sets of data to calculate the calibration model parameters. When the initial values of $[\ u_0\ \ v_0\ \ \omega_0\ \ \gamma_0\ \ \alpha_0\ \ \beta_0\ ]$, $[\ x_0\ \ y_0\ \ k_1\ ]$, and $[\ f_x\ \ f_y\ ]$ are [0 0 0 0 76 310] (unit:degree), [724 471 0.1063] pixels, and [15.6 mm 15.6 mm], respectively, the system parameters are [−0.1625 −0.178 0.10614 0.0345 77.19 310.49] (unit:degree), [719.03 470 −0.0009] pixels, and [15.614 mm 15.65 mm].

After the model is solved, it is necessary to evaluate the accuracy of the model parameters. First, the reliability of the results of the model is analyzed theoretically. The image centroid coordinates are used to represent the optical axis vector of the mirror as the X-axis, and the calculated solar vector value is considered the Y-axis in the fitting analysis. The linear fitting correlation coefficients of the two groups of values are considered to perform the reliability analysis of the evaluation model solution results. The fitting results of the two groups of data are shown in Figure 8.

Table 1. Data to solve the model.

Time	hh:mm:ss	9:06:53	9:09:09	9:11:46	9:14:33	9:17:14	9:21:06	9:24:12	9:27:23
Encoder angle	Pitch	103.667	103.271	102.524	102.524	103.579	104.7	105.952	107.029
value/(°)	Azimuth	167.563	167.256	167.278	167.256	167.585	167.278	166.663	166.355
Sun	Altitude	29.041	29.397	29.804	30.234	30.644	31.227	31.688	32.156
position/(°)	Azimuth	131.971	132.452	133.012	133.615	134.202	135.059	135.756	136.48
Sun cen-	Pixel x	206.383	216.259	244.145	273.59	317.017	345.299	351.167	372.528
troid/(pixel)	Pixel y	324.101	285.591	225.422	205.229	242.376	273.299	314.541	349.653

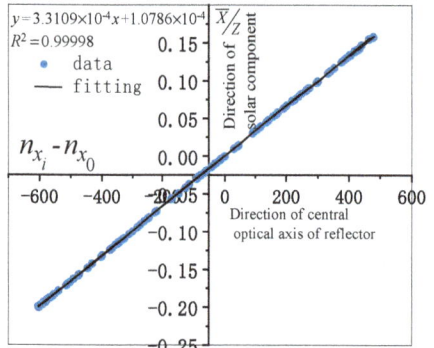
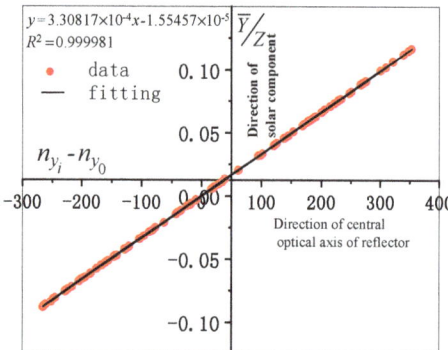

Figure 8. Fitting of the image centroid and solar vector.

It can be seen from the fitting results in Figure 8 that the image centroid coordinate represents the mirror normal direction consistent with the solar vector, and the fitting correlation coefficient R^2 is greater than 0.99998, which indicates a high linear correlation. Therefore, the reliability of the model results can be analyzed considering the theoretical data. Second, we analyze the error of the system calculation model. The system error caused by multipoint data optimization is used to analyze the pixel difference caused by the camera observation and angle difference caused by the encoder elevation and azimuth. The pixel, pitch, and azimuth error distributions corresponding to the systematic error distribution generated by the solution model are shown in Figure 9.

The error distribution data in Figure 9 show that the pixel error corresponds to the system model solution error, and the pixel average error and standard deviation in the X-axis direction are 1.253 pixels and 1.014 pixels, respectively. The average error and standard deviation in the Y-axis direction are 0.61 pixels and 0.45 pixels, respectively. The average error and standard deviation of the azimuth axis are 0.024° and 0.019°, respectively. The average error and standard deviation in the pitch axis direction are 0.012° and 0.0085°, respectively. According to the standard deviation data, these results are within the allowable error range. Therefore, from the theoretical error data, the reliability of the calculation model results is further verified.

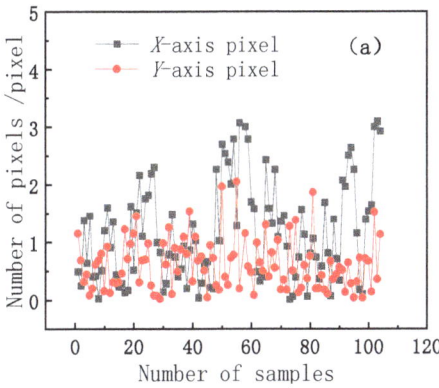
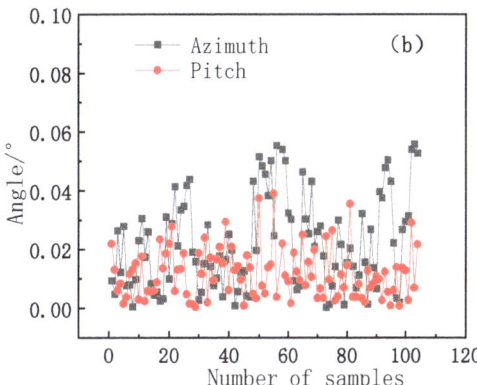

Figure 9. (a) Error distribution of the solution model corresponding to X and Y pixels; (b) pitch and azimuth angle error.

5.3. Model Experiment Verification

In this step, we further verify the accuracy of the model parameters. Through the experiment, using the model inverse solution algorithm after calibration, the corresponding encoder pitch and azimuth target positions corresponding to the sun at different times are inversely solved, and the motor is driven to the target position. Finally, the accuracy of the model is verified by the actual observation of the camera. Part of the test data of the validated model is shown in Table 2, where only 8 sets of data are presented.

Table 2. Data used to solve the model.

Sun position/(°)	Altitude	29.811	32.342	34.575	37.129	39.697	39.58	37.73	30.35
	Azimuth	141.002	145.701	150.732	158.317	172.518	188.583	199.285	217.961
Encoder target angle of inverse solution/(°)	Pitch	106.787	109.314	111.555	114.126	116.719	116.653	114.829	107.534
	Azimuth	169.585	164.905	159.895	152.336	138.186	122.168	111.467	92.747
Encoder measurement angle/(°)	Pitch	106.831	109.292	111.577	114.17	116.741	116.697	114.807	107.512
	Azimuth	169.629	164.927	159.873	152.292	138.164	122.212	111.489	92.703
Error target and measurement/(°)	Pitch	0.044	−0.022	0.022	0.044	0.022	0.044	−0.022	−0.022
	Azimuth	0.044	0.022	−0.022	−0.044	−0.022	0.044	0.022	−0.044

The first row in Table 2 indicates the solar altitude and azimuth angles when the central light axis of the reflector is aligned with the sun at different times. The second row indicates the target positions of the pitch and azimuth encoders, as calculated with the model inverse solution algorithm after calibration. The third row indicates the actual position measurement values of the encoder. The device considers the data presented in the second row as the target position, rotates the motor to the target position, and uses the encoder to detect the actual position as the feedback signal to further ensure the motion control accuracy of the turntable. The fourth row of data is the difference between the third row of data and the second row of data, which represents the pitch and azimuth control deviation. Figure 10 shows that the standard deviations of the pitch and azimuth angle control errors are 0.0176° and 0.0305°, respectively. The comparison and analysis of the pitch and azimuth encoder test data indicate that the model calculations are consistent with the measured values. The error range is approximately 0.04°, and the accuracy is better than 0.1°, which satisfies the verification requirements of the calibration model. The accuracy of the model is thus preliminarily verified by analyzing the motion control accuracy of the turntable and through actual observations by the solar observer.

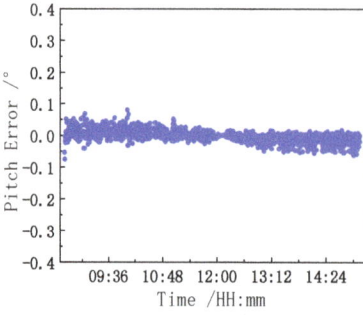
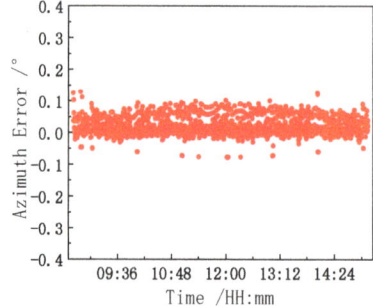

Figure 10. Pitch and azimuth control error.

Through the inverse calibration model, the motor is driven and controlled, and the model is preliminarily verified. To further verify the accuracy of the model parameters, by considering the actual observation of the camera after calibration, the solar image is tracked and collected, and the centroid coordinates of the solar image are used for verification. The centroid coordinates of the solar image at different times are compared with the camera main point coordinates to reflect the deviation degree of the center light axis of the reflector pointing toward the sun. The root mean square error (RMSE) of the two groups of data is calculated by Equation (21) to quantitatively evaluate the correctness of the model solving parameters and the tracking control accuracy of the system.

$$\sigma(\theta) = \sqrt{\frac{\sum (x - x_0)^2}{n - 1}} \quad \sigma(\varphi) = \sqrt{\frac{\sum (y - y_0)^2}{n - 1}} \quad (21)$$

Here, $\sigma(\theta)$ and $\sigma(\varphi)$ are the RMSEs of the pitch and azimuth respectively; x_0 and y_0 are the coordinates of the principal point of the camera after calibration; x and y are the image centroid coordinates.

The centroid test data of the experimental verification model are presented in Table 3, where only 8 sets of data are listed.

Table 3. Partial centroid test data of the validated model.

Sun centroid/(pixel)	Pixel x	722.355	723.784	723.463	721.453	721.31	720.256	720.454	719.536
	Pixel y	474.071	472.398	473.061	469.46	472.441	469.79	471.393	470.585
Camera main point/(pixel)	Pixel x	719.000	719.000	719.000	719.000	719.000	719.000	719.000	719.000
	Pixel y	470.000	470.000	470.000	470.000	470.000	470.000	470.000	470.000
Error centroid and main point/(pixel)	Pixel x	3.355	4.784	4.463	2.453	2.31	1.256	1.454	0.536
	Pixel y	4.071	2.398	3.061	-0.54	2.441	-0.21	1.393	0.585

The first row in Table 3 indicates the measured image centroid coordinates when the reflector centroid axis is aligned with the sun according to the target value of the inverse calibration model. The second row of data pertains to the use of a checkerboard to calibrate the camera's main point coordinates. The third row shows the deviation between the measured image centroid and camera main point. The two sets of data and deviations are shown in Figure 11.

According to the two sets of data in Figure 9a,b, it can be determined by formula (21) that the RMSE values of the X- and Y-axis pixels are 2.0995 pixels and 0.8689 pixels, respectively, and the corresponding pixel angle resolution errors are 0.0377° and 0.0144°. The synthetic angular resolution error is calculated by formula (22) combined with the standard uncertainty formula [30], and the synthetic angular resolution error is 0.0403°.

$$u_c = \sqrt{\sum_{i=1}^{N} u_i^2} \tag{22}$$

Here, u_i is the component of error uncertainty.

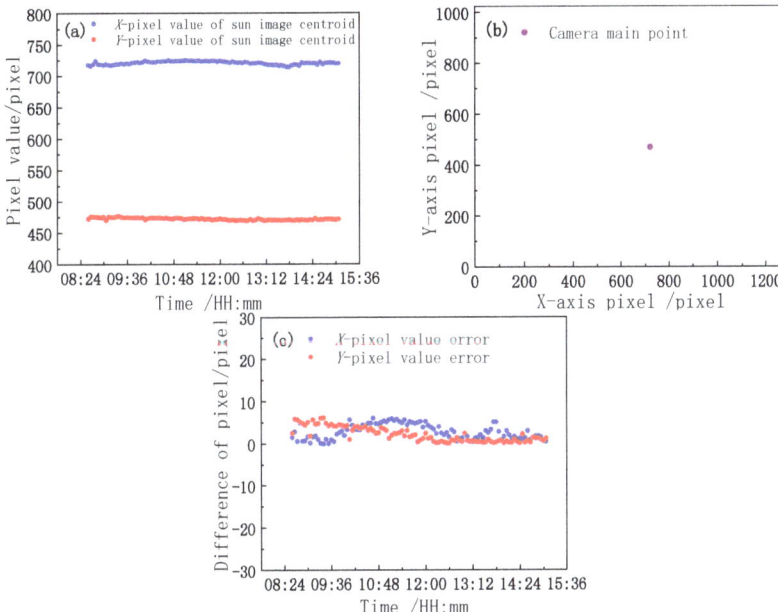

Figure 11. (a) Measured image centroid; (b) camera main point; (c) deviation.

It can be determined from the above analysis data that a small deviation exists between the centroid coordinates of the solar image obtained by the actual observation of the camera as the observation value and the coordinates of the main point of the camera as the real value. Nevertheless, the two sets of data are consistent, which demonstrates the accuracy of the calibration model. At the same time, the tracking control accuracy of the system is also measured through the RMSE. Because the tracking accuracy of the system represents the normal pointing accuracy of the mirror, the tracking control accuracy of the system is also the pointing accuracy of the system.

5.4. Accuracy Analysis of System Tracking

Through the experimental verification and analysis of the calibration model, the accuracy of system tracking using the model is evaluated. The tracking accuracy of the system mainly includes the motion control accuracy, external image processing algorithm accuracy and calibration model calculation accuracy. The accuracy of the motion control pertains to the accuracy (0.0003°) of the solar position calculated with the astronomical algorithm [31] and detection accuracy of the encoder (0.02°). The accuracy of the external image processing algorithm pertains to the accuracy of the image centroid extraction algorithm (0.032°) [32–36], average reprojection error of the camera calibration (0.1299 pixels), interference of the solar image noise and accuracy of the calibration model calculation. The uncertainty sources affecting the tracking accuracy of the system are presented in Table 4. The system tracking accuracy summarizes all the factors. The RMSE of the solar image obtained by the actual observation of the camera as the observation value and camera principal point coordinate as the real value is comprehensively evaluated as 0.0403°, and the tracking accuracy is noted to be better than 0.1°, which meets the requirements of the comprehensive pointing control accuracy of the system [37–40].

Table 4. Uncertainty analysis of system calibration.

Uncertainty Sources in System Tracking Control		Uncertainty and Error
Internal source	Motion control error	0.02°
	Astronomical algorithm accuracy	0.0003°
	Encoder detection accuracy	0.02°
External source	Image processing algorithm accuracy	0.030°
	Image centroid extraction algorithm accuracy	0.0204°
	Camera calibration average reprojection error	0.0021°
	Solar image noise interference	0.0219°
	Calibration model solution accuracy	0.180°
	Comprehensive evaluation accuracy	0.0403°

According to the data in Table 4, the uncertainty of the system calibration is approximately 0.0403°. That is, the tracking control accuracy of the system is 0.0403°, which is greatly improved compared with the tracking accuracy of the tracking equipment in the solar photovoltaic industry and the tracking accuracy of foreign point light sources [24,41–46]. This finding demonstrates the effectiveness of the calibration model in this paper.

Overall, the motion control error, encoder detection accuracy and image centroid extraction algorithm accuracy are the main error sources in the system control accuracy. Therefore, it is necessary to enhance the detection accuracy of the encoder, overcome the interference caused by the mechanical transmission error and unbalanced force in the motion processes, and optimize the image quality and image centroid extraction algorithm. Moreover, by enhancing the accuracy of the calibration camera and reducing the influence of the error caused by the model, the tracking accuracy of the system can be further increased to enhance the comprehensive pointing accuracy of the system and more effectively realize radiometric calibration and MTF detection of high-spatial resolution satellites.

6. Conclusions

A high-precision automatic geometric calibration modeling method for a point light turntable is proposed. Based on the principle of geometric calibration modeling, a high-precision automatic calibration model is established. By analyzing the reliability of the experimental data and solving the model, the feasibility and effectiveness of the method are demonstrated theoretically and experimentally. This approach can overcome the problem of the low precision of normal and single-point calibration, which limits the enhancement of the pointing accuracy. Moreover, the approach can reduce the calibration time, accelerate the calibration progress and increase the work efficiency, which facilitates high-frequency and high-efficiency networking automation to carry out the calibration of point light sources with different energy levels and increase the pointing accuracy of the system, achieve high-precision control of the central optical axis of the point light source reflector to point toward the target position, and reflect the light spot toward the satellite entrance pupil. Finally, this work lays a foundation for the high-precision, high-frequency, operational on-orbit radiometric calibration and MTF detection of high-resolution satellites. In addition, this system modeling method provides a theoretical basis for heliostat and solar photovoltaic equipment calibration.

Author Contributions: The contributions of the authors to this paper are as follows: Conceptualization, R.L., L.Z. and W.X.; methodology, R.L. and L.Z.; software, X.L.; investigation, R.L. and W.X.; data curation, R.L. and C.H.; writing—original draft preparation, R.L.; writing—review and editing, R.L. and X.W.; mechanical structure, J.L.; supervision, L.Z. and X.W. All the authors have read and agreed to the published version of the manuscript.

Funding: This research received no external funding.

Acknowledgments: The authors appreciate the editor and anonymous reviewers for their constructive comments and suggestions regarding this study. Moreover, the first author thanks his tutor for his constructive guidance and all the colleagues in the research group for their support.

Conflicts of Interest: The authors declare no conflict of interest.

References

1. Wu, Y.; Lu, J.; Lin, R.; Dang, X. Research on the Standard of High-resolution Earth Observation Satellite Optical Remote Sensor. *Aerosp. Stand.* **2018**, *1*, 4–9.
2. Gu, X.; Tian, G.; Yu, T.; Li, X.; Gao, H.; Xie, Y. *Principle and Method of Radiometric Calibration for Aerospace Optical Remote Sensor*, 1st ed.; Sci. Press: Beijing, China, 2013; pp. 1–68.
3. Zheng, X. High-Accuracy Radiometric Calibration of Satellite Optical Remote Sensors. *Spacecr. Recovery Remote Sens.* **2011**, *32*, 36–43.
4. Li, Z.; Zheng, X.; Tang, L.; Li, R.; Xi, H. An Advanced Technology of Absolute Radiometric Calibration for Optical Remote Sensors. *J. Remote Sens.* **2007**, *11*, 581–588.
5. Qiao, Y.; Zheng, X.; Wang, X.; Zhang, L.; Yi, W.; Wang, L. Whole-Process Radiometric Calibration of Optical Remote Sensors. *J. Remote Sens.* **2006**, *10*, 616–623.
6. Zheng, X. Some respects for development of radiometric calibration techniques for optical remote sensing satellite. *J. Atmos. Environ. Opt.* **2014**, *9*, 2–8.
7. Xu, W.; Zhang, L.; Si, X.; Li, X.; Yang, B.; Shen, Z. On-orbit modulation transfer function detection of high resolution optical satellite sensor based on reflected point sources. *Acta Opt. Sin.* **2017**, *37*, 0728001.
8. Wang, M.; Zhou, S.; He, M.; Cheng, X. Characterization and Calibration Method of Satellite Sensor Radiometric Calibration Site. *Geomat. Spat. Inf. Technol.* **2015**, *7*, 24–27.
9. Li, R.; Zhang, L.; Xu, W.; Si, X.; Li, J.; Hu, Y.; Wang, X.; Wang, J. High-precision calibration modeling of point-light-source turntable based on solar vector. *Acta Opt. Sin.* **2019**, *39*, 1–12.
10. Li, R.; Zhang, L.; Li, X.; Wang, W.; Li, J.; Hao, J. Specular Normal Calibration technology of Point Source. *Acta Photonica Sin.* **2020**, *49*, 04120021-12.
11. Schiller, S.; Silny, J. The specular array radiometric calibration (SPARC) method: A new approach for absolute vicarious calibration in the solar reflective spectrum. *Proc. SPIE* **2010**, 78130E. [CrossRef]
12. Xu, W.; Zhang, L.; Chen, H.; Li, X.; Yang, B.; Wang, J. In-flight radiometric calibration of high resolution optical satellite sensor using reflected point sources. *Acta Opt. Sin.* **2017**, *37*, 0328001.
13. Xu, W.; Zhang, L.; Li, X.; Si, X.; Xu, Y. On-Orbit Radiometric Calibration of High-Resolution Optical Remote Sensing Satellite Based on Sub-Pixel Targets. *Acta Opt. Sin.* **2019**, *39*, 15–22.
14. Schiller, S. Application of The Specular Array Radiometric Calibration (SPARC) Method for The Vicarious Calibration of Landsat Sensors. In Proceedings of the ASPRS 2016 Annual Conference and co-located JACIE Workshop, Fort Worth, TX, USA, 11–15 April 2016; Volume 581, pp. 2–3.
15. Comprehensive Vicarious Calibration and Characterization of a Small Satellite Constellation Using the Specular Array Radiometric Calibration (SPARC) Method. 2017. Available online: https://digitalcommons.usu.edu/smallsat/2017/all2017/108/ (accessed on 23 March 2021).
16. Leger, D.; Viallefont, F.; Hillairet, E.; Meygret, A. In-flight refocusing and MTF assessment of SPOT5 HRG and HRS cameras. In Proceedings of the SPIE 4881, Sensors, Systems, and Next-Generation Satellites VI, Crete, Greece, 23–27 September 2002; Volume 4881.
17. Helder, D.; Choi, T.; Rangaswamy, M. In-flight characterization of spatial quality using point spread functions. Post-Launch Calibration of Satellite Sensors. In Proceedings of the International Workshop on Radiometric and Geometric Calibration, Gulfport, MS, USA, 2–5 December 2003; Morain, S.A., Budge, A.M., Eds.; CRC Press: Boca Raton, FL, USA, 2003.
18. Improving Spatial and Temporal Coverage in Earth Observations through Inter-Sensor Data Harmonization. 2019. Available online: https://digitalcommons.usu.edu/calcon/ (accessed on 23 March 2021).
19. Schiller, S.; Silny, J. Measuring Atmospheric Optical Depth Directly from Satellite Imagery. In Proceedings of the JACIE 2011 Civil Commercial Imagery Evaluation Workshop, Boulder, CO, USA, 29–31 March 2011.
20. Schiller, S.; Silny, J. Specular Array Radiometric Calibration (SPARC). In Proceedings of the In-Flight Sensor Performance Assessment SAEF/SMUG Conference, SI Stonegate Facility, Chantilly, VA, USA, 26 June 2012.
21. Schiller, S.; Silny, J. In-Flight Performance Assessment of Imaging Systems Using the Specular Array Radiometric Calibration (SPARC) Method. In Proceedings of the 11th Annual Joint Agency Commercial Imagery Evaluation (JACIE) Workshop, Fairfax, VA, USA, 18 April 2012.
22. Xu, W. On-orbit MTF Estimation of High Resolution Satellite Optical Sensor. Ph.D. Thesis, Chinese Academy of Sciences, Hefei, China, 2011.
23. Feng, H.; Zhang, L.; Li, J. High-precision pointing technology with reflected point light source based on CCD camera. *Acta Opt. Sin.* **2018**, *38*, 0528004. [CrossRef]

24. Optimizing Product Quality and Cost for Small Satellite Earth-Imaging Constellations by Incorporating Automated Vicarious Calibration in the Processing Pipeline, Small Satellite Conference. 2018. Available online: https://digitalcommons.usu.edu/smallsat/2018/all2018/356/ (accessed on 23 March 2021).
25. Wang, S. *Photogrammetry*, 1st ed.; Surveying and Mapping Press: Beijing, China, 2013; pp. 1–68.
26. Zhu, L. *Fundamentals of Photogrammetry*, 1st ed.; Surveying and Mapping Press: Beijing, China, 2018; pp. 25–41.
27. Zhang, Z. A flexible new technique for camera calibration. *IEEE Trans. Pattern Anal. Mach. Intell.* **2000**, *22*, 1330–1334. [CrossRef]
28. Wang, J.; Wang, W.; Ma, Z. Hybrid-model based camera distortion iterative calibration method. *Bull. Surv. Mapp.* **2019**, *4*, 103–106.
29. Cui, E. *Research on Key Techniques of High-Precision Binocular Visual Measurement System*; University of Chinese Academy of Sciences: Changchun, China, 2018.
30. Fei, Y. *Error Theory and Data Processing*, 7th ed.; China Machine Press: Beijing, China, 2017; pp. 88–91.
31. Solar Position Algorithm for Solar Radiation Applications Technical Report. 2008. Available online: https://www.nrel.gov/docs/fy08osti/34302.pdf (accessed on 23 March 2021).
32. Yang, P.; Xie, L.; Liu, J. Zernike moment based high-accuracy sun image centroid algorithm. *J. Astronaut.* **2011**, *32*, 1963–1969.
33. Zhan, Y.; Zheng, Y.; Zhang, C.; Ma, G.; Luo, Y. Image centroid algorithms for sun sensors with super wide field of view. *Acta Geod. Cartogr. Sin.* **2015**, *44*, 1078–1084.
34. Tu, B.; Han, K.; Wang, H.; Bai, J.; Jin, Z. Design of Digital Sun Sensor with Large Field. *Chin. J. Sens. Actuators* **2011**, *24*, 336–341.
35. Chen, Y.; Feng, Y.; Wei, L.; Zhao, H.; Zhu, Z. Experiment Research on Subpixel Location Error of the Facula Centroid. *Opto-Electron. Eng.* **2010**, *37*, 80–84.
36. Gao, S.; Zhao, M.; Zhang, L.; Zhou, Y. Improved algorithm about subpixel edge detection of image based on zernike orthogonal moments. *Acta Autom. Sin.* **2008**, *34*, 1163–1168. [CrossRef]
37. Yuan, X.; Yu, J. Calibration of constant angular error for high resolution remotely sensed imagery. *Acta Geodaetica et Cartographica Sinica* **2008**, *37*, 36–41.
38. Tu, X.; Xu, M.; Liu, L. The geometric calibration of airborne three-line-scanner ADS40. *Acta Geodaetica et Cartographica Sinica* **2011**, *40*, 78–83.
39. Liu, Y.; Jin, G.; He, H. Research on model of pointing error of a three-axis simulation turntable. *J. Harbin Inst. Technol.* **2005**, *37*, 701–704.
40. Xue, L.; Chen, T.; Xu, T.; Liu, Y.; Li, B. High-precision calculation for attitude angles of fast steering mirror. *Opt. Precis. Eng.* **2016**, *24*, 2000–2009.
41. Awasthi, A.; Shukla, A.; Manohar, S.; Zarchi, H. Development of a machine vision dual-axis solar tracking system. *Sol. Energy* **2018**, *169*, 136–143.
42. Song, J.; Zhu, Y.; Zhou, J.; Zhang, Y. Daylighting system via fibers based on two-stage sun-tracking model. *Sol. Energy* **2014**, *108*, 331–339. [CrossRef]
43. Fathabadi, H. Comparative study between two novel sensorless and sensor based dualaxis solar trackers. *Sol. Energy* **2016**, *138*, 67–76. [CrossRef]
44. Awasthi, A.; Shukla, A.; Manohar, S.; Dondariya, C.; Shukla, K.; Porwal, D.; Richhariya, G. Review on sun tracking technology in solar PV system. *Sol. Energy* **2020**, *6*, 392–405. [CrossRef]
45. Wang, X.; Hu, H.; Wang, B. Dual-Mode Solar Tracking Controller. *Electron. Opt. Control* **2012**, *19*, 80–83. [CrossRef]
46. Fan, H.; Kang, T.; Zhang, P.; Liang, W.; Huang, R. Research on dual-mode tracking system of solar azimuth. *Mod. Electron. Tech.* **2020**, *43*, 51–55.

Article
Surface Reconstruction Assessment in Photogrammetric Applications

Erica Nocerino [1,*], Elisavet Konstantina Stathopoulou [2,3], Simone Rigon [2] and Fabio Remondino [2]

1. LIS UMR 7020, Aix-Marseille Université, CNRS, ENSAM, Université De Toulon, Domaine Universitaire de Saint-Jérôme, Bâtiment Polytech, Avenue Escadrille Normandie-Niemen, 13397 Marseille, France
2. 3D Optical Metrology (3DOM) Unit, Bruno Kessler Foundation (FBK), 38123 Trento, Italy; estathopoulou@fbk.eu (E.K.S.); srigon@fbk.eu (S.R.); remondino@fbk.eu (F.R.)
3. Laboratory of Photogrammetry, National Technical University of Athens (NTUA), 15780 Athens, Greece
* Correspondence: erica.nocerino@univ-amu.fr

Received: 10 September 2020; Accepted: 12 October 2020; Published: 16 October 2020

Abstract: The image-based 3D reconstruction pipeline aims to generate complete digital representations of the recorded scene, often in the form of 3D surfaces. These surfaces or mesh models are required to be highly detailed as well as accurate enough, especially for metric applications. Surface generation can be considered as a problem integrated in the complete 3D reconstruction workflow and thus visibility information (pixel similarity and image orientation) is leveraged in the meshing procedure contributing to an optimal photo-consistent mesh. Other methods tackle the problem as an independent and subsequent step, generating a mesh model starting from a dense 3D point cloud or even using depth maps, discarding input image information. Out of the vast number of approaches for 3D surface generation, in this study, we considered three state of the art methods. Experiments were performed on benchmark and proprietary datasets of varying nature, scale, shape, image resolution and network designs. Several evaluation metrics were introduced and considered to present qualitative and quantitative assessment of the results.

Keywords: surface reconstruction; mesh model; 3D reconstruction; visibility constraints; volumetric methods; dense point cloud; multiple view stereo (MVS); dense image matching (DIM); photogrammetry; computer vision

1. Introduction

The 3D reconstruction of the physical shape or geometry of either single objects or complex scenes is a topic of interest in countless application scenarios, varying from more industrial analyses [1], cultural heritage related studies [2,3], environmental mapping [4,5] and city modeling [6,7] to the latest autonomous driving and navigation applications [8]. Polygonal meshes in the form of triangular or quadrilateral faces are typically used to represent the digital surface of such objects or scenes in the 3D space.

The employed technique used to acquire the input data highly affects the quality of the final surface reconstruction. Among a large variety of active and passive optical sensors and methods, image-based 3D reconstruction is frequently used due to its easiness, portability, efficiency and reliability. In particular, dense image matching (DIM) is the process of calculating the 3D coordinates of each pixel visible in at least two images, thus generating a dense representation of the scene. In photogrammetry, DIM follows the image orientation, triangulation and camera calibration steps commonly calculated within the bundle adjustment (BA) process [9].

Equivalent to this, in the computer vision community, the task of reconstructing a dense 3D representation of the scene from a collection of images is known as multi-view stereo (MVS) [10], typically performed as a subsequent step to the Structure from Motion (SfM) procedure.

Traditionally, the final output of DIM is a 3D point cloud (or, in mapping applications, 2.5D digital elevation model, DEM) of the scene [9,11]. The surface or mesh reconstruction is usually applied to the point cloud resulting from the DIM, without any further checks on the images and their orientations.

On the other hand, MVS encompasses distinct 3D reconstruction methods that may deliver different output products in the form of depth maps, point clouds, volume scalar-fields and meshes [12]. Such different scene representations have been initially developed for visualization and graphics applications, each of them being optimized for different purposes, following also the evolution in hardware and computational power. A certain class of MVS approaches generates a refined mesh model using photo-consistency (i.e., pixel color similarity) measures and the so-called visibility information (i.e., the image orientations and thus the 2D–3D projections) [12,13].

Paper's Motivation and Aim

This study aimed to investigate the surface reconstruction problem for image-based 3D reconstruction scenarios. The paper builds upon the following considerations:

- In a traditional photogrammetric pipeline, the meshing step interpolates a surface over the input 3D points. This is usually disjointed from the 3D point cloud generation DIM but can potentially leverage and take advantage of additional information from the previous steps of the workflow, i.e., visibility constraints and photo-consistency measures which are generally not considered in popular meshing algorithms as Poisson [14].
- Dense point clouds can be heavily affected by poor image quality or textureless areas, resulting in high frequency noise, holes and uneven point density. These issues can be propagated during the mesh generation process.
- Volumetric approaches for surface reconstruction based on depth maps are well-established, time-efficient methods for depth sensors, also known as RGB-D [15], and might be a valid approach also for pure image-based approaches.

The aim of this work was thus to evaluate whether the integration of visibility information (image orientation) and photo-consistency and during the meshing process can potentially lead to an improvement of the mesh quality (and successive products). For this reason, three diverse surface reconstruction approaches were considered and evaluated on diverse datasets (Figure 1):

- *Method 1*: Surface generation and refinement are incorporated in the 3D reconstruction pipeline. The mesh is generated after depth maps and dense point clouds are estimated and is subsequently refined considering visibility information (i.e., image orientation) to optimize a photo-consistency score over the reconstructed surface [13,16].
- *Method 2*: Surface generation is disjoint from the image-based 3D reconstruction procedure. The dense point cloud, as obtained from Method 1, is converted to a mesh model without the use of any visibility constraints or photo-consistency checks [14,17].
- *Method 3*: Given the image poses, a mesh model is generated from the depth maps produced in Method 1, employing a volume integration approach [15,18,19]. Again, in this method, visibility and photo consistency information are not taken into consideration while reconstructing the surface.

The results of the considered approaches were evaluated using several metrics, including accuracy, completeness and roughness. On the contrary, the computational time was not considered a key factor for this investigation.

The rest of the article is divided as follows. Section 2 reviews the main concepts and steps of DIM and MVS. Section 3 provides an overview of the available DIM/MVS benchmark datasets, examining their suitability for the present study; surface reconstruction and assessment criteria are also addressed. The considered surface reconstruction methods are then introduced in Section 4. The employed datasets, carefully chosen to cover a wide range of image scale, image resolution and application scenarios (from close range to aerial photogrammetry), and the adopted comparative metrics are presented in Section 5, followed by a discussion of the obtained results in Section 6.

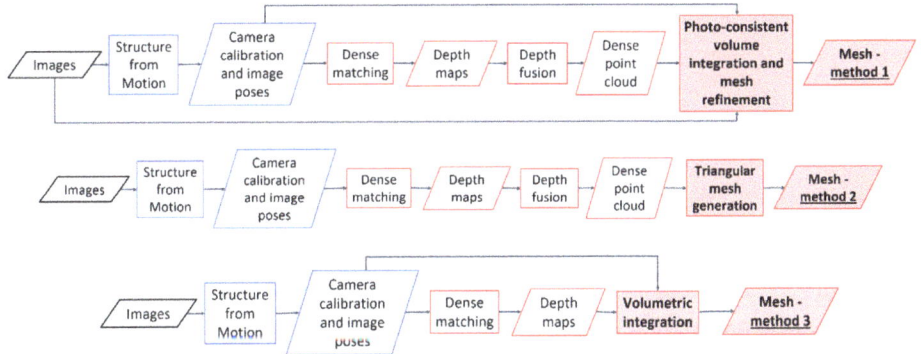

Figure 1. The three surface generation approaches investigated in the paper.

2. On DIM and MVS

Matching is a general term used to define approaches for finding correspondences between two images, sets of features or surfaces [20]. In photogrammetry, image matching indicates the (automatic) procedure of identifying and uniquely matching corresponding (homologous, conjugate) features (points, patterns and edges) between two (i.e., stereo) or more (i.e., multi-view) overlapping images. In computer vision, the analogous step is the so-called "stereo correspondence" problem [21].

Image matching can be sparse or dense, stereo or multi-view. In sparse matching, detectors and descriptors are usually employed to extract and characterize a set of sparse and potentially matching image features; their local appearance is then used to search and match corresponding locations in other images. Some approaches first extract only highly reliable features and then use them as seeds to grow additional matches [10]. Sparse matching algorithms are an integral part of automatic image orientation procedures implemented in SfM algorithms.

In dense image matching (DIM), a huge number of correspondences (up to pixel-to-pixel) between image pairs (dense stereo matching) or multiple views is established. The dense correspondence problem is still a crucial and active research topic for applications where dense and detailed 3D data generation are needed. It is more challenging than the sparse correspondence problem, since it requires inferring correspondences also in textureless, reflective and challenging areas [10]. Szeliski [10] identified four main steps that are usually implemented in dense correspondence algorithms: (1) matching cost computation; (2) cost (support) aggregation; (3) disparity computation and optimization; and (4) disparity refinement. Based on the various implementations of the aforementioned fundamental steps, diverse methods have been proposed. Several approaches have been developed to measure the agreement between the pixels and find the best match, from local to semi-global [22] and global methods, from area or patch-based [23,24] to feature-based [25] or a combination of them [26]. The most important used criterion to find corresponding pixels is known as photo-consistency, which estimates the similarity of two (or more) pixels between two images [12]. Examples of photo-consistency metrics are the Sum of Squared Differences (SSD), Sum of Absolute Differences (SAD), Normalized Cross Correlation (NCC) and Mutual Information (MI) [12].

The term dense stereo matching refers to the subclass of dense correspondence methods focusing on establishing correspondences between pixels in a stereo pair of images [20]. When three or more overlapping images are involved in the reconstruction process, the dense matching problem is defined as multi-view, multi-view stereo or multiple view. The ultimate goal of MVS is to reconstruct a complete and potentially globally consistent 3D representation of the scene from a collection of images acquired from known positions [10,27].

Examples of MVS algorithms are surface-based stereo, voxel coloring, depth map merging, level set evolution, silhouette and stereo fusion, multi-view image matching, volumetric graph cut and carved visual hulls [10]. An exhaustive taxonomy of DIM and MVS methods is extremely complex and, thus, several classification schemes have been proposed up to now. For instance, Seitz et al. [27] divided MVS algorithms based on six criteria: scene representation, photo-consistency measure, visibility model, shape prior, reconstruction algorithm and initialization requirements. Aanæs et al. [28] divided MVS approaches into two main categories: point cloud based methods (e.g., [23,29–32]) and volume-based methods (e.g., [33–35]).

In this paper, we adapt the categorization proposed by Furukawa and Hernández [12], focusing on the different output of the MVS procedure in terms of scene's representation. MVS starts with the search of corresponding pixels in the images in order to transform these dense correspondences into depth maps and/or point clouds. Visibility and occlusions estimation can be integrated in the matching process and are usually performed in coarse-to-fine manner as the dense reconstruction progresses to optimize the photo-consistency computation [12]. Based on the photo-consistency result (i.e., once the corresponding pixels have been identified in the images), depth maps are reconstructed for each image used as reference and matched with its visual neighbors. The resulting depth maps are then merged to produce the final 3D point cloud. Alternatively, when corresponding pixels with the highest photo-consistency score are found in two or more images, they are directly converted into 3D coordinates using collinearity in order to generate a dense point cloud.

Subsequently, the mesh generation follows. Oriented (i.e., with normals) and unoriented (i.e., without normals) point clouds can be converted into mesh models using several algorithms, such as Poisson surface reconstruction [14]. Alternatively, more sophisticated optimization techniques or volumetric surface reconstruction approaches have also been largely investigated [13]. Some of them require dense point clouds as an intermediate step while generating the surface model [36,37]. Other volumetric methods, such as the so-called Truncated Signed Distance Field algorithm (TSDF), use straightaway depth maps and generate a surface by dividing the 3D space into 3D voxel cells where each voxel is labeled with a distance [38]. However, while surface reconstruction from depth maps is quite common when using RGB-D sensors [15,39], in image-based 3D applications of metric accuracy it is still not fully exploited. Indeed, point clouds are the most common and requested product of a photogrammetric project while mesh models are generally produced mainly for rendering and visualization purposes.

Finally, a mesh refinement step can be undertaken [12]. This requires that images are considered again to verify the photo-consistency, this time over the reconstructed mesh surface. The vertices are moved to optimize their location, individually or all together. In the optimization process, a regularization term can influence the smoothness of the final mesh and, when available, silhouettes can be included as an additional consistency measure.

3. Benchmarks and Assessment of Surface Reconstruction Approaches

The current section is divided in two parts. In the first, existing benchmarks and evaluation methods adopted in photogrammetry and computer vision are reviewed, showing that they mainly focus on dense point clouds. The second part addresses the quality metrics developed in computer graphics for the assessment of surface reconstruction approaches. Some of these metrics were adopted in the comparative evaluation presented in Section 6.

The use of benchmarks is a common practice in the scientific community for the purpose of assessing and comparing different techniques, methods and algorithms. They collect data characterized by relevant features and evaluated according to significant metrics. A benchmark is usually composed of three elements: (1) input data to apply the investigated method; (2) reference or ground truth data against which the achieved results are compared; and (3) assessment criteria for the evaluation procedure.

3.1. DIM/MVS Benchmarks

Bakuła et al. [40] and Özdemir et al. [41] provided overviews of benchmarking initiatives proposed in photogrammetry and remote sensing. Knapitsch et al. [42] and Schops et al. [43] discussed and proposed benchmarks specifically focusing on MVS. Each of the available benchmarks has unique features, which cover different scene characteristics, from small objects in laboratory conditions, such as in Middlebury MVS [27,44] and DTU Robot Image Data Sets [28,45] or 3DOMcity benchmark [41,46], to more and complex scenes both indoor and outdoor (Strecha [47], ETH3D [43,48]; Tanks and Temples [42,49]; ISPRS-EuroSDR benchmark on High Density Aerial Image Matching [50,51]. Image resolutions also vary, from very small (0.2 Mpx Middlebury MVS) to medium (6 Mpx Strecha) up to high (24 Mpx 3DOMcity and ETH3D) and very high resolution aerial images (136 Mpx of the ISPRS-EuroSDR benchmark). Frames extracted from videos of different quality are also available (ETH3D and Tanks and Temples).

Ground truth data for benchmarking MVS methods are usually acquired with laser scanner systems and used in the form of a point cloud. Strecha and Middlebury MVS convert the point cloud into a triangle mesh, yet the reference models are not publicly available from Middlebury MVS. In some cases, the ground truth is available only for "training" data, while additional scenes are provided for the evaluation (ETH3D and Tanks and Temples).

Most of the evaluation procedures impose a resampling or regularization of the MVS 3D data to be evaluated and, if the submitted result is a mesh, a conversion into point cloud (DTU Robot Image Data Sets, ETH3D, Tanks and Temples) is performed.

The investigated methods are often evaluated by submitting the obtained results online (e.g., Middlebury MVS and 3DOMcity); however, open source code is also made available for offline testing and training (DTU Robot Image Data Sets, ETH3D and Tanks and Temples).

The assessment protocol requires the reference/ground truth (GT) and tested/submitted data (D) to be aligned, i.e., co-registered in the same reference system. This may be accomplished in different ways: (i) using the provided image interior and exterior orientation parameters (Middlebury MVS and ETH3D); (ii) computing a 7-Degrees of Freedom (DoF) spatial transformation through absolute orientation of the image exterior orientation parameters (3DOMcity and Tanks and Temples); and (iii) with an iterative closest point (ICP) refinement between the reference and test data (Tanks and Temples).

The common metrics used in the evaluation are accuracy and completeness (Middlebury MVS, DTU Robot Image Data Set and 3DOMcity), also defined, respectively, as precision and recall (ETH3D and Tanks and Temples). Both criteria entail the computation of the distance between the two models. For the accuracy assessment the distance is computed from the submitted data (D) to the ground truth (GT). For the completeness evaluation, it is the opposite, i.e., from GT to D. The computed distances can be signed (Middlebury MVS and 3DOMcity) or unsigned (Tanks and Temples). A threshold distance is usually adopted to find the fraction or percentage of points falling within the allowable threshold, which is decided according to the data density and noise. As additional accuracy parameters, DTU Robot Image Data Set and 3DOMcity characterize the distance distributions with statistics, such as the mean and median values, also performing some outlier removal. ETH3D and Tanks and Temples combine the accuracy/precision p and completeness/recall r values into a single score, i.e., their harmonic mean ($F1$ in ETH3D and F in Tanks and Temples), computed as: $(2 \cdot p \cdot r)/(p + r)$.

3.2. Surface Reconstruction and Assessment Criteria

Although aiming at the quality assessment of MVS approaches, the benchmarks described in the previous section mainly focus on dense point clouds. However, the surface reconstruction problem is also relevant in computer graphics. A survey on surface reconstruction methods from point clouds in computer graphics was provided by [52], who also distinguished the different evaluation criteria in geometric accuracy, topological accuracy, structure recovery and reproducibility. An analogy can be established between the quality assessment in photogrammetry and computer vision (Section 3.1) and the geometric accuracy in this context, which also requires the comparison with a ground truth. The Hausdorff distance (i.e., the maximum of the distances of all points of one mesh to the other [53]), mean and root mean square distance [54] or error in normals are frequently used geometric error measures. Metro [55] is a very popular tool for measuring the (geometric) difference between a reference mesh and its simplified version.

When dealing with polygonal mesh surfaces, while geometry mainly refers to the position of vertices, topology refers to the connectivity, or graph, of the mesh elements, i.e., vertices, edges and triangles [56]. Visual exemplifications of connected components, manifolds, self-intersections and boundaries are shown in Figure 2.

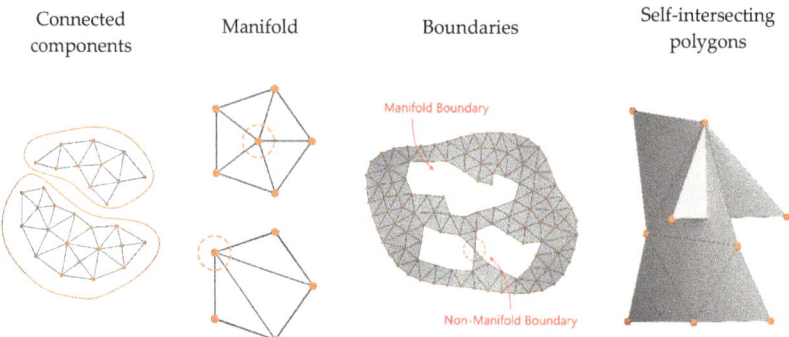

Figure 2. Examples of polygonal mesh topology properties.

With the aim of overcoming the issue of a "real" ground truth, i.e., the lack of the computational representation of the reference surface, Berger et al. [57] proposed a benchmark for the assessment of surface reconstruction algorithms from point clouds where point cloud data are simulated as being acquired with a laser scanner. Implicit surfaces, i.e., continuous and smoothly blended surfaces [58] of different complexity, are used as initial geometric shapes for sampling and are adopted as reference models for a quantitative evaluation. Both geometric and topological error measures are reported.

Crucial in computer graphics is the assessment of the mesh quality degradation resulting from simplification, resampling and other operations, for example compression and watermarking, which alter the original mesh not only from a geometric/topological point of view but also introduce visual distortions. Mesh visual quality (MVQ) assessment is, indeed, adopted as a criterion to design and optimize mesh processing algorithms [59]. While in photogrammetry the quality of a produced model is usually assessed in terms of its accuracy, precision and resolution, in computer graphics the term quality indicates *"the visual impact of the artefacts introduced by computer graphics"* algorithms [60]. It has been shown that metrics frequently adopted for assessing the geometric accuracy of the mesh (i.e., Hausdorff, mean and root mean square distances) do not correlate well with the human perception of surface quality and therefore quality scores consistent with the perception of human observers have been introduced [60–62]. These perceptually driven quality metrics, which try to mimic the human visual system (HVS), are based on roughness, local curvature, saliency, contrast and structural properties of the mesh and require a reference mesh to estimate the introduced degradation. Roughness infers the geometric non-smoothness [63] of a surface and can be computed as either a local or a global property: while local roughness may provide the high frequency behavior of the mesh vertices in a local region, global roughness is an indication of the average low frequency surface characteristic [64]. Curvature is also adopted as a measure to indicate structure and noise, well correlating with visual experience [65]. Databases comprising reference and distorted meshes are publicly available for assessing MVQ degradation introduced by geometrical processing [66] (subjective quality assessment of 3D models). However, since MVQ metrics are computationally expensive, the 3D models contained in these databases are of lower resolution (up to 50k vertices and 100k triangles) than typical models for photogrammetric applications. Lately, machine learning tools have been applied to estimate visual saliency and roughness of meshes without any reference model [67].

4. Investigated Surface Generation Methods

Three surface generation methods are hereafter considered. All three approaches require images with known camera parameters (interior and exterior) as input. Therefore, to avoid bias, all three methods share the same interior and exterior camera parameters and undistorted images, as well as depth values and dense point clouds estimations.

SIFT feature points [68] are first extracted, then matched with cascade hashing [69] and finally the camera poses are computed along with the sparse point cloud within an incremental bundle adjustment, as implemented within OpenMVG library [70]. Distortions are removed from the images before the dense matching step, which generates depth maps and a dense point cloud. Based on PatchMatch stereo [71,72], Shen [24] introduced a patch-based stereo approach where the depth of each pixel is calculated using random assignment and spatial propagation. OpenMVS is an open-source library that closely follows this idea while applying some optimization steps for more efficiency and is thus broadly used in 3D reconstruction research [42,73,74]. First, the best neighboring views are selected based on viewing direction criteria, and potential stereo pairs are formed. Rough depth maps are generated based on the sparse point clouds and iteratively refined using photo-consistency (zero mean normalized cross correlation ZNCC). Estimated depth maps are subsequently filtered taking into consideration visibility constraints while enforcing consistency among neighboring views. Finally, overlapping depth maps are merged to generate the fused dense 3D point cloud of the scene by minimizing redundancies and eliminating occluded areas.

The three employed surface generation methods are reported in detail in the following sections.

4.1. Photo-Consistent Volume Integration and Mesh Refinement (Method 1, M1)

The mesh reconstruction method exploiting photo-consistency and image visibility information is based on the approach introduced by Jancosek and Pajdla [13] as implemented by OpenMVS. The 3D space is initially discretized in tetrahedra using Delaunay tetrahedralization starting from the dense points and free space is modeled from the visibility information of the input 3D points. The final surface results as the interface between the free and the full space (graph cut optimization) while respecting visibility constraints, i.e., the image orientation and the projection of the 3D points back to the 2D image plane. Several mesh optimization steps can be performed to obtain an optimal mesh result, being pure geometric, such as smoothing, non-manifold and spike removal, or photo-consistent. Surface curvature, as expressed by point normals, is also taken into consideration during mesh reconstruction and thus complex regions are represented with high density triangles, while smoother areas may be wrapped into triangles of larger edges [75]. Photo-consistent refinement algorithms are generally efficient enough to produce detailed surfaces even from a rough input. In this method, an extra step of mesh refinement solution based on the idea described in [16] is implemented, by adding a photometric consistency score along with the geometric regularization term weighted by a scalar regularization weight. Mesh texturing is also enabled in this method, by assigning a best view to each face and generating a texture atlas, as described in [76]. In our experiments, following the OpenMVS implementation, we performed the extra mesh refinement step in order to take full advantage of the visibility information.

4.2. Surface Generation from Point Cloud (Method 2, M2)

Poisson surface reconstruction from oriented (i.e., with normals) point clouds [14] is a well-known and commonly adopted meshing approach. It creates a watertight surface, solving the reconstruction problem as a solution of a Poisson equation. An indicator function is computed with value one inside the surface and zero outside. In this work, M2 is based on the screened Poisson formulation [77], as implemented in CloudCompare [78]. The term "to screen" is adopted by the authors to indicate the screening term associated with the Poisson equation. The screening term reduces the over-smoothing of the data by introducing a soft constraint that forces the reconstruction to follow the input points. The volume occupied by the orientated points is partitioned using an adaptive octree structure, whose depth d (or level) can be decided accordingly. Selecting a depth implies constructing a voxel grid with a resolution no better than $2^d \times 2^d \times 2^d$. The octree level is automatically adapted to the original point sampling density, with the selected reconstruction depth being an indication of the maximum achievable mesh resolution. Beside the depth value, another critical parameter is the samples per node. It defines the number of points included in each voxel grid or node: the more noisy are the input data, the higher should be the number of points falling in each node of the octree, which may result in a loss of geometric details. If the original points have color information, as in our experiments, the RGB values are interpolated and transferred to the vertices of the generated mesh.

4.3. TSDF Volume Integration (Method 3, M3)

The Truncated Signed Distance Field (TSDF) volume integration is a volumetric reconstruction method broadly used while working with low-cost RGB-D sensors and real-time scenarios. It became a standard method since Newcombe et al. [15] used it in the KinectFusion project followed by various extensions and optimizations thereafter [79–82]. TSDF methods divide the 3D space (volume) into a discretized set of voxels and fuse distance information into them and is optimized for reconstruction speed. It is commonly combined with the marching cubes algorithm [83] (to generate a mesh, using the voxel grid created by TSDF and creating triangles on the edges. In more detail, SDF functions yield the shortest distance to any surface for every 3D point: depending on the sign, a point can be inside (negative) or outside (positive) the object boundaries, with the surface boundaries lying exactly on the zero crossing. On the other hand, in TSDF methods, a truncation threshold is added to omit everything outside this range. The standard method, although efficient under certain scenarios, has some default fundamental limitations as the voxel size itself defines the resolution of the final mesh and anything below this threshold cannot be reconstructed or erroneous results are produced when slanted surfaces are present, requiring alternative optimization solutions (e.g., [82,84]). In this work, M3 uses the TSDF implemented in the Intel Open3D library [19]. The resulting mesh may consist of a large number of unnecessary polygons, so further optimization steps may be performed: (1) merge the vertices of the mesh that are within a certain tolerance; (2) eliminate all edges and vertices that are non-manifold; (3) divide the mesh into clusters; and (4) eliminate all clusters with an area less than a certain value.

5. Datasets and Evaluation Metrics

The purpose of this study, namely the understanding and quantification of the potential benefits of surface reconstruction methods fully integrated into the photogrammetric pipeline, requires the usage of available benchmark data and metrics (Section 3), while pushing towards the identification of additional test cases and evaluation measures.

5.1. Datasets

Covering a broad range of application scenarios was the highest priority while choosing the evaluation datasets summarized in Table 1. Some of them are derived from existing benchmarks while others originate from original projects realized by FBK/3DOM.

Table 1. Case studies and related characteristics.

Dataset	Type of Scene	Type of Acquisition	Num of Images/Total Mpx	Scene Size/Mean Image GSD	Ground Truth	Evaluation Criteria
FBK/AVT	Urban	Aerial nadir—single shots	4/120	$(1 \times 1 \times 0.1)$ km^3/10 cm	-	Profiles, plane fitting, topology correctness
Strecha Fountain	Building facade	Terrestrial—single shots	11/66	$(5 \times 4 \times 5)$ m^3/ 30 mm	Mesh from laser scanner	Accuracy, completeness, roughness, profiles, topology correctness
FBK/3DOModena	Building facade	Terrestrial—single shots	14/320	$(11 \times 3 \times 9)$ m^3/ 20 mm	Mesh from laser scanner	Accuracy, completeness, roughness, profiles, topology correctness
Tanks and Temples—Ignatius	Statue	Terrestrial—video	263/535	$(2 \times 2 \times 3)$ m^3/ 30 mm	Mesh from laser scanner	Accuracy, completeness, roughness, profiles, topology correctness

Table 1. Cont.

Dataset	Type of Scene	Type of Acquisition	Num of Images/Total Mpx	Scene Size/Mean Image GSD	Ground Truth	Evaluation Criteria
FBK/3DOM wooden ornament	Asset	Terrestrial—single shots	32/740	$(305 \times 95 \times 25)$ mm^3/0.6 mm	Mesh from structured light scanner	Accuracy, completeness, roughness, profiles, topology correctness

Table 2 reports the selected parameters for the three investigated methods and the obtained final mesh resolution.

Table 2. Main parameters and final mesh resolution of the investigated methods.

		FBK/AVT	Strecha Fountain	Tanks and Temples—Ignatius	FBK/3DOM Modena	FBK/3DOM Wooden Ornament
Image Resolution for Depth Maps and Dense Point Cloud Generation		$\frac{1}{4}$	Full	Full	$\frac{1}{4}$	Full
M1	Regularity weight	0.4	0.4	0.4	0.4	0.4
	Resolution (mm)	80.0	1.3	1.2	3.0	0.04
M2	Voxel grid size (mm)	160.0	0.7	0.3	1.5	0.02
	Samples per node	20	1.5	20	1.5	20
	Resolution (mm)	160.0	0.3	0.2	0.7	0.01
M3	Voxel grid size (mm)	500.0	3.8	15	6.2	1.4
	Resolution (mm)	160.0	2.7	5.0	3.0	0.07

5.2. Evaluation Approach and Criteria

To enable the evaluation approach, a series of steps was undertaken. First, for the datasets for which the ground truth is available in the form of a point cloud, a surface was reconstructed using the same approach as in Section 4.2, preserving the original point cloud resolution. The meshing result was evaluated by computing the distance between the original point cloud and the derived mesh. Only the vertices that fall within a defined threshold (i.e., three times the average point cloud resolution) were retained. Moreover, interpolated triangles with a side length greater than about ten times the average mesh resolution and small disconnected components were eliminated from the mesh models generated by the three methods described in Section 4. Finally, a common datum was defined for the reference and evaluated meshes. The co-registration between the reference mesh and surfaces to be compared (called "data") was performed in a two-step procedure: (i) an absolute orientation through reference points or image exterior orientation parameters where available; and (ii) 7-DoF spatial transformation refinement through ICP between the photogrammetric dense point cloud and the reference mesh.

The following metrics were used to evaluate the results:

- *Accuracy* was evaluated as the signed Euclidean distance between the vertices of the (photogrammetric) data mesh and the (scanner) reference mesh. The signed Euclidean distance was chosen instead of the Hausdorff distance to highlight any possible systematic error. For this, both CloudCompare and Meshlab [85] implementations were tested, providing equivalent results. The following values were computed: mean, standard deviation (STDV), median and normalized maximum absolute deviation from the median (NMAD = 1.4826 × MAD), root mean square (RMS) and outliers percentage.

- *Completeness* was defined as the signed Euclidean distance between the (scanner) reference mesh and the (photogrammetric) data mesh. The percentage of vertices of photogrammetric data mesh falling within the defined threshold (in%) was adopted as a measure for completeness.
- *F-score* was defined as in [42] (see Section 3.1).
- Local *roughness* was computed as the absolute distance between the mesh vertex and the best fitting plane estimated on its nearest neighbors within a defined kernel size. The method implemented in CloudCompare was adopted. Adapting the standard parameters generally used to quantify the roughness [86], mean and RMS roughness values are reported to describe the local behavior of the vertices in their local region (i.e., within the selected kernel). The kernel size was carefully selected according to surface resolution.
- Local *noise* was assessed on selected planar regions where the plane fitting RMS was computed.
- *Sections* were extracted from the meshes and the mean and RMS signed distance values from data to reference are reported.
- Local *curvature variation*, expressed as normal change rate, was computed over a kernel size, i.e., the radius defining the neighbor vertices around each point where the curvature was estimated. As for the roughness metric, the kernel size was decided according to the surface resolution and size of the geometric elements (3D edges). The normal change rate is shown as a color map to highlight high geometric details (e.g., 3D edges), which appear as sharp green to red contours, and high frequency noise, shown as scattered green to red areas. The method implemented in CloudCompare was here adopted.
- The *topology* of each generated surface is evaluated in terms of the percentage of self-intersecting triangles over the total number of faces.

Given the above, the accuracy, completeness and F-score provide insight on the global geometric correctness of the reconstructed mesh, or in other words its closeness to the reference model. At the same time, the roughness and fitting of planar areas are a measure of the high frequency noise generated in the meshing process, while the normal change rate mainly shows the ability of reproducing geometric elements, such as 3D edges and contours. Finally, the percentage of self-intersecting triangles is an indication of the level of topological errors produced by the surface generation approach.

6. Results and Discussion

In this section, the results of the performed analyses are discussed. Firstly, the dataset without ground truth data is presented, reporting evaluations in terms of profiles, normal change rate maps and plane fitting (Section 6.1). Then, mesh results with datasets featuring a ground truth mesh are presented (Section 6.2).

6.1. Evaluation without a Reference Mesh: The Aerial Case Study

For the aerial dataset, no reference model is available. Thus, the quantitative evaluation is reported in terms of plane fitting RMS in two different areas, P1 and P2 in Figure 3a. The difference in the surface reconstruction approaches is also qualitatively shown in the section profiles S1 and S2 (Figure 3b) and normal rate change (Figure 3c).

High noise and discrepancies can be observed while comparing the three methods. All methods present topological errors, with non-manifold vertices as well as self-intersecting faces for M2 and M3. M1 appears less noisy, in terms of both plane fitting RMS (Table 3) and normal change rate (Figure 3c). The section profiles also show a less bumpy pattern.

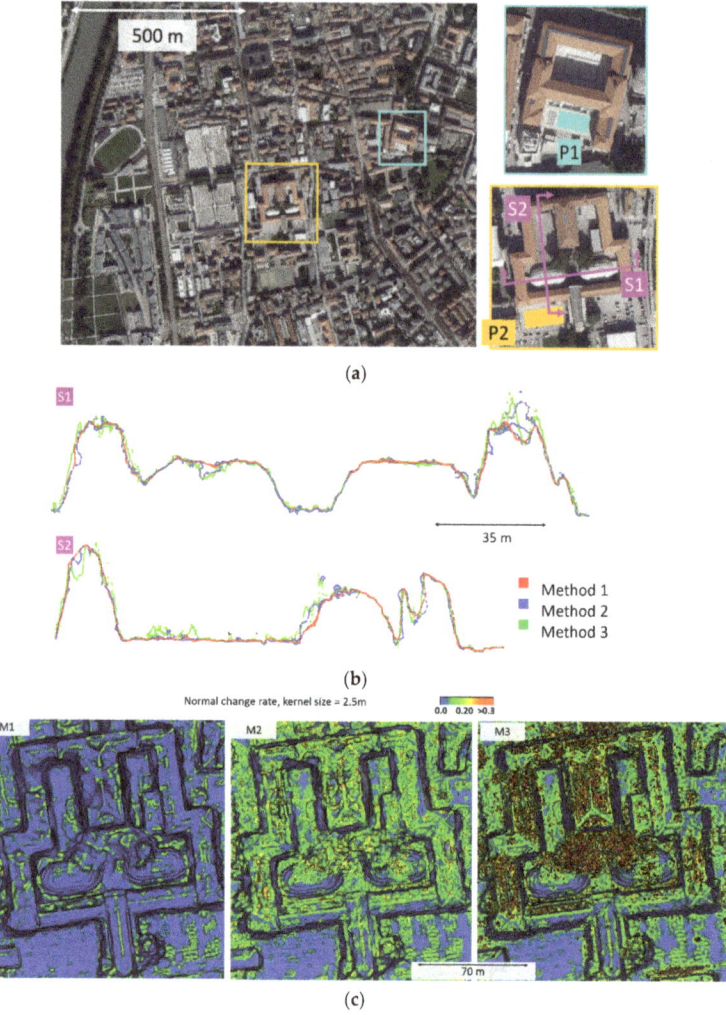

Figure 3. (a) Orthographic view of the urban area, with details of the extracted areas for the plane fitting analysis (P1 and P2) and sections (S1 and S2); (b) profiles of the extracted sections; and (c) normal change rate maps on a building.

Table 3. Quantitative and topological analyses for the aerial dataset: plane fitting and percentage of self-intersecting faces.

Method	Plane Fitting RMS (m)		Percent of Self-Intersecting Faces
	P1	P2	
M1	0.352	0.602	-
M2	0.391	0.606	0.01%
M3	0.385	0.547	0.5%

To evaluate the suitability of the investigated approaches for specific photogrammetric applications that require orthophotos as the final outcome, orthographic views of the mesh models are shown

(Figure 4). M1, which integrates the texturing step downstream the photogrammetric pipeline, provides a result visually more comparable to a standard orthophoto. The visual appearance is qualitatively better than the other two views, which are derived from color-vertex meshes. However, artifacts in the building edges due to geometric defects in the mesh model can be observed.

Figure 4. Orthographic view of the textured mesh from M1 and color vertex surfaces from M2 and M3.

6.2. Evaluation with a Reference Mesh

Figure 5 shows the surface models for the datasets where a reference surface model is available. The related analyses are summarized in Tables 4–7 (see Section 5.2 for the definition of the metrics) and visually shown in Figures 6–9 for the datasets *Fountain*, *Modena*, *Ignatius* and *Wooden ornament*, respectively. The roughness map provides information on the geometric non-smoothness of the surface. The normal change rate map highlights high geometric details (e.g., 3D edges), which appear in the images as sharp green to red contours, and high frequency noise, shown as scattered green to red areas.

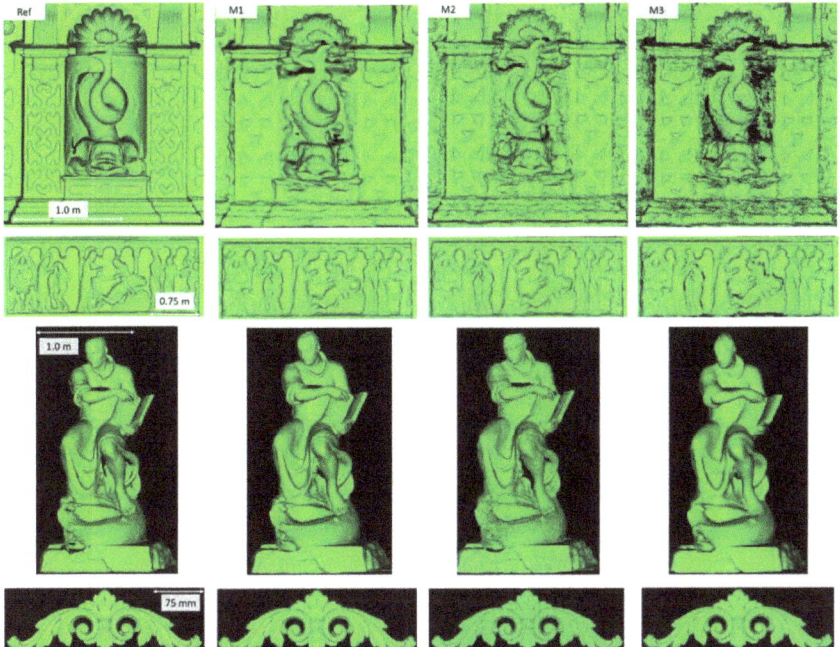

Figure 5. Shaded surface models of evaluated datasets. From left to right: Reference, M1, M2 and M3. From top to bottom: *Fountain*, *Modena's* bas-relief, *Ignatius* and *Wooden ornament*.

Table 4. Quantitative and topological analyses for the *Fountain* dataset. Values are in mm. Threshold and kernel values are set equal to 9.0 and 10 mm, respectively. Mean and RMS values in the Sections columns are double as we considered two sections (Figure 6).

Method	Accuracy						Completeness		F-Score	Roughness		Sections		% of Self-Intersecting Faces
	MEAN	STDV	RMS	MEDIAN	NMAD	OUT%	IN%			MEAN	RMS	MEAN	RMS	
M1	3.3	11.0	11.5	1.7	6.8	3.7	73.4		0.886	1.0	1.3	6.5	14.9	-
												0.4	15.9	
M2	2.1	12.6	12.7	0.4	6.7	4.1	73.6		0.898	1.7	2.2	6.9	16.2	0.03%
												0.1	15.0	
M3	3.6	27.9	28.1	2.1	8.6	7.8	71.1		0.827	1.8	2.3	8.7	18.3	0.07%
												−0.2	27.7	

Table 5. Quantitative and topological analyses for the *Modena* dataset. Values are in mm. Threshold and kernel values are set equal to 4.5 and 20 mm, respectively. Mean and RMS values in the Sections columns are double as we considered two sections (Figure 7).

Method	Accuracy						Completeness	F-Score	Roughness		Sections		% of Self-Intersecting Faces
	MEAN	STDV	RMS	MEDIAN	NMAD	OUT%	IN%		MEAN	RMS	MEAN	RMS	
M1	5.4	21.2	21.9	−0.1	5.6	9.8	54.3	0.779	1.0	1.4	−0.1	6.0	-
											1.3	8.4	
M2	5.0	17.5	18.2	0.8	6.4	7.5	52.7	0.758	1.5	2.0	1.0	8.0	-
											2.0	9.0	
M3	7.1	20.9	22.1	1.3	7.1	9.1	52.7	0.736	2.1	2.8	0.9	7.2	-
											1.7	9.0	

Table 6. Quantitative and topological analyses for the *Ignatius* dataset. Values are in mm. Threshold and kernel values are set equal to 3 and 10 mm, respectively. Mean and RMS values in the Sections columns are double as we considered two sections (Figure 8).

Method	Accuracy						Completeness	F-Score	Roughness		Sections		% of Self-Intersecting Faces
	MEAN	STDV	RMS	MEDIAN	NMAD	OUT%	IN%		MEAN	RMS	MEAN	RMS	
M1	1.6	3.2	3.6	1.4	2.7	1.5	65.3	0.820	0.4	0.5	3.4	7.3	-
											0.5	2.7	
M2	1.7	7.4	7.6	1.0	3.0	2.7	85.7	0.798	1.2	1.5	2.5	7.8	0.02%
											−0.2	3.1	
M3	4.9	11.4	12.4	3.0	3.0	9.2	58.8	0.643	3.6 *	3.3 *	5.4	9.7	-
											2.4	4.3	

* Kernel size equal to 50 mm.

Table 7. Quantitative and topological analyses for the *Wooden ornament* dataset. Values are in mm. Threshold and kernel values are set equal to 0.225 and 0.5 mm, respectively. Mean and RMS values in the Sections columns are double as we considered two sections (Figure 9).

Method	Accuracy					Completeness		F-Score	Roughness		Sections		% of Self-Intersecting Faces
	MEAN	STDV	RMS	MEDIAN	NMAD	OUT%	IN%		MEAN	RMS	MEAN	RMS	
M1	0.06	0.23	0.24	**0.03**	**0.11**	5.5	**99.9**	0.80	**0.02**	**0.03**	0.05	**0.12**	-
M2	**0.05**	**0.13**	**0.14**	0.04	0.11	**1.9**	90.8	**0.93**	0.06	0.07	**0.05**	0.14	0.002%
											0.04	**0.10**	
											0.07	0.17	
M3	0.36	1.34	1.4	0.03	0.14	10.1	77.9	0.86	0.04	0.05	0.09	0.13	0.06%
											0.61		

The accuracy, completeness and F-score values (Tables 4–7) reveal that the three investigated approaches perform similarly, with M1 and M2 usually outperforming M3.

M1 also exhibits the best metrics in terms of roughness for all the datasets. The visual inspection and metric values for the section profiles point out that all three methods tend to over-smooth the geometric details compared to the ground truth and that the sections from M1 are usually less noisy. All the investigated mesh models present non-manifold vertices, and, other than the *Modena*'s bas-relief (Table 5), M2 (Tables 4, 6 and 7) and M3 (Tables 4 and 7) are also characterized by self-intersecting faces.

The normal change rate maps convey additional insight on the different performances of the three surface generation methods. It is evident that none of the approaches can reproduce the geometric details of the reference mesh. However, M3 and especially M2 are more affected by high frequency noise, easily distinguishable in the green to red spots spread over the models (Figures 6 and 8 from the *Fountain* and *Wooden ornament* datasets). For *Ignatius*, due to the significantly lower resolution of M3, a different kernel size is adopted for the normal change rate estimation, clearly implying an over-smoothed geometry with respect to the other approaches. In the *Modena*'s bas-relief, the normal change rate does not highlight significant differences among the investigated methods.

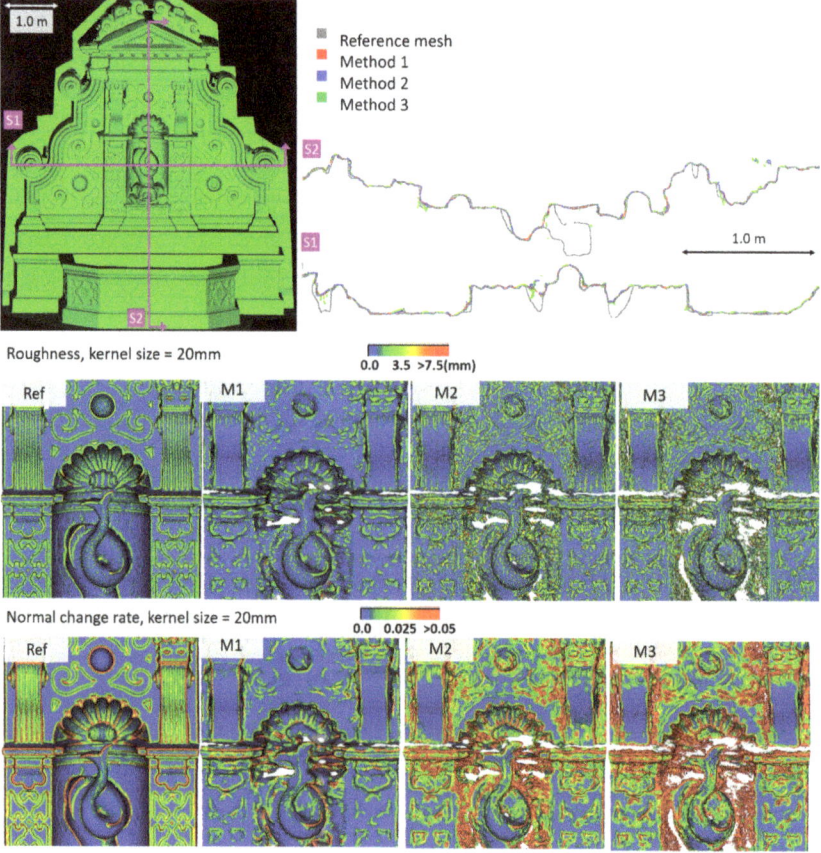

Figure 6. Section profiles (**top**); roughness (**middle**); and normal change rate maps (**bottom**) for the *Fountain* dataset.

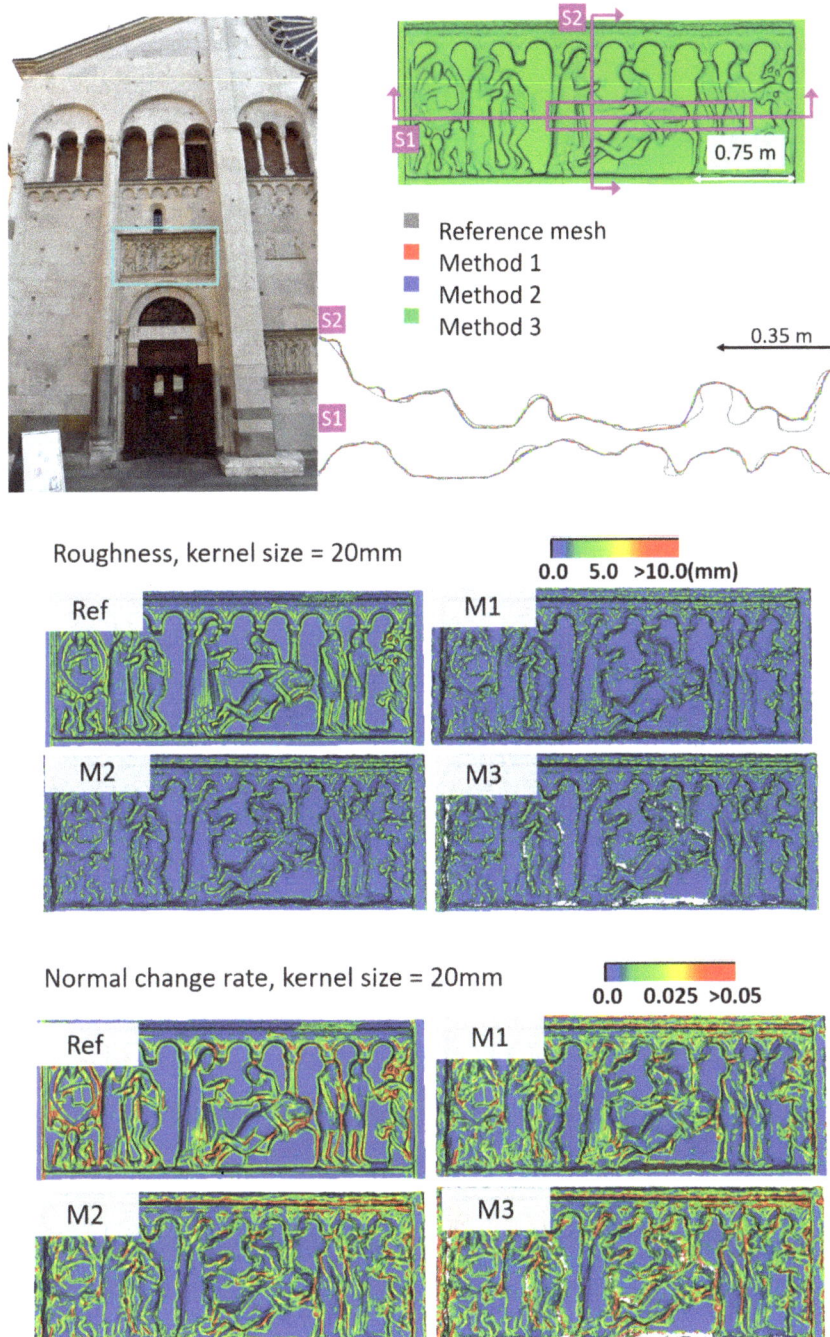

Figure 7. Section profiles (**top**); roughness (**middle**); and normal change rate maps (**bottom**) for the *Modena* dataset.

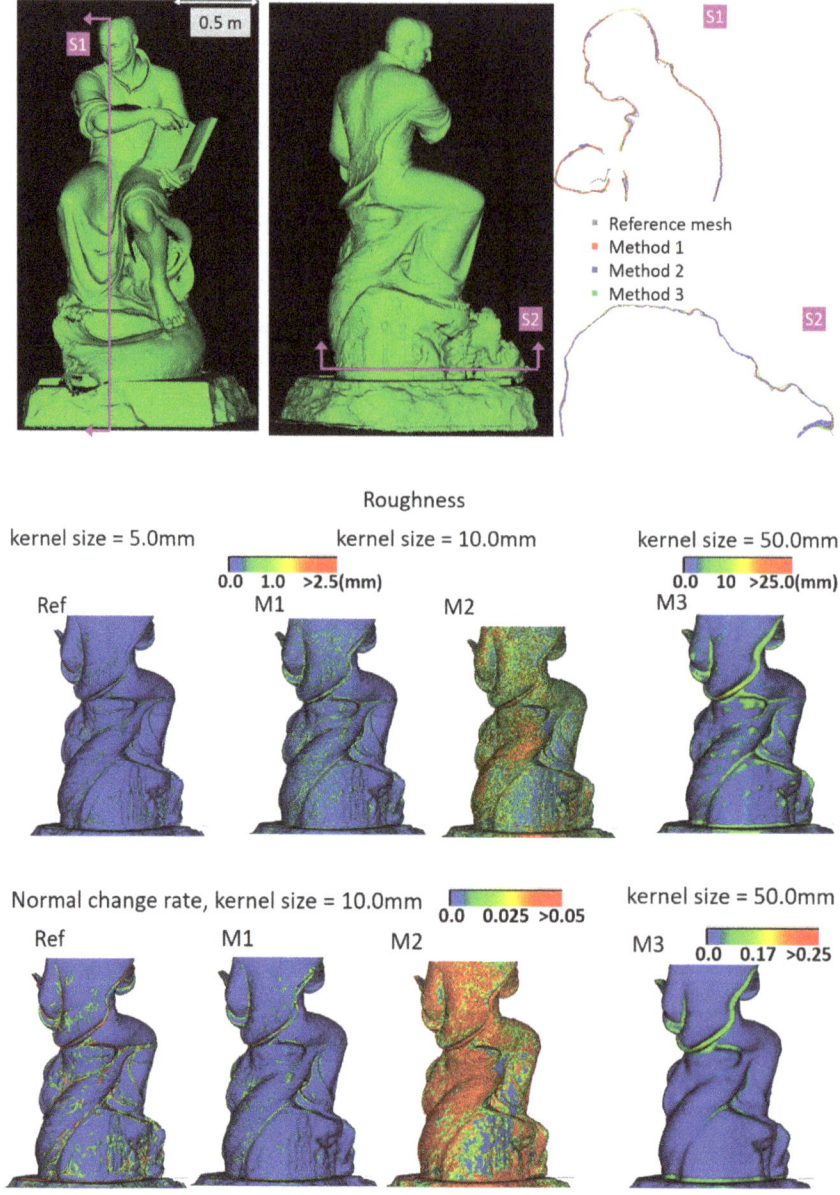

Figure 8. Section profiles (**top**); roughness (**middle**); and normal change rate maps (**bottom**) for the *Ignatius* dataset.

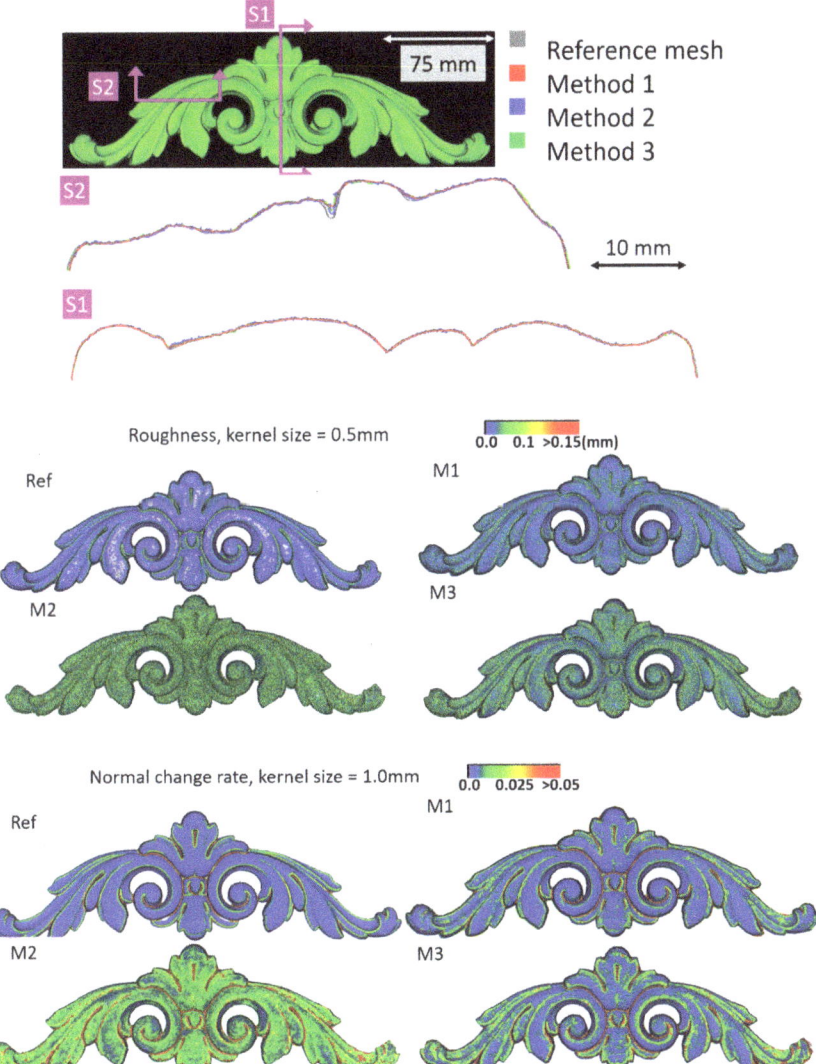

Figure 9. Section profiles (**top**); roughness (**middle**); and normal change rate maps (**bottom**) for the *Wooden ornament* dataset.

From the analysis of the roughness and normal change rate maps, it can be deduced that the methods do not show significant differences, when the starting data (dense point cloud or depth maps) are not heavily affected by high frequency noise (*Modena* dataset, Figure 7). When noise characterizes the intermediate MVS results such as for the *Fountain* (Figure 6), *Ignatius* (Figure 8) and *Wooden ornament* (Figure 9) datasets, M1 generally produces less noisy surfaces while preserving better the geometric details.

7. Conclusions

Surface or mesh reconstruction is a cross-disciplinary topic and an important step in a 3D modeling procedure. It can be fully integrated into the image-based pipeline as the final output of the MVS step or applied separately from the main workflow, which implies the use of popular surface reconstruction algorithms such as Poisson.

We investigated three different approaches of surface reconstruction in the context of photogrammetric applications: (1) the mesh generation step, incorporated in the reconstruction pipeline, takes into account photo-consistency and visibility information (M1); (2) the surface reconstruction is "outsourced" from the main reconstruction workflow and does not exploit visibility constraints or photo-consistency checks (M2); and (3) provided the image orientation parameters, the mesh is generated by integrating the depth maps with a volumetric method, without any visibility or photo-consistency information (M3). The comparative analysis aimed at quantifying the improvement of approaches fully integrated into the 3D reconstruction procedure and leveraging geometric and photo-consistency constraints, against methods disjointed from the dense reconstruction procedure that do not further exploit image content information or the results from bundle adjustment.

We first revised the concepts and steps of MVS and reviewed existing benchmarks, highlighting their limitations in the context of this work. Many of the publicly available data do not provide reference data in the form of mesh models and the employed assessment criteria are usually narrowed to the global geometric correctness through accuracy and completeness scores, ignoring other important features such as the reproduction of fine geometric details or noise level. An overview of surface assessment criteria adopted in computer graphics was also provided, with a focus on those considered in this study to quantify the reconstruction noise and geometric details.

The three considered methods were introduced, and the selected datasets and evaluation metrics were described. Drawing a definite conclusion was out of the scope of the paper. The results of the investigation show that, in experiments with a reference model, M1 and M2 performed similarly in terms of accuracy and completeness. However, the surface generation method integrated into the image-based reconstruction workflow (M1) generally outperformed the other two approaches in recovering geometric details and reducing the noise in all the considered case studies, regardless of the characteristics of the given images (scale, resolution, texture, etc.).

Although relevant for some applications, especially in real time, computational efficiency was not included in this evaluation, because the main interest was to test the best achievable quality even at the expense of long calculation times. However, it should be mentioned that M3 proved to be computationally more efficient than M1 and M2, i.e., on average 5–10 times faster than M1 and 2.5–5 time faster than M2, differences that can get larger as the complexity of the dataset increases in terms of resolution and noise. The presented study also emphasized the lack of benchmarks and assessment criteria specifically addressing the surface reconstruction problem for applications where metric accuracy matters.

Our future work will include the expansion of the current investigation to further MVS approaches and integrating perceptual evaluation metrics into rigorous accuracy assessment procedures. The robustness of the different methods to possible variations in the interior and exterior orientation parameters will also be examined. Moreover, we plan to further investigate the inclusion of visibility and semantic constraints in the 3D reconstruction pipeline towards the optimization of the final products.

Author Contributions: Conceptualization, E.N. and F.R.; methodology, E.N., E.K.S., S.R., and F.R.; software, E.N., E.K.S., and S.R.; validation, E.N., E.K.S., and F.R.; formal analysis, E.N., E.K.S., and S.R.; investigation, E.N., E.K.S., and S.R.; resources, F.R.; data curation, E.N., E.K.S., and S.R.; writing—original draft preparation, E.N. and E.K.S.; writing—review and editing, E.N., E.K.S., F.R., and S.R.; visualization, E.N. and S.R.; supervision, E.N. and F.R.; and project administration, F.R. All authors have read and agreed to the published version of the manuscript.

Funding: This research received no external funding.

Conflicts of Interest: The authors declare no conflict of interest.

References

1. Yogeswaran, A.; Payeur, P. *3D Surface Analysis for Automated Detection of Deformations on Automotive Body Panels*; New Advances in Vehicular Technology and Automotive Engineering: Ijeka, Croatia, 2012.
2. Nicolae, C.; Nocerino, E.; Menna, F.; Remondino, F. Photogrammetry applied to problematic artefacts. *Int. Arch. Photogram. Remote Sens. Spat. Inf. Sci.* **2014**, *40*, 451. [CrossRef]
3. Scopigno, R.; Cignoni, P.; Pietroni, N.; Callieri, M.; Dellepiane, M. Digital fabrication techniques for cultural heritage: A survey. In *Computer Graphics Forum*; Wiley Online Library: Hoboken, NJ, USA, 2017; pp. 6–21.
4. Starek, M.J.; Mitasova, H.; Hardin, E.; Weaver, K.; Overton, M.; Harmon, R.S. Modeling and analysis of landscape evolution using airborne, terrestrial, and laboratory laser scanning. *Geosphere* **2011**, *7*, 1340–1356. [CrossRef]
5. Barbarella, M.; Fiani, M.; Lugli, A. Landslide monitoring using multitemporal terrestrial laser scanning for ground displacement analysis. *Geomat. Nat. Hazards Risk* **2015**, *6*, 398–418. [CrossRef]
6. Haala, N.; Rothermel, M. Image-Based 3D Data Capture in Urban Scenarios. In Proceedings of the Photogrammetric Week 2015, Stuttgart, Germany, 7–11 September 2015; pp. 119–130.
7. Toschi, I.; Ramos, M.M.; Nocerino, E.; Menna, F.; Remondino, F.; Moe, K.; Poli, D.; Legat, K.; Fassi, F. Oblique photogrammetry supporting 3D urban reconstruction of complex scenarios. *Int. Arch. Photogram. Remote Sens. Spat. Inf. Sci.* **2017**, *42*, 519–526. [CrossRef]
8. Krombach, N.; Droeschel, D.; Houben, S.; Behnke, S. Feature-based visual odometry prior for real-time semi-dense stereo SLAM. *Robot. Auton. Syst.* **2018**, *109*, 38–58. [CrossRef]
9. Remondino, F.; Spera, M.G.; Nocerino, E.; Menna, F.; Nex, F. State of the art in high density image matching. *Photogram. Rec.* **2014**, *29*, 144–166. [CrossRef]
10. Szeliski, R. *Computer Vision: Algorithms and Applications*; Springer Science & Business Media: Berlin, Germany, 2010.
11. Ressl, C. Assessing the Accuracy of Dense Image Matching (or Benchmarking DIM). Keynote Presentation. In Proceedings of the ISPRS Technical Commission II Symposium, Riva del Garda, Italy, 2018. Available online: https://www.isprs.org/tc2-symposium2018/images/ISPRS-Keynote_Ressl.pdf (accessed on 15 October 2020).
12. Furukawa, Y.; Hernández, C. Multi-view stereo: A tutorial. *Found. Trends Comput. Graph. Vis.* **2015**, *9*, 1–48. [CrossRef]
13. Jancosek, M.; Pajdla, T. Exploiting visibility information in surface reconstruction to preserve weakly supported surfaces. *Int. Sch. Res. Not.* **2014**. [CrossRef]
14. Kazhdan, M.; Bolitho, M.; Hoppe, H. Poisson surface reconstruction. In Proceedings of the Fourth Eurographics Symposium on Geometry Processing, Cagliari, Sardinia, 26–28 June 2006; Volume 7.
15. Newcombe, R.A.; Izadi, S.; Hilliges, O.; Molyneaux, D.; Kim, D.; Davison, A.J.; Kohi, P.; Shotton, J.; Hodges, S.; Fitzgibbon, A. KinectFusion: Real-time dense surface mapping and tracking. In Proceedings of the 2011 10th IEEE International Symposium on Mixed and Augmented Reality, Basel, Switzerland, 26–29 October 2011; pp. 127–136.
16. Vu, H.H.; Labatut, P.; Pons, J.P.; Keriven, R. High accuracy and visibility-consistent dense multiview stereo. *IEEE Trans. Pattern Anal. Mach. Intell.* **2011**, *34*, 889–901. [CrossRef]
17. Jamin, C.; Alliez, P.; Yvinec, M.; Boissonnat, J.D. CGALmesh: A generic framework for Delaunay mesh generation. *ACM Trans. Math. Softw. TOMS* **2015**, *41*, 1–24. [CrossRef]
18. Curless, B.; Levoy, M. A Volumetric Method for Building Complex Models from Range Images. In Proceedings of the 23rd Annual Conference on Computer Graphics and Interactive Techniques August 1996, New Orleans, LA, USA, 4–9 August 1996.
19. Zhou, Q.Y.; Park, J.; Koltun, V. Open3D: A modern library for 3D data processing. *arXiv* **2018**, arXiv:1801.09847. Available online: https://arxiv.org/abs/1801.09847 (accessed on 14 October 2020).
20. Fisher, R.B.; Breckon, T.P.; Dawson-Howe, K.; Fitzgibbon, A.; Robertson, C.; Trucco, E.; Williams, C.K. *Dictionary of Computer Vision and Image Processing*; HOVA MART LLC: Bayonne, NJ, USA, 2013.
21. Granshaw, S.I. Photogrammetric terminology. *Photogrammet. Rec.* **2016**, *31*, 210–252. [CrossRef]
22. Hirschmuller, H. Accurate and efficient stereo processing by semi-global matching and mutual information. In Proceedings of the 2005 IEEE Computer Society Conference on Computer Vision and Pattern Recognition (CVPR'05), San Diego, CA, USA, 20–25 June 2005; Volume 2, pp. 807–814.

23. Furukawa, Y.; Ponce, J. Accurate, dense, and robust multiview stereopsis. *IEEE Trans. Pattern Anal. Mach. Intell.* **2009**, *32*, 1362–1376. [CrossRef]
24. Shen, S. Accurate multiple view 3D reconstruction using patch-based stereo for large-scale scenes. *IEEE Trans. Image Process.* **2013**, *22*, 1901–1914. [CrossRef] [PubMed]
25. Remondino, F.; Zhang, L. Surface reconstruction algorithms for detailed close-range object modeling. *Int. Arch. Photogram. Remote Sens. Spat. Inf. Sci.* **2006**, *36*, 117–123.
26. Remondino, F.; El-Hakim, S.; Gruen, A.; Zhang, L. Development and performance analysis of image matching for detailed surface reconstruction of heritage objects. *IEEE Signal. Process. Mag.* **2008**, *25*, 55–65. [CrossRef]
27. Seitz, S.M.; Curless, B.; Diebel, J.; Scharstein, D.; Szeliski, R. A comparison and evaluation of multi-view stereo reconstruction algorithms. In Proceedings of the 2006 IEEE Computer Society Conference on Computer Vision and Pattern Recognition (CVPR'06), New York, NY, USA, 17–22 June 2006; Volume 1, pp. 519–528.
28. Aanæs, H.; Jensen, R.R.; Vogiatzis, G.; Tola, E.; Dahl, A.B. Large-scale data for multiple-view stereopsis. *Int. J. Comput. Vis.* **2016**, *120*, 153–168. [CrossRef]
29. Campbell, N.D.; Vogiatzis, G.; Hernández, C.; Cipolla, R. Using multiple hypotheses to improve depth-maps for multi-view stereo. In *European Conference on Computer Vision*; Springer: Berlin/Heidelberg, Germany, 2008; pp. 766–779.
30. Goesele, M.; Curless, B.; Seitz, S.M. Multi-view stereo revisited. In Proceedings of the 2006 IEEE Computer Society Conference on Computer Vision and Pattern Recognition (CVPR'06), New York, NY, USA, 17–22 June 2006; Volume 2, pp. 2402–2409.
31. Hiep, V.H.; Keriven, R.; Labatut, P.; Pons, J.P. Towards high-resolution large-scale multi-view stereo. In Proceedings of the 2009 IEEE Conference on Computer Vision and Pattern Recognition, Miami, FL, USA, 20–25 June 2009; pp. 1430–1437.
32. Tola, E.; Strecha, C.; Fua, P. Efficient large-scale multi-view stereo for ultra high-resolution image sets. *Mach. Vis. Appl.* **2012**, *23*, 903–920. [CrossRef]
33. Hernández, C.; Vogiatzis, G.; Cipolla, R. Probabilistic visibility for multi-view stereo. In Proceedings of the 2007 IEEE Conference on Computer Vision and Pattern Recognition, Minneapolis, MN, USA, 17–22 June 2007; pp. 1–8.
34. Kolev, K.; Brox, T.; Cremers, D. Fast joint estimation of silhouettes and dense 3D geometry from multiple images. *IEEE Trans. Pattern Anal. Mach. Intell.* **2012**, *34*, 493–505. [CrossRef]
35. Liu, S.; Cooper, D.B. A complete statistical inverse ray tracing approach to multi-view stereo. In Proceedings of the IEEE CVPR 2011, Providence, RI, USA, 20–25 June 2011; pp. 913–920.
36. Zach, C.; Pock, T.; Bischof, H. A globally optimal algorithm for robust tv-l 1 range image integration. In Proceedings of the 2007 IEEE 11th International Conference on Computer Vision, Rio de Janeiro, Brazil, 14–21 October 2007; pp. 1–8.
37. Kuhn, A.; Mayer, H.; Hirschmüller, H.; Scharstein, D. A TV prior for high-quality local multi-view stereo reconstruction. In Proceedings of the IEEE 2014 2nd International Conference on 3D Vision, Tokyo, Japan, 8–11 December 2014; Volume 1, pp. 65–72.
38. Werner, D.; Al-Hamadi, A.; Werner, P. Truncated signed distance function: Experiments on voxel size. In *International Conference Image Analysis and Recognition*; Springer: Cham, Switzerland, 2014; pp. 357–364.
39. Proença, P.F.; Gao, Y. Probabilistic RGB-D odometry based on points, lines and planes under depth uncertainty. *Robot. Auton. Syst.* **2018**, *104*, 25–39. [CrossRef]
40. Bakuła, K.; Mills, J.P.; Remondino, F. A Review of Benchmarking in Photogrammetry and Remote Sensing. In Proceedings of the International Archives of the Photogrammetry, Remote Sensing & Spatial Information Sciences, Warsaw, Poland, 16–17 September 2019.
41. Özdemir, E.; Toschi, I.; Remondino, F. A multi-purpose benchmark for photogrammetric urban 3D reconstruction in a controlled environment. *Int. Arch. Photogramm. Remote Sens. Spat. Inf. Sci.* **2019**, *4212*, 53–60. [CrossRef]
42. Knapitsch, A.; Park, J.; Zhou, Q.Y.; Koltun, V. Tanks and temples: Benchmarking large-scale scene reconstruction. *ACM Trans. Graph. ToG* **2017**, *36*, 1–3. [CrossRef]
43. Schops, T.; Schonberger, J.L.; Galliani, S.; Sattler, T.; Schindler, K.; Pollefeys, M.; Geiger, A. A multi-view stereo benchmark with high-resolution images and multi-camera videos. In Proceedings of the IEEE Conference on Computer Vision and Pattern Recognition 2017, Honolulu, HI, USA, 21–26 July 2017.
44. Middlebury, M.V.S. Available online: https://vision.middlebury.edu/mview/ (accessed on 2 September 2020).

45. DTU Robot Image Data Sets. Available online: http://roboimagedata.compute.dtu.dk/ (accessed on 2 September 2020).
46. 3DOMcity Benchmark. Available online: https://3dom.fbk.eu/3domcity-benchmark. (accessed on 2 September 2020).
47. Strecha, C.; Von Hansen, W.; Van Gool, L.; Fua, P.; Thoennessen, U. On benchmarking camera calibration and multi-view stereo for high resolution imagery. In Proceedings of the 2008 IEEE Conference on Computer Vision and Pattern Recognition, Anchorage, AK, USA, 23–28 June 2008; pp. 1–8.
48. ETH3D. Available online: https://www.eth3d.net/ (accessed on 2 September 2020).
49. Tanks and Temples. Available online: https://www.tanksandtemples.org/ (accessed on 2 September 2020).
50. Cavegn, S.; Haala, N.; Nebiker, S.; Rothermel, M.; Tutzauer, P. Benchmarking high density image matching for oblique airborne imagery. In Proceedings of the 2014 ISPRS Technical Commission III Symposium, Zurich, Switzerland, 5–7 September 2014; pp. 45–52.
51. ISPRS-EuroSDR. Benchmark on High Density Aerial Image Matching. Available online: https://ifpwww.ifp.uni-stuttgart.de/ISPRS-EuroSDR/ImageMatching/default.aspx (accessed on 2 September 2020).
52. Berger, M.; Tagliasacchi, A.; Seversky, L.M.; Alliez, P.; Guennebaud, G.; Levine, J.A.; Sharf, A.; Silva, C.T. A survey of surface reconstruction from point clouds. *Comput. Graph. Forum* **2017**, *36*, 301–329. [CrossRef]
53. Guthe, M.; Borodin, P.; Klein, R. Fast and Accurate Hausdorff Distance Calculation between Meshes. In Proceedings of the Conference Proceedings WSCG'2005, Plzen, Czech Republic, 31 January–4 February 2005; ISBN 80-903100 7 9.
54. Aspert, N.; Santa-Cruz, D.; Ebrahimi, T. Mesh: Measuring errors between surfaces using the hausdorff distance. In Proceedings of the IEEE International Conference on Multimedia and Expo, Lausanne, Switzerland, 26–29 August 2002; Volume 1, pp. 705–708.
55. Cignoni, P.; Rocchini, C.; Scopigno, R. Metro: Measuring error on simplified surfaces. In *Computer Graphics Forum*; Blackwell Publishers: Oxford, UK; Boston, MA, USA, 1998; Volume 17, pp. 167–174.
56. O'Gwynn, B.D. A Topological Approach to Shape Analysis and Alignment. Ph.D. Dissertation, The University of Alabama at Birmingham, Birmingham, AL, USA, 2011.
57. Berger, M.; Levine, J.A.; Nonato, L.G.; Taubin, G.; Silva, C.T. A benchmark for surface reconstruction. *ACM Trans. Graph.* **2013**, *32*, 1–7. [CrossRef]
58. Opalach, A.; Maddock, S.C. An Overview of Implicit Surfaces. Available online: https://www.researchgate.net/publication/2615486_An_Overview_of_Implicit_Surfaces (accessed on 15 October 2020).
59. Dong, L.; Fang, Y.; Lin, W.; Seah, H.S. Perceptual quality assessment for 3D triangle mesh based on curvature. *IEEE Trans. Multimed.* **2015**, *17*, 2174–2184. [CrossRef]
60. Lavoué, G.; Mantiuk, R. Quality assessment in computer graphics. In *Visual Signal Quality Assessment*; Springer: Cham, Switzerland, 2015; pp. 243–286.
61. Corsini, M.; Larabi, M.C.; Lavoué, G.; Petřík, O.; Váša, L.; Wang, K. Perceptual metrics for static and dynamic triangle meshes. In *Computer Graphics Forum*; Blackwell Publishing Ltd.: Oxford, UK, 2013; Volume 32, pp. 101–125.
62. Abouelaziz, I.; Chetouani, A.; El Hassouni, M.; Cherifi, H. Mesh visual quality assessment Metrics: A Comparison Study. In Proceedings of the IEEE 2017 13th International Conference on Signal.-Image Technology & Internet-Based Systems (SITIS), Jaipur, India, 4–7 December 2017; pp. 283–288.
63. Moreau, N.; Roudet, C.; Gentil, C. Study and Comparison of Surface Roughness Measurements. In Proceedings of the Journées du Groupe de Travail en Modélisation Géométrique, Lyon, France, 27 March 2014.
64. Kushunapally, R.; Razdan, A.; Bridges, N. Roughness as a shape measure. *Comput. Aided Des. Appl.* **2007**, *4*, 295–310. [CrossRef]
65. Yildiz, Z.C.; Capin, T. A perceptual quality metric for dynamic triangle meshes. *EURASIP J. Image Video Process.* **2017**, *2017*, 12. [CrossRef]
66. Subjective Quality Assessment of 3D Models. Available online: https://perso.liris.cnrs.fr/guillaume.lavoue/data/datasets.html (accessed on 2 September 2020).
67. Abouelaziz, I.; Chetouani, A.; El Hassouni, M.; Latecki, L.J.; Cherifi, H. No-reference mesh visual quality assessment via ensemble of convolutional neural networks and compact multi-linear pooling. *Pattern Recognit.* **2020**, *100*, 107174. [CrossRef]

68. Lowe, D.G. Distinctive image features from scale-invariant key points. *Int. J. Comput. Vis.* **2004**, *60*, 91–110. [CrossRef]
69. Cheng, J.; Leng, C.; Wu, J.; Cui, H.; Lu, H. Fast and Accurate Image Matching with Cascade Hashing for 3d Reconstruction. In Proceedings of the IEEE Conference on Computer Vision and Pattern Recognition, Columbus, OH, USA, 24–27 June 2014; pp. 1–8.
70. Moulon, P.; Monasse, P.; Perrot, R.; Marlet, R. Openmvg: Open multiple view geometry. In *International Workshop on Reproducible Research in Pattern Recognition*; Lecture Notes in Computer Science; Springer: Berlin, Germany, 2016.
71. Barnes, C.; Shechtman, E.; Finkelstein, A.; Goldman, D.B. PatchMatch: A randomized correspondence algorithm for structural image editing. *ACM Trans. Graph.* **2009**, *28*. [CrossRef]
72. Bleyer, M.; Rhemann, C.; Rother, C. PatchMatch Stereo-Stereo Matching with Slanted Support Windows. *InBmvc* **2011**, *11*, 1–11.
73. Stathopoulou, E.K.; Welponer, M.; Remondino, F. Open-Source Image-Based 3d Reconstruction Pipelines: Review, Comparison and Evaluation. *Int. Arch. Photogram. Remote Sens. Spat. Inf. Sci.* **2019**, *42*, 331–338. [CrossRef]
74. Stathopoulou, E.K.; Remondino, F. Multi-view stereo with semantic priors. *Int. Arch. Photogram. Remote Sens. Spat. Inf. Sci.* **2019**, *4215*, 1157–1162. [CrossRef]
75. Zaharescu, A.; Boyer, E.; Horaud, R. Transformesh: A topology-adaptive mesh-based approach to surface evolution. In *Asian Conference on Computer Vision*; Springer: Berlin/Heidelberg, Germany, 2007; pp. 166–175.
76. Waechter, M.; Moehrle, N.; Goesele, M. Let there be color! Large-Scale Texturing of 3D Reconstructions. Available online: https://www.gcc.tu-darmstadt.de/media/gcc/papers/Waechter-2014-LTB.pdf (accessed on 14 October 2020).
77. Kazhdan, M.; Hoppe, H. Screened Poisson surface reconstruction. *ACM Trans. Graph. ToG.* **2013**, *32*, 1–3. [CrossRef]
78. CloudCompare. Available online: https://www.danielgm.net/cc/ (accessed on 2 September 2020).
79. Bondarev, E.; Heredia, F.; Favier, R.; Ma, L.; de With, P.H. On photo-realistic 3D reconstruction of large-scale and arbitrary-shaped environments. In Proceedings of the 2013 IEEE 10th Consumer Communications and Networking Conference (CCNC), Las Vegas, NV, USA, 11–14 January 2013; pp. 621–624.
80. Li, F.; Du, Y.; Liu, R. Truncated signed distance function volume integration based on voxel-level optimization for 3D reconstruction. *Electron. Imaging.* **2016**, *21*, 1–6. [CrossRef]
81. Whelan, T.; Johannsson, H.; Kaess, M.; Leonard, J.J.; McDonald, J. Robust real-time visual odometry for dense RGB-D mapping. In Proceedings of the 2013 IEEE International Conference on Robotics and Automation, Karlsruhe, Germany, 6–10 May 2013; pp. 5724–5731.
82. Splietker, M.; Behnke, S. Directional TSDF: Modeling Surface Orientation for Coherent Meshes. *arXiv* **2019**, arXiv:1908.05146.
83. Lorensen, W.E.; Cline, H.E. Marching cubes: A high resolution 3D surface construction algorithm. *ACM Siggraph Comput. Graph.* **1987**, *21*, 163–169. [CrossRef]
84. Dong, W.; Shi, J.; Tang, W.; Wang, X.; Zha, H. An efficient volumetric mesh representation for real-time scene reconstruction using spatial hashing. In Proceedings of the 2018 IEEE International Conference on Robotics and Automation (ICRA), Brisbane, QLD, Australia, 21–25 May 2018; pp. 6323–6330.
85. Meshlab. Available online: https://www.meshlab.net/ (accessed on 2 September 2020).
86. Santos, P.M.; Júlio, E.N. A state-of-the-art review on roughness quantification methods for concrete surfaces. *Construct. Build. Mater.* **2013**, *38*, 912–923. [CrossRef]

Publisher's Note: MDPI stays neutral with regard to jurisdictional claims in published maps and institutional affiliations.

© 2020 by the authors. Licensee MDPI, Basel, Switzerland. This article is an open access article distributed under the terms and conditions of the Creative Commons Attribution (CC BY) license (http://creativecommons.org/licenses/by/4.0/).

Article

Rough or Noisy? Metrics for Noise Estimation in SfM Reconstructions

Ivan Nikolov * and Claus Madsen

Department of Architecture, Design and Media Technology, Aalborg University, Rendsburggade 14, DK-9000 Aalborg, Denmark; cbm@create.aau.dk
* Correspondence: iani@create.aau.dk

Received: 6 August 2020; Accepted: 1 October 2020; Published: 8 October 2020

Abstract: Structure from Motion (SfM) can produce highly detailed 3D reconstructions, but distinguishing real surface roughness from reconstruction noise and geometric inaccuracies has always been a difficult problem to solve. Existing SfM commercial solutions achieve noise removal by a combination of aggressive global smoothing and the reconstructed texture for smaller details, which is a subpar solution when the results are used for surface inspection. Other noise estimation and removal algorithms do not take advantage of all the additional data connected with SfM. We propose a number of geometrical and statistical metrics for noise assessment, based on both the reconstructed object and the capturing camera setup. We test the correlation of each of the metrics to the presence of noise on reconstructed surfaces and demonstrate that classical supervised learning methods, trained with these metrics can be used to distinguish between noise and roughness with an accuracy above 85%, with an additional 5–6% performance coming from the capturing setup metrics. Our proposed solution can easily be integrated into existing SfM workflows as it does not require more image data or additional sensors. Finally, as part of the testing we create an image dataset for SfM from a number of objects with varying shapes and sizes, which are available online together with ground truth annotations.

Keywords: Structure from Motion (SfM); 3D reconstruction; noise estimation; point clouds; roughness

1. Introduction

Structure from Motion (SfM) is widely used for visualization and inspection purposes in the building [1–3], manufacturing [4] and energy industries [5], as well as for geology [6–8] and cultural preservation [9–11]. Because of the reliance of SfM on 2D image data, it is prone to geometric noise and topological defects, if optimal image capturing conditions are not met (Figure 1). This has prompted a number of benchmarks [12–14] on the accuracy and robustness of SfM solutions, as well as on the best possible lighting conditions, camera positions, image density and captured object surface characteristics. The problem of determining if noise is present on a 3D reconstructed mesh and differentiating between noise and the inherent roughness that surfaces and objects have is not a trivial one. Because topological defects and noise on the surface of SfM reconstruction are caused by a combination of sub-optimal capturing conditions, the surface properties of the scanned object and the camera used to capture the 2D, they cannot easily be quantified.

The main contribution of this paper is the exploration, development and evaluation of a number of metrics for determining if the underlying 3D reconstructed surface is noisy or rough. An overview of the idea proposed in this paper is shown in Figure 2. The proposed metrics are chosen based on the known weaknesses of SfM solutions, as well as on the underlying principals used in many of the state of the art mesh simplification, quality assessment and denoising algorithms, given in the next section. For testing the proposed metrics, we have created a image dataset from a number of number of different objects. This dataset, together with the ground truth noise annotations for testing are available online (Dataset: dx.doi.org/10.17632/xtv5y29xvz.2).

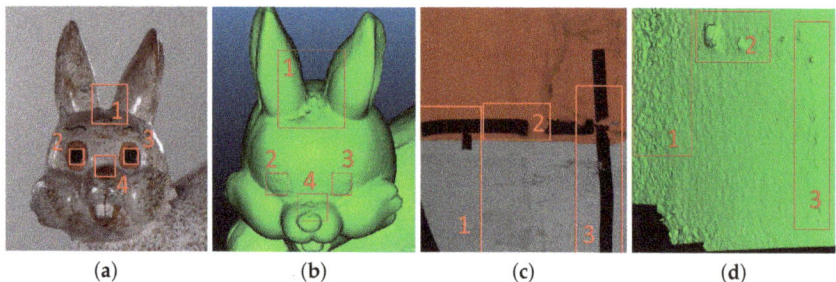

Figure 1. Illustration of Structure from Motion (SfM) reconstruction geometrical errors, which need to be distinguished from real surface roughness. Noise parts are shown in red. The problematic areas in (**a**,**c**), lead to geometrical errors in the reconstruction as seen in (**b**,**d**).

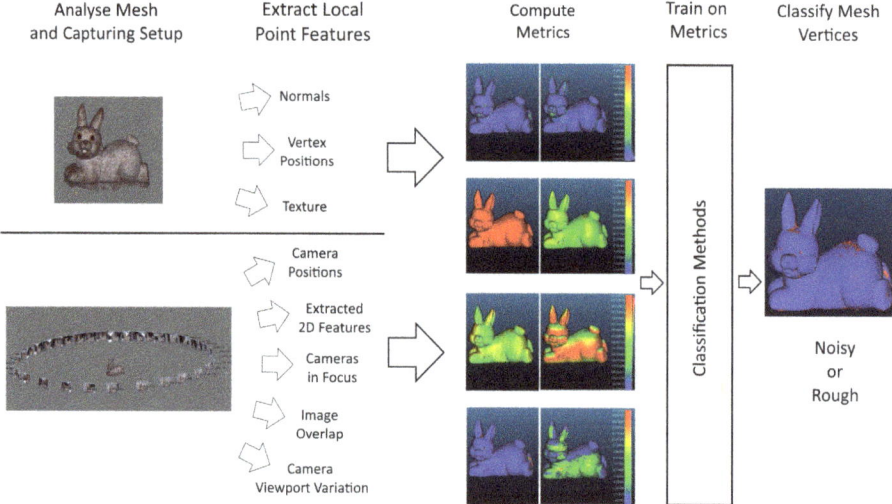

Figure 2. Overview of the proposed idea for using metrics extract from the mesh and capturing setup used for SfM reconstruction, to determine if the underlying surface is noisy or rough.

2. State of the Art

Most of the commercial SfM solutions rely on global or isotropic smoothing algorithms. These algorithms remove noise, but smooth out smaller details. Reconstruction solutions like Metashape [15], ContextCapture [16], Reality Capture [17], etc. use this approach, with additional options for mesh surface refinement. Such global denoising algorithms are also presented by [18–20].

Local feature or anisotropic algorithms analyze the underlying mesh geometry and normals to distinguish noisy areas from high surface roughness areas and preserve smaller details. The research from [21] uses a pre-filtering step and a L_1-median normal filtering, while [22] uses filtered facet normal descriptors and training of a neural network for calculating regression functions. Other research is focused on classifying normal regions and using isotropic neighbourhoods [23] or iterative estimation of normals and vertex movement [24,25].

Another important factor for detecting noise is the geometric visibility of roughness, especially on complex surfaces. There are multiple proposed solutions by [26–29], using local visibility features, curvature calculation and normals to detect parts of meshes with low or high roughness. These methods are used both for detecting noise on smooth meshes, but also for introducing watermarking to meshes without distorting their appearance.

Most of the described mesh denoising algorithms are not focused directly on SfM reconstructions and thus they do not use a lot of the information which can be taken from SfM production pipelines. In this paper we propose noise estimation metrics, which can predict noise risk and be used to distinguish noise caused by sub-optimal SfM reconstructions from the inherent roughness of the reconstructed objects. These metrics combine knowledge taken directly from 3D meshes reconstructed using SfM, with information taken from their textures, as well as from the camera setup used to capture the images used for reconstructing the object, such as camera positions, orientations, focal length and internal parameters. No external sensors or additional captured data are required for any of the presented metrics. With this our main contributions in this paper can be summarized:

- We present a number of metrics that can be easily calculated as part of the normal SfM workflow;
- We explore the correlation between each metric and the presence of noise on reconstructed objects;
- We train classical supervised learning methods using combinations of these metrics and demonstrate how to verify their accuracy;
- To verify the robustness of the metrics, we test them on objects with varying surface textures, shapes and sizes;
- We provide the captured database of images used to create the SfM reconstructions, together with the manually annotated ground truth data as part of the paper. This way others can use it for comparison and testing noise estimation and removal implementations.

3. Methodology

As part of this paper we propose nine metrics for detecting noise on SfM reconstructed meshes. These can be divided into two groups—metrics based on findings in the areas of mesh visual quality and roughness detection, and ones based on the SfM reconstruction weaknesses to sub-optimal capturing conditions. A total of five main observational hypotheses are made for the appearance of noise and geometric inaccuracies in SfM reconstructions and for each, one or more metrics are chosen as a way to describe each one. The observations are given in the numbered list below, with corresponding metrics shown in Table 1. In the next sections, each of the metrics will be explained in detail.

1. Noise manifests as either clumped together high frequency vertices or flat patches and holes—when the initial feature detection and matching methods in the SfM pipeline do not produce enough correct matches, the produced 3D surfaces can end up with overlapping or missing parts. These manifest in geometrical surface errors, as seen in Figure 3a;
2. SfM noise normally comes from smooth, monochrome colored surfaces—monochrome surfaces normally lack robust features like edges and angles, while smooth and transparent surfaces, produce reflections, which change with the view direction, making correct feature matching impossible (Figure 3b);
3. Noise is present on parts of the object that have not been seen from enough camera positions—SfM needs to gather information of the object from multiple directions, to provide a correct geometrical representation of the micro and macro shape of the surfaces. Not enough camera variation can lead to 3D surface "guessing" and deformed patches. An example of this can be seen in Figure 3c, where one object obscures another surface from being seen by the cameras resulting in noise;
4. Noise is present on parts of the object that have been seen from enough camera positions, but were not in focus—surface features need to be extracted and matched, but if parts of the object are blurred and out of focus, not enough information can be extracted from them. This

is visualized in Figure 3d, where the back of the object becomes out of focus, resulting in not enough features captured;

5. Noise is present on parts of the object that have been seen from enough camera positions, but those positions were not diverse enough—if all the capturing positions are from the same direction, not enough information can be extracted for the shape of the surface. This can be seen in Figure 3e, where multiple images are taken from a surface, but none of them have enough angular diversity in vertical direction, resulting in the reconstruction of the bottom of the surface being noisy.

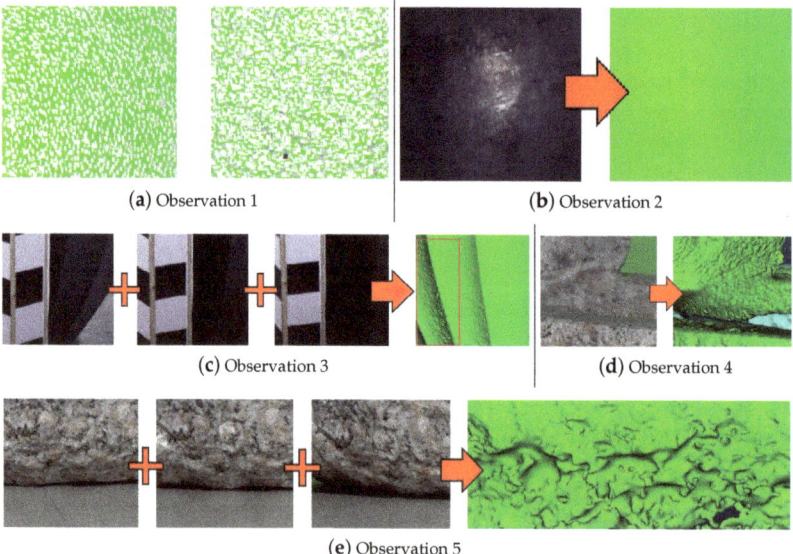

Figure 3. Examples of the five main observational hypotheses, used as a basis for the chosen mesh-based and capturing setup-based metrics.

Table 1. The five observational hypotheses and the chosen metrics, used to describe them. The different metrics are either based only on the reconstructed mesh itself or on the capturing setup—camera positions, intrinsic parameters, etc.

Observation	Metrics	Type
1	Local Roughness from Gaussian Curvature ($LRGC_m$) Difference of Normals (DON_m)	Mesh-based
2	Vertex Local Spatial Density (VD_m) Vertex Local Intensity Entropy (VIE_m)	Mesh-based
3	Number of Cameras Seeing Each Vertex (NCV_s) Projected 2D Features (PF_s)	Capturing Setup-based
4	Vertices in Focus (ViF_s)	Capturing Setup-based
5	Vertices Seen from Parallel Cameras (VPC_s) Vertex Area of Visibility (VAV_s)	Capturing Setup-based

A visualization of each of the metrics on the surface of a reconstructed mesh is given in Figure 4. In the subsections below we will focus on each of the metrics' theoretical basis, extraction methods, interpretation, etc. For easier readability each of the metric abbreviations will have a subscript of m for mesh-based or s for capturing setup-based. Before computing each metric, the reconstructed object is scaled to absolute real-world scale. Once all the metrics have been presented, they will be analyzed to determine their level of correlation. This will be presented in the Results Section 5.

3.1. General Mesh-Based Metrics

In this subsection we will cover, the metrics extracted directly from the 3D reconstructed mesh. They are based on the vertex positions, normals and vertex color. These metrics are based on observational hypotheses 1 and 2, presented in Section 3.

3.1.1. Local Roughness from Gaussian Curvature (LRGC$_m$)

Rationale: *Noise on the SfM surface appears as a geometric disturbance, which creates high roughness areas on otherwise smooth surface patches.*

The first calculated metric is the mesh's local roughness, depending on a metric closely related to Gaussian curvature. The metric was first proposed by [27], in their paper for mesh quality assessment. Local curvature is widely used for visual quality assessment and denoising, as a characteristic describing the local changes of the surface. Their proposed algorithm first calculates the Gaussian curvature like metric (GC) in an area around each vertex, essentially describing how much the area deviates from a planar surface. This is done using Equation (1), where $N_i^{(F)}$ is all the neighbour faces around a point i and α_j is the angle between the current vertex and the one which is incident to it.

$$GC_i = \left| 2\pi - \sum_{j \in N_i^{(F)}} \alpha_j \right| \tag{1}$$

Once the local curvature is calculated, a Laplacian matrix of the angles between the connected neighbours and each vertex is derived. Finally the local roughness metric LRGC is defined as a weighted difference between the Gaussian curvatures of each vertex and its neighbours, weighted according to the calculated Laplacian matrix. This is shown in Equation (2), where D_{ij} is the Laplacian matrix and $N_i^{(V)}$ is all the vertices in the neighbourhood of the current one. An in-depth explanation of the method can be seen in [27].

$$LRGC_i = \left| GC_i - \frac{\sum_{j \in N_i^{(V)}} (D_{ij} \cdot GC_j)}{\sum_{j \in N_i^{(V)}} D_{ij}} \right| \tag{2}$$

This metric is robust to curved surfaces and gives gradual and smooth values. The method gives a scale independent surface roughness measure. An example of the metric can be seen in Figure 4a, where higher values denote higher roughness and higher risk of noise.

3.1.2. Difference of Normals (DON$_m$)

Rationale: *Noise on SfM surfaces appears as high frequency surface changes, especially on the edges of the mesh and surrounding holes in it.*

The metric is proposed by [30] and is used for surface roughness detection, point cloud segmentation, obstacle detection, etc. It is a scale dependent local value, sensitive to specific resolutions of roughness. Two radii r_1 and r_2 of different sizes are chosen around each vertex. The normals of the area below the neighbourhood for each radius are computed and their difference gives the final metric. Equation (3) is used for calculating the difference of normals, where $\hat{n}(p,r)$ is the normal of the surface under each of the radii for every vertex i and $r_1 < r_2$. Get the final measure, the magnitude of this vector is calculated, which is between $[0,1]$.

$$DON_i = \left| \frac{\hat{n}(p_i, r_1) - \hat{n}(p_i, r_2)}{2} \right| \tag{3}$$

In their work, [30] demonstrate that high frequency areas contain smaller details in point clouds. SfM noise is normally represented as high frequency signal in clustered areas on the surface of the reconstruction. This is why we focus on capturing very high frequency surface changes. After looking at the scale of the input data, the larger radius is set heuristically to 2% of the size of each object, while the smaller radius is set to ten times smaller factor, as suggested in [30]. This makes it independent from the scale of the object. With these input parameters, the difference of normals is especially sensitive to roughness at the edges of objects and allows it to provide a more focused additional roughness metric to $LRGC_m$ metric. The calculated metric is visualized in Figure 4b, where higher values denote higher difference between the local normals and higher risk of noise.

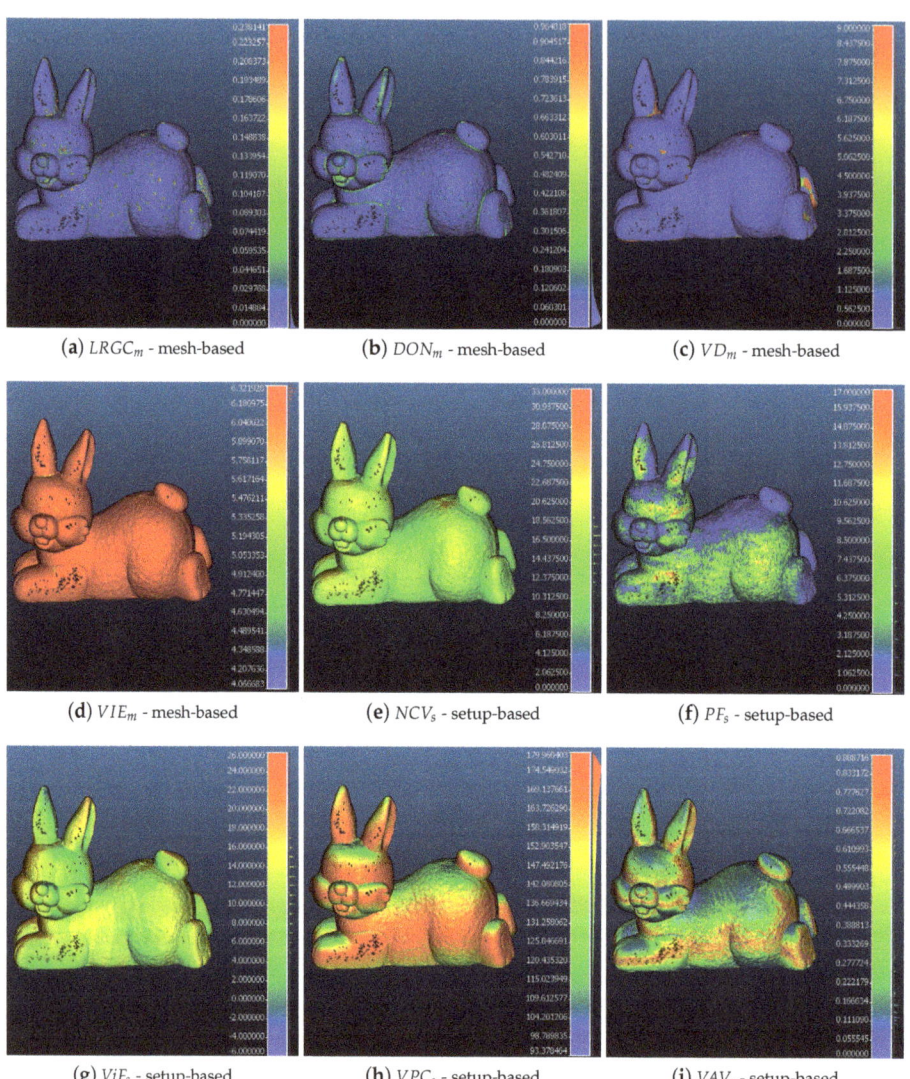

Figure 4. Visualization of all the proposed metrics as heat maps. For Local Roughness from Gaussian Curvature (LRGC), DON_m, VD_m, higher values (indicated with red color) indicate higher risk of noise, while for VIE_m, NCV_s, PF_s, ViF_s, VPC_s and VAV_s—higher values, indicate lower risk of noise.

3.1.3. Vertex Local Spatial Density (VD$_m$)

Rationale: *When surface errors occur in SfM reconstructions, the resultant reconstruction contains areas of high vertex density, even on supposedly smooth real world object areas.*

This metric is based on point cloud segmentation methods like the one proposed by [31], using area of interest spatial neighbourhood grouping like K nearest neighbours. This metric is calculated by first computing a number of progressively larger search radii, connected to the overall size of the reconstructed object. The size is chosen heuristically and is in the interval $R_{VD} = [0.1\% : 0.5\%]$ from the size of the object, as this is seen as the vertex density that best explains the possibility of noise. The mesh global maximum of neighbours for each of the radii is calculated. A percentage of these maximum values is taken and used as a threshold in the subsequent calculations. The lower this percentage is the less the local spatial density can be before it is viewed as problematic. For this paper the percentage is set to 60%.

For each vertex the number of neighbours is captured for each of the radii. If the number is above the threshold, a score is given for that vertex. The more instances get a number higher than the threshold, the higher the final score for that vertex. This is shown in Equation (4), where $N_i^{(r_j)}$ is the set of all neighbours for the current radius, $N_{max}^{(r_j)}$ is the maximum set of all neighbours, DC is the density coefficient in percentage and s is the score. This way a vertex density score scaled to the global density of the object on multiple size levels is achieved. This makes the metric invariant to the scale of the object and it can be comparable between objects of different sizes. The calculated density metric is shown in Figure 4c, where higher values indicate parts of higher vertex density and higher risk of noise.

$$VD_i = \sum_{r_j \in R_{VD}} s(j) \quad , \text{for } s(j) = \begin{cases} 1, & \text{if } N_i^{(r_j)} \geq DC \cdot N_{max}^{(r_j)} \\ 0, & \text{otherwise} \end{cases} \quad (4)$$

3.1.4. Vertex Local Intensity Entropy (VIE$_m$)

Rationale: *SfM reconstruction tends to produce errors and noise when the object surface is featureless and monochrome [32].*

The intensity for each vertex is calculated from the texture RGB data. These intensities are then used to calculate the local entropy of the mesh. Color has been used for mesh and depth map denoising [25,33] and it is shown to give good results. We choose to use entropy [34], as it can be more easily calculated locally on a point cloud, compared to other edge detection algorithms and can give a measure of the surface color intensity change. To calculate the entropy H we use Equation (5), where P_i is probability of the occurrence of the specific intensity level at vertex p_i and N is the maximum number of possible intensity values equal to 256. The visualization of the entropy is given in Figure 4d, where higher values indicate higher entropy and more varied surface color, with lower risk of noise.

$$H = -\sum_{i=0}^{N} P_i \log_2 P_i \quad (5)$$

3.2. Capturing Setup-Based Metrics

The following metrics are unique for SfM meshes, as they are extracted from the camera capturing setup and utilize the position, orientation, view density of the cameras, etc. The main factors for selecting these metrics, are the dependencies demonstrated by [14,35,36], between the quality of the capturing setup and the resultant reconstruction. To calculate these metrics a Unity implementation is created for positioning the reconstruction and calculated camera positions, as well as reprojecting the necessary data. We use the Unity engine, because of the easy programming pipeline using C#, fast ray cast computation and the possibility to visualize and compute large 3D model relatively fast and easy.

An overview of the used development pipeline is given in Section 4. These metrics are based on the hypothesis observations 3, 4 and 5.

3.2.1. Number of Cameras Seeing Each Vertex (NCV_s)

Rationale: *To create a good SfM reconstruction, a high amount of overlap between images is required [9], [11], which means that vertices "seen" by many cameras have a lower risk to contain noise.*

To compute this metric, all the pixels of each of the calculated cameras are projected to the reconstructed mesh. The metric is calculated by projecting the captured images from the calculated camera positions towards the reconstructed mesh. Each vertex is scored depending on the amount of image pixels projected onto it, meaning that the higher the score the more cameras have "seen" the vertex. The visualization of the metric is shown in Figure 4e.

This metric gives an overview of how certain we are, whether the data created by the SfM system is representative of the real world object. If not enough photos are taken from certain parts of the real life objects, there is a bigger chance that the reconstruction of these parts will contain noise or holes. The following metrics will expand on the information captured by this metric.

3.2.2. Projected 2D Features (PF_s)

Rationale: *To create the SfM reconstruction, 2D feature points are extracted from each image. These features are matched between images and used in the triangulation of the sparse point cloud and the reprojection of camera positions [37]. By projecting these points to the mesh, areas of higher certainty can be found, by exploiting the fact that areas not containing any found and matched features, will produce lower quality reconstructions*

We look at the 2D features extracted in the triangulation and camera position calculation step of the SfM pipeline. In this step features are extracted from each image and matched between them. In most SfM solutions, these 2D feature descriptors are not disclosed, but they are mostly variations of SURF [38] or free alternatives like FAST [39] and ORB [40]. An example image with captured feature points can be seen in Figure 5, where it can be seen that smooth areas like the eyes and nose of the bunny statue have much less features. For each camera position, the already calculated feature descriptor points are extracted. A radius around each point is set and the points under that area are projected to the 3D reconstructed model. For each 3D point the metric as aggregated depending on how many of these matched feature point areas are projected onto it.

Figure 5. An image used as input to the SfM solution and calculated feature points. A radius is set around each of the features and all points that are in the area are projected to the reconstructed mesh.

The higher the value of this metric for each vertex, the more 2D features were projected onto it. Figure 4f shows this metric. As these 2D features are used in the reconstruction itself it is hypothesized that a high metric will have less noise.

3.2.3. Vertices in Focus (ViF$_s$)

Rationale: *Structure from Motion matches points between images for creating the initial sparse point cloud and camera position and orientation calculation. If parts of the object are captured out of focus, these points would have blurring on them. This can increase the possibility for reconstruction noise to be present in these parts.*

To calculate the metric, first the near N_p and far F_p focal plains are calculated for each camera using the formulas presented in Equation (6). There H_f is the hyperfocal distance, which is the distance between the camera and the closest surface, which is in focus, when the lens is focused on infinity, while the CoC is the circle of confusion calculated according to [41]. The focal length F and aperture A are known from the EXIF data contained in the images and the distance to the object D is calculated from the camera to the closest surface of the reconstruction. Because the object is scaled before capturing the metrics, the measured distances between cameras and the object should be in correct units.

$$N_p = \frac{H_f \cdot D}{H_f + (D - F)}, \quad F_p = \frac{H_f \cdot D}{H_f - (D - F)} \tag{6a}$$

$$H_f = \frac{F^2}{A CoC}, \quad CoC = \frac{F}{1720} \tag{6b}$$

A ray is cast from each pixel of the camera, to the corresponding face from the reconstructed model and the distance between the two is calculated. Vertices of faces outside of the focal planes are scored with -1 for cameras which have seen them, while ones that are inside the focal planes are scored with 1. A lower score indicates more out of focus cameras having seen the vertex and a higher chance of it being noisy. The metric can be seen in Figure 4g, where the lower the value, the more times it has been out of focus and the higher risk for noise.

3.2.4. Vertices Seen from Parallel Cameras (VPC$_s$)

Rationale: *Even if multiple images have captured the surface of the object, if all of them "see" it from large angles, without at least one central image to connect them, there is a possibility of SfM calculation error [42].*

This metric is captured by computing the angle between each normal and the forward direction of each of the calculated cameras that can "see" the vertex. This is achieved by using Equation (7), where α_m is the calculated angle between the normal N_i of vertex v_i and the camera forward direction vector C_f for each camera seeing the vertex $[0, i]$. Two 3D vectors are parallel, if the angle between them is either 180 or 0 degrees, but the camera has to be able to see the vertex, so an angle of 0 degrees is not likely. The closer at least one angle is to 180 degrees, the less chance there is of noise. Figure 4h shows this metric.

$$\alpha_i = \arccos \frac{C_f \cdot N_i}{|C_f \cdot N_i|} \tag{7a}$$

$$\alpha_{max} = \max_{\{1:i\}} \alpha_i \tag{7b}$$

3.2.5. Vertex Area of Visibility (VAV$_s$)

Rationale: *To capture a surface's shape, SfM requires images from multiple positions and angles, so all parts of the topology are visible. If only little variation is given in the imaging positions, the resultant mesh can exhibit noise patches, surface deformations and holes [42].*

The metric requires the calculation of the area in space, from which each vertex is seen. We assume that the object surface is visible from every camera point of view. To model this metric, first a hemisphere is placed on the position of each vertex, oriented depending on the underlying normal. A hemisphere is chosen, as the assumption is that the cameras need to be able to physically see surface and the presence of self-occlusion. A ray is cast from each camera that "sees" the vertex. The points of intersection between each ray and the hemisphere are calculated and their 3D coordinates are saved. An example of this can be seen in Figure 6, with the camera position pulled closer and the hemisphere colored for easier visualization.

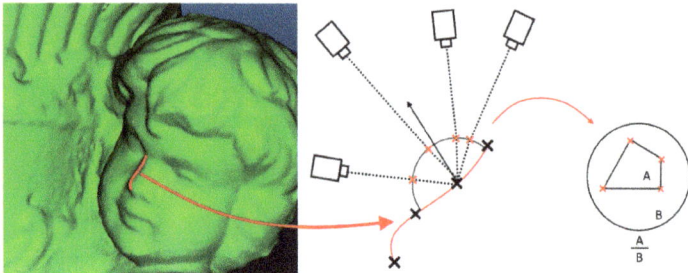

Figure 6. Visualization of the calculated hemisphere positioned above each vertex in the mesh and the camera position, together with the intersection points. The distance from the camera to the vertex position is made in a smaller scale for easier visualization. Once all the intersection points are found the area between them is calculated and the ratio between it and the whole area is used for the metric.

We then project the points in 2D, to avoid working with spherical geometry. The Lambert azimuthal equal-area projection, is chosen as it represents correctly the area in all regions of the sphere. For the projection Equation (8) is used, where (x, y, z) are the Cartesian coordinates of the points on the sphere and (X, Y) are the projected ones. The metrics is calculated as a ratio between the area of the projected points and the whole area. An example of the metric can be seen in Figure 4i, where the higher the values are, the higher the area of visibility is and the lower the risk of noise. This means that even if a lot of cameras have seen the point, if their angular coverage from different positions is not large enough this would be penalized.

$$X = \sqrt{\frac{2}{1-z}}x, \quad Y = \sqrt{\frac{2}{1-z}}y \tag{8}$$

4. Implementation

In this section a short overview of the implementation pipeline is given. The different processing environments for extracting each of the metrics are given in Figure 7. The initial data of the reconstructed mesh, the camera positions and orientations and extracted feature points are taken directly from the SfM software. For our current implementation Agisoft Metashape [15] is used, but the same data can be extracted from many of the commercial and open source SfM applications. In our case Metashape uses a Python based API for automation of the SfM pipeline, which can be also used to extract the required data and parse it in a structure, used for metric extraction. For the purely mesh-based metrics only the reconstruction itself is used and the processing is done directly in Python. For extracting data and manipulating the 3D data, the library open3D [43] is used in. The extracted features are manipulated and the areas around them calculated, by using OpenCV [44] for Python. The capturing setup-based metrics are calculated through the use of the Unity game engine [45]. The engine uses C#, with specific optimizations for vector and GPU computations. Normally used for making games and interactive experiences, we use the powerful 3D features of the engine, the camera settings and the fast and easy ray calculating capabilities. The data from the Metashape Python API in these cases is saved to a custom format containing all the mesh data—vertices, faces, normals, color information, as well as camera positions and orientation. For these metrics, the EXIF data from each image is also used, for calculating the proper field of view and depth of field of each of the cameras. The setup-based metrics are calculated per mesh vertex, by casting rays from each pixel of the camera positions to the reconstructed surfaces. An example view of the Unity implementation is given in Figure 8a, where the reconstruction together with the calculated camera positions and their forward direction vectors are given. The projected points on the mesh are used to calculate the NCV metric and show which parts of the object are seen by the particular camera. The input photo and the equivalent view from the Unity camera are given in Figure 8b,c.

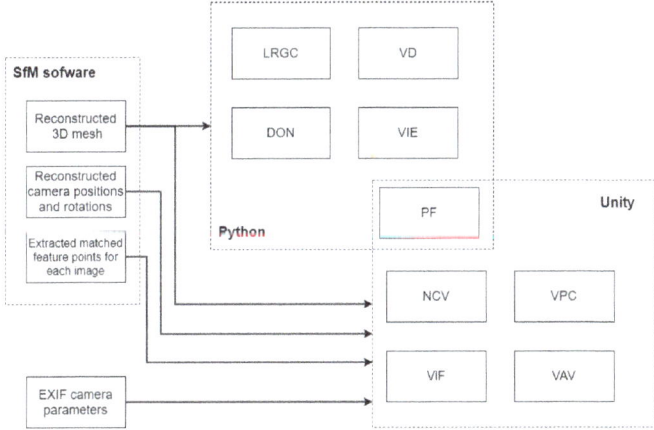

Figure 7. Overview of the implementation pipeline, showing what input and programming environments are used to calculate each of the metrics. The mesh-based metrics are directly computed in Python, while the capturing-setup based ones use a combination between Python and the Unity game engine.

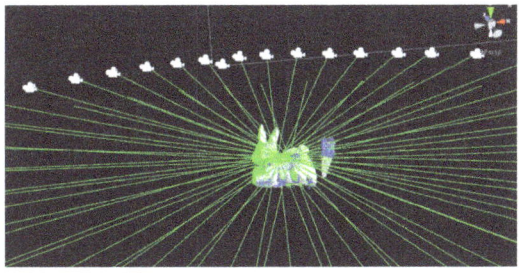

(**a**) View from NCV metric calculation

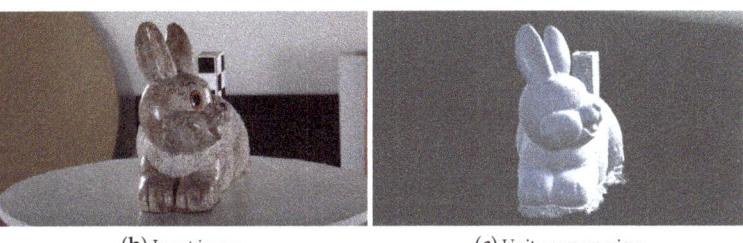

(**b**) Input image (**c**) Unity camera view

Figure 8. Views from the Unity implementation used for the capturing setup-based metric extraction.

5. Testing and Results

Testing the proposed metrics was done in a number of steps. First the correlation between the different metrics was calculated. This gave an initial idea if any of them gave redundant information, too similar to the others. The second step was to create a dataset of images and SfM reconstructions. These objects had varied sizes, shapes, roughness levels and were made from different materials with different textures. We then manually annotated each one of the reconstructions on a vertex level—as noise and not noise. This annotation was used as ground truth for testing the accuracy of the proposed metrics.

We then separated the reconstructed objects into testing and training data and used the metrics together with the annotated data to train a number of supervised learning classification methods. The accuracy of the proposed metrics could then be evaluated for segmentation of the testing data into noise and not noise vertices.

To evaluate if all metrics were useful for detecting noise, we first calculated the correlation between the appearance of noise and each of the metrics. We then used that information to retrain the best performing supervised classification method on different subsets of the metrics and evaluate the resultant accuracy.

Finally, we also evaluated the proposed solution in a wider industrially relevant context, by using a reconstruction of a wind turbine blade for testing and evaluating the results from it.

5.1. Data Gathering

To ensure the robustness of the proposed metrics, objects with different shape, size, roughness and color, as well as material were used. All the objects are shown in Figure 9. Special care was taken to create a diverse set of objects, to lower the possibility of bias in the proposed metrics. Some of the ways the dataset could be separated:

1. By size of the objects—we had objects ranging from 150 mm (cups shown in Figure 9i,j, etc.) to 800 mm (the black vase Figure 9d and sea vase Figure 9f), together with the wind turbine blade segment, which was more than 1500 mm long;
2. By material—we had objects made from stone, ceramics, plastic, clay, wood and metal. This guaranteed that we could have varying surface properties like reflectivity, texture and color variation;
3. By shape complexity—we had objects with simple shapes and repeated patterns like the different cups and vases, as well as objects complex shapes, with all the possible problems that could arise from that—self-occlusion (Figure 9c) or thin and narrow regions (Figure 9g,h).

A Canon 5Ds DSLR camera was used for capturing images of the objects. The resolution was set to 8688 × 5792 and a zoom lens with a variable focal length of 30–105 mm was used. The zoom lens was used, so the focal length can be easily changed depending on the size of the object. The focal length was set at the start of the capturing process for each object and kept the same throughout, only being changed if needed, once a new object is selected. This was done to guarantee that the captured object was always in frame and most parts of it also in focus. The focal length was changed depending on the size of the object. For the initial and subset tests 36 images were taken in a circle around each object in one horizontal band. The camera was setup to such a height, so it stayed perpendicular to the side of the objects. The research by [14], shows that this one vertical band capturing setup ensures that the objects can be reconstructed, but there is a possibility of geometrical noise on their surfaces. For the industrial context test 2 × 17 images in vertically stacked horizontal bands were used, because of the larger size of the wind turbine blade, compared to the objects used in the initial and subset. This way the front of the blade can be captured and reconstructed. All the objects were reconstructed using Agisoft Metashape and all the required data—camera positions, orientations, internal parameters, etc., were extracted from the program workflow, as explained in Section 4. To make them more manageable to work with the reconstructions were sub-sampled to around 50k vertices. The actual number depended on the size and complexity of the shape of the object.

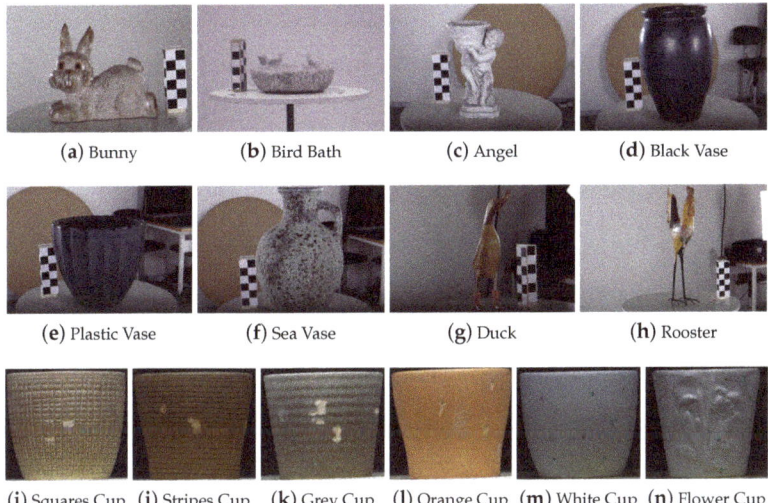

Figure 9. Objects selected for the robustness test. These objects have widely varying shape, size, roughness profiles and materials.

The processing times of the reconstructions was between 15 and 20 min, with extracting the two types of metrics using the Python and Unity processing pipeline added around 10 min more. The processing time for the capturing setup ones was heavily dependent on the number of used images and the resolution of the captured images. The mesh-based metrics' processing time depends on the number of vertices in the input reconstructions.

For testing the proposed solution and training the classification methods, a roughness/noise ground truth was created for all the used objects. The ground truth was made manually by annotating all the reconstructed meshes and masking all vertices of surfaces containing noise or any other topological defects. The software used for annotation of the mesh vertices was also developed in Unity (Figure 10) and at the end of the process the information for each vertex for each of the objects was saved into an array of values—showing 0 for clear surfaces and 1 for noise and geometrical defects. This annotated data were also used for testing the correlation between the appearance of noise and the different metrics.

5.2. Correlation Analysis

The correlation between the different independent metrics needed to be tested, to ensure that highly correlated ones were removed, as they did not give any new information and could introduce uncertainty and interfere the detection of the noise. In addition, the correlation between the metrics and the appearance of noise was also analyzed. To compute the correlation between the metrics a correlation matrix was calculated using the Pearson correlation coefficient [46]. The matrix is shown in Figure 11.

Figure 10. View from the annotation tool used for creating the roughness versus noise ground truth for each of the meshes. The vertices painted red are set as reconstruction noise.

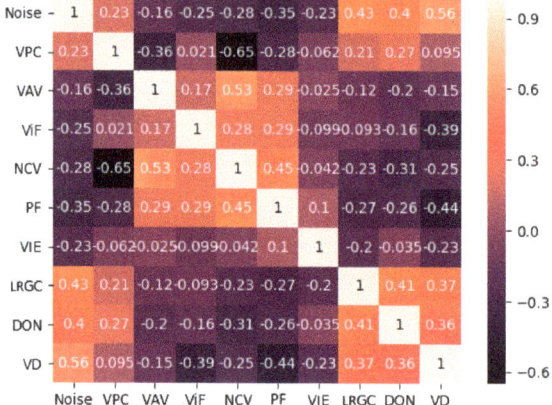

Figure 11. Correlation matrix of the used metrics, together with the dependent variable. For easier visualization the metrics are shown with their coded names—VPC_s: vertices seen from parallel camera, VAV_s: vertex area visibility, ViF_s: vertices in focus, NCV_s: number of cameras seeing each vertex, PF_s: projected 2D features, VIE_m: vertex local color entropy, $LRGC_m$: local roughness from Gaussian curvature, DON_m: difference of normals and VD_m: vertex local spatial density.

We chose to consider a cutoff between metric correlation higher than 0.5 and with the dependent variable lower than 0.1. From the correlation matrix it can be seen that one of the metrics had a high correlation with the others—the number of cameras seeing each vertex (NCV_s). Because this metric was quite generic and much of the information that it carried was present in the vertices seen from parallel camera (VPC_s), with correlation of 0.65 and the vertex area visibility (VAV_s), with correlation of 0.53, as well as projected 2D features (PF_s) metric, we chose not to include NCV_s in the final set of metrics.

The correlation between the independent variable metrics and the dependent variable, which in our case was the presence of noise and geometric inaccuracies, was further explored. From the correlation matrix in Figure 11, we could deduce that three mesh roughness metrics $LRGC_m$, DON_m and VD had the highest correlation with the presence of noise. This was expected as these metrics were directly connected to the topology of the mesh. From the capturing setup-based metrics the most correlated ones to the presence of noise were PF_s, NCV_s, ViF_s, but NCV_s was removed from

5.3. Initial Testing

For the initial test we used all the proposed metrics, except NCV_s. Further testing of subsets of metrics will be given in Section 5.4. The metrics were used to train a number of supervised learning classification methods—support vector machines (SVM), K-nearest neighbours (KNN), naive Bayes (NB), decision trees (DT), as well as more complex ensemble methods—random forests (RF) and AdaBoost (AB). The implementations were taken from Scikit-learn [47]. The hyperparameter used for each classifier are given in Table 2. Because of the limited number of test objects, we used a cross validation, where we trained on all but one and tested on it. We did this for each of the objects. Because the two classes—noise and not-noise were not balanced, an oversampling strategy was deployed when pre-processing the training data. The oversampling was done using Synthetic Minority Over-Sampling Technique (SMOTE) [48].

Table 2. Used hyperparameters for the tested classification methods—support vector machines (SVM), K-nearest neighbours (KNN), naive Bayes (NB), decision trees (DT), random forests (RF) and AdaBoost (AB).

Method	Parameters
SVM	C = 8, kernel = linear, gamma = scale
RF	n_estimators=150, max_depth=10, min_sample_split = 3
AB	n_estimators=150, learning_rate = 0.5
KNN	n_neighbors = 5, weights = uniform, algorithm = auto
NB	default parameters
DT	criterion= entropy, max_depth=10, min_sample_split = 2

Because of the imbalanced dataset, we focused not only on the accuracy, but on the precision, recall and F1-score, which are shown in Table 3. The table presents the average of all calculated performance factors for all the tested objects. From these, the AdaBoost classifer provided the best results, depending on the combination of the calculated factors.

Table 3. Average results from the 14 objects and the chosen classical classifiers—support vector machines (SVM), K-nearest neighbours (KNN), naive Bayes (NB), decision trees (DT), random forests (RF) and AdaBoost (AB).

Method	ACC	Precision	Recall	F_1
SVM	0.816	0.569	0.842	0.679
RF	0.824	0.580	0.879	0.699
AB	0.851	0.630	0.844	0.742
KNN	0.812	0.568	0.789	0.660
NB	0.809	0.558	0.832	0.668
DT	0.824	0.578	0.885	0.699

All the tested classifiers gave satisfactory results, with high recall, which indicated that it classified noise vertices as such. On the other hand they also classified non-noise vertices as noise, which was shown by the low levels of precision. This shows that metrics could be useful for signalling to possible areas under risk of noise and could be a part of a semi-automatic SfM noise estimation pipeline, where a user then verifies the results. For an easier visualization of the performance of the achieved noise risk assessment, the pseudo-colored visualizations of the annotated and classified noise vertices are also given in Figure 12. Looking closer at these visualizations, some problems can be seen in the classified noise from rough objects like the bird bath (Figure 12h) and the sea vase (Figure 12l), where the noise and roughness had a very closely related appearance. The same can be seen on objects like the bunny (Figure 12g) and the angel statue (Figure 12i), where the small rougher surface patches could sometimes closely resemble noise, especially close to self areas of self-occlusion, because of their more complex shapes.

Further complicating the non-trivial task were the manually annotated areas. For example, in the case of the two white cups (Figure 12e,f) the overall low reconstruction accuracy meant that there was noise with different levels of severity. Where the cutoff between acceptable surface and noise was could become very arbitrary, without classifying the whole surface as noisy. One way to alleviate this was to have multiple people annotate the same objects and get an average annotation. This will be further explored in the Conclusion and Future Work Section 6.

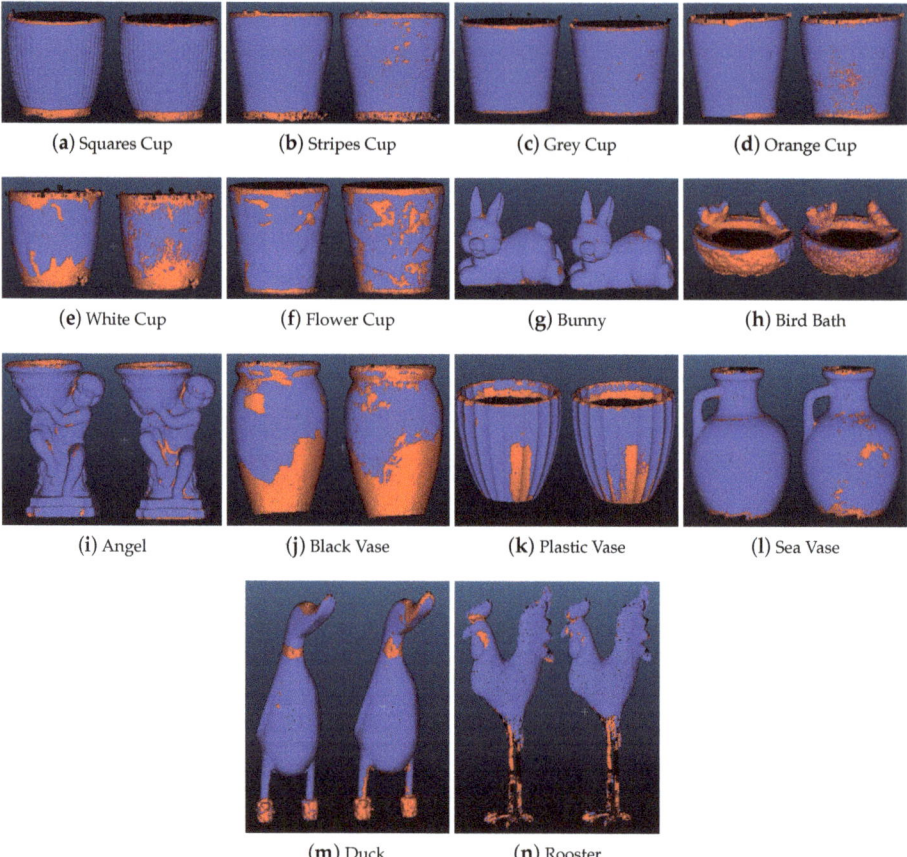

Figure 12. The annotated ground truth vertices on the left and the same classified vertices using our proposed method on the right. The noise vertices are colored red, while the non-noise ones are blue.

5.4. Subset Testing

The calculated results in the previous section were based on all metrics except NCV_s. To test how much influence each of the metrics had on the calculated performance, a number of subset tests were performed. A total of five main tests were set up as shown in Table 4. Because both the $LRGC_m$ and DON_m are used in the literature for point cloud classification, they were used separately, as a baseline naive first test for detecting noise on SfM reconstructions. The second test checked if NCV_s would have negative influence on the results, because of its high correlation with VPC_s and VAV_s

metrics. All other metrics were used for this test scenario. Using the information gathered in Section 5.2, the $LRGC_m$, DON_m and VD_m were set as main metrics, because of their high correlation with the presence of noise. The third scenario tested how important are the mesh and capturing setup-based metrics for the performance of noise estimation. The fourth test took the three designated main metrics and created five subsets, but adding each of the capturing setup-based metrics, to see how important they were separately. The final test again took the main metrics and combined them with the other ones, which were either more correlated or less correlated to the noise.

Table 4. The four main subset test scenarios. Each of the scenarios is designed to test the impact of different metrics or combination of metrics on the final results.

Testing Scenario	Description
1	$LRGC_m$ and DON_m separately
2	All metrics, with and without the most correlated metric—NCV_s
3	Mesh-based versus capturing setup-based metrics
4	Each capturing setup-based metric's impact on the results
5	Impact on the results from different combinations of setup-based metrics

The best performing classification method from the initial test was chosen for this scenario—AdaBoost. It was retrained with the different subsets of metrics and the results are given in Table 5. Again the average of the calculated performance factors using the left one out strategy for cross validation. For visualization purposes the resultant detected noise from each subset for one of the test objects is shown in Figure 13, together with the ground truth annotated noise.

Table 5. Results from testing different subsets of the proposed metrics. Each of the subsets is used to train the best performing classification method from the first testing scenario AdaBoost. Different subsets are created to test the posed question in Table 4.

Subsets	ACC	Precision	Recall	F1	Testing Scenario
Only $LRGC_m$	0.723	0.492	0.652	0.574	1
Only DON_m	0.686	0.407	0.788	0.537	1
All, without NCV_s	**0.889**	**0.674**	0.863	**0.756**	2
All, with NCV_s	0.852	0.635	0.848	0.725	2
$LRGC_m$, VD_m, DON_m	0.828	0.592	0.833	0.692	3
$LRGC_m$, VD_m, DON_m, VIE_m	0.837	0.611	0.822	0.701	3
VPC_s, VAV_s, ViF_s, PF_s	0.707	0.425	0.753	0.544	3
$LRGC_m$, VD_m, DON_m, PF_s	0.840	0.615	0.829	0.706	4
$LRGC_m$, VD_m, DON_m, ViF_s	0.838	0.615	0.809	0.699	4
$LRGC_m$, VD_m, DON_m, VAV_s	0.837	0.612	0.811	0.698	4
$LRGC_m$, VD_m, DON_m, VPC_s	0.839	0.614	0.824	0.704	4
$LRGC_m$, VD_m, DON_m, NCV_s	0.831	0.603	0.799	0.701	4
$LRGC_m$, VD_m, DON_m, PF_s, ViF_s	0.814	0.565	**0.869**	0.683	5
$LRGC_m$, VD_m, DON_m, VIE_m, VPC_s, VAV_s	0.839	0.615	0.822	0.703	5

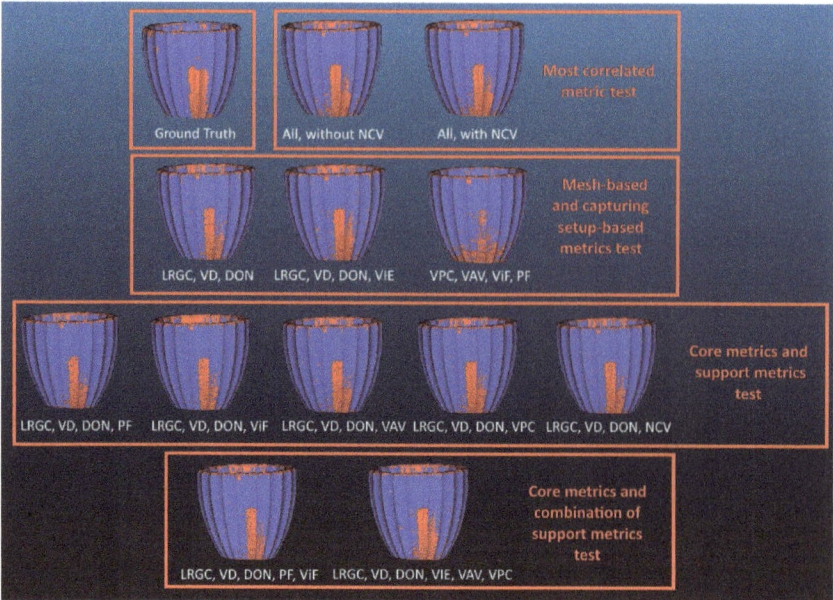

Figure 13. Visualization of the noise estimation results, using different subsets of metrics, together with the ground truth annotation. The different testing scenarios are separated for easier comparison.

The naive approaches to using only the $LRGC_m$ and DON_m yielded overall lower results, showing that only analyzing the roughness profile of the reconstruction could not completely separate noise from real world surface roughness. The results also showed that, as expected, the mesh-based metrics gave the highest effect on the performance of the classification method, meaning that they were the most useful in discriminating between noise and surface roughness. The texture metric VIE_m helped boost the overall accuracy and precision of the detection. This can be seen in Figure 13, with a lot less random noise vertices, compared to the purely $LRGC_m$, VD_m, DON_m trained detector. The capturing setup-based metrics on their own were too vague to properly discern between noise and surface roughness, as seen from the lower overall accuracy. When introducing them to the mesh-based metrics, it could be seen that they also boosted the overall performance when segmenting the noise from the roughness. Overall different combinations of the metrics could be useful in different situations, depending if it was more important to detect more of the noise correctly, but also mis-classified some of the roughness as noise, or vice-versa. The combination between the mesh-based metrics with the different capturing-setup metrics also showed that depending on the structure of the objects different capturing metrics could be useful. Larger objects benefited more from the ViF_s and VPC_s metrics, while smaller objects benefited more from VAV_s and VPC_s metrics. The PF_s metric was the one that always gave positive impact to the performance, as it was directly connected to the captured 2D feature points.

5.5. Industrial Context Test

The final test was made to give a wider industrial application context to the proposed metrics. We wanted to test if the described metrics could be used on data from different areas. This would also provide a better understanding on the generalization capabilities of the proposed metrics. We chose to test on wind turbine blade data, as this is an industrial inspection area which has began to use SfM for capturing information more and more and research is focused on ensuring the high quality of the reconstructions [49]. In addition, wind turbine blade data are hard to acquire, because of the requirements by blade manufacturers, that blades in use are not normally imaged. If the proposed metrics can be used to train noise recognition methods on generic data and then can be used no wind turbine blade surface reconstructions, it would make researching and benchmarking SfM results from blades surfaces much more easily accessible.

For the test, a decommissioned wind turbine blade segment was selected (Figure 14a). To ensure that the blade had different types of surface roughness and damaged areas, it was additionally sandblasted. The image capture was done in an outdoor environment. Because the object was considerably larger than the ones used in the previous tests and normally the leading edge and sides of blades are inspected, a different image capture pattern was selected. Two vertical bands of 17 images in a semi-circle pattern are captured, leading to 34 images in total. The best performing classifier was chosen from the first two tests—AdaBoost.

We chose also the best performing combination of metrics—all except NCV_s. All the reconstructions used in the previous testing scenarios were used as training data for AdaBoost. To evaluate the performance of the metrics on the blade, ground truth noise and roughness annotations were also made for it. The calculated classification results had an accuracy of 0.843, while the precision was 0.786 and recall was 0.877. For this test the precision-recall curve was also calculated for giving a better idea of the performance of the trained model using the proposed metrics (Figure 14b). We chose to use it instead of a ROC curve, on the basis of the unbalanced dataset. This way the calculated results were going to be less skewed and "optimistic" [50]. The area under the curve (AUC) of the precision-recall curve is 0.877. Finally, the pseudo-colored visualization of the classified and annotated vertices for the wind turbine blade model are given in Figure 14c. Overall the metrics provided acceptable results, by capturing all the problem areas around the top, bottom and back of the object, without misclassifying the real damaged areas of the edge of the blade. This showed that a transfer learning effect could be used, where the training could be done on more easily accessible generic 3D reconstruction objects and how noise was seen on them, and then the trained classifier could be used on specialized input data like wind turbine blades, with high level of accuracy.

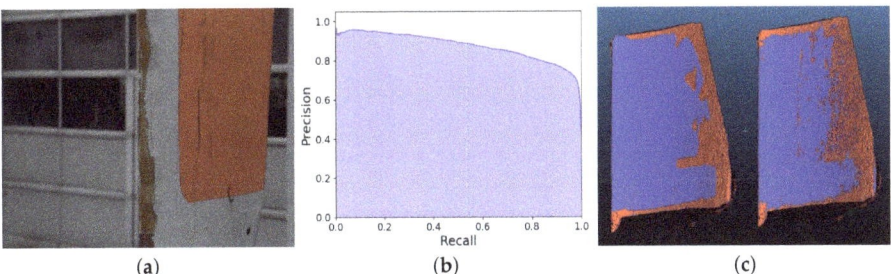

Figure 14. The wind turbine blade used for the second testing scenario (**a**), together with the precision-recall curve of the classification model (**b**) and the visualized annotation compared to classified vertices (**c**). Red vertices are noise, blue are non-noise.

6. Conclusions and Future Work

The problem of detecting noise and geometric disturbances of 3D reconstructed meshes resulting from SfM is a non-trivial one. In these meshes noise and regular surface roughness can exhibit the same characteristics, making it difficult for detecting noise without miss classifying the roughness. This is why in this paper we present a number of metrics based on both the mesh surface and on the capturing setup. This combination of metrics is chosen, as it has been observed from the state of the art in SfM testing and benchmarking, that the appearance of geometrical errors and noise on the reconstructions is highly correlated to the quality of the capturing setup, the used camera and the number of images taken. By combining these metrics and analysing their performance we are trying to address a gap in the knowledge of SfM results and how they can be used in applications like industrial inspection and surface roughness estimation. In addition, none of the proposed metrics require external sensor data and can be easily integrated in normal SfM production pipeline.

To test the metrics a dataset of images is captured from a number of objects with different shapes, sizes, textures and materials. These objects are then reconstructed and the metrics are captured from them. The amount of correlation between the metrics and between the metrics and the presence of noise is computed and is seen that only one of the metrics—the NCV_s is highly correlated to the others. A number of classical supervised learning classification methods are trained on the metrics, together with ground truth manually annotated data. The results from classifying the meshes as noisy and not noisy vertices are shown to be usable, with the metrics generally giving a good overview which parts of the meshes contain noise, with some noise miss-classified as roughness. On the other hand surface patches, which contain real life damages are correctly classified as not noise. The captured dataset of images, together with the ground truth annotations will be available online for use for training and testing purposes.

Different combinations of the proposed metrics are also tested, to see how individual metrics influence the performance of detecting noise. We demonstrate that a naive approach of just using the roughness of the surface of the reconstruction does not yield high quality results, with an overall accuracy between 0.68 to 0.72. The results could be dramatically improved by introducing a combination of all the mesh-based metrics proposed in the paper, pushing the accuracy to 0.85. The mesh-based metrics manage to describe the rough parts of objects, but tend to be less discriminative between the parts with high roughness and the ones with geometrical errors. The use of capturing setup-based metrics is shown to be helpful in discerning between the two, as they pinpoint areas of the reconstructed surface, that have been reconstructed under sub-optimal conditions. Combining them with the mesh-based metrics yield at least another 5–6% increase in the performance of the noise estimation, depending on which mesh-based metrics, they are combined with.

Finally we test the larger context of the proposed metrics for detecting noise on 3D reconstructions, which have significant difference from the data used for capturing the training metrics. This way such robustness can be tested. A wind turbine blade is selected, as their inspection has become of particular research interest. The blade also has a different size, shape and material from all the other tested objects. We demonstrate that we can achieve usable results, without miss-classifying any surface damage as reconstruction noise. This result also shows that the proposed metrics can be used as a form of transfer learning, where a noise detector can be trained on generic widely available data and then used on specialized data, which does not contain a large enough dataset, like wind turbine blade surfaces. The produced results of 0.843 accuracy 0.786 precision and 0.877 recall, show that the same level of quality of noise estimation can be achieved for wind turbine blades, which can be seen as an extended general applicability of the presented research.

The next step in verifying the results of the publication, would be comparing the reconstructed meshes to ground truth of the object, captured with a high resolution scanner. The difference between the two can be used, as a more objective noise ground truth, which can be then used to compare to the estimated noise risk. A look into global deformations in the overall shape of the reconstructed objects, as well as self-occlusions and fractal parts of the objects, can also be used to further introduce more metrics for assessing the risk of noise. Finally, one can also look even more into the influence of the camera specifications on the possibility of noise, such as the use of fixed focus lens versus an automatic focus one, as well as the use of rolling versus a global shutter.

Our future work would build on the results from this paper, by comparing them to both traditional mesh denoising algorithms and newer point cloud and mesh classification methods using convolutional and deep neural networks. For this a larger dataset of SfM object reconstruction is being build, so enough data are present. Finally, it is deemed interesting to look into detecting the illumination levels of the environment and see if they can be used as reliable indicators, as the role of the capturing setup lighting in the presence of noise, requires more research.

Author Contributions: Conceptualization, I.N. and C.M.; methodology, I.N.; software, I.N.; validation, I.N.; formal analysis, I.N.; investigation, I.N.; resources, I.N.; data curation, I.N.; writing—original draft preparation, I.N.; writing—review and editing, C.M.; visualization, I.N.; supervision, C.M.; project administration, C.M.; funding acquisition, C.M. All authors have read and agreed to the published version of the manuscript.

Funding: This work has received funding by the Energy Technology Development and Demonstration Program 374 (project number 64015-0046) under the Danish Energy Agency.

Conflicts of Interest: The authors declare no conflict of interest and that non of the funders have had any role in shaping the results and direction of the paper.

Abbreviations

The following abbreviations are used in this manuscript:

SfM	Structure from Motion
LRGC	Local Roughness from Gaussian Curvature
DON	Difference of Normals
VD	Vertex Local Spatial Density
VIE	Vertex Local Intensity Entropy
NCV	Number of Cameras Seeing Each Vertex
PF	Projected 2D features
ViF	Vertices in Focus
VPC	Vertices Seen from Parallel Cameras
VAV	Vetex Area of Visibility
GC	Gaussian Curvature
SVM	Support Vector Machines
KNN	K-nearest Neighbours
NB	Naive Bayes
DT	Decision Trees
RF	Random Forest
AB	AdaBoost

References

1. Siebert, S.; Teizer, J. Mobile 3D mapping for surveying earthwork projects using an Unmanned Aerial Vehicle (UAV) system. *Autom. Constr.* **2014**, *41*, 1–14. [CrossRef]
2. Tuttas, S.; Braun, A.; Borrmann, A.; Stilla, U. Acquisition and consecutive registration of photogrammetric point clouds for construction progress monitoring using a 4D BIM. *PFG J. Photogramm. Remote Sens. Geoinf. Sci.* **2017**, *85*, 3–15. [CrossRef]
3. Chaiyasarn, K.; Kim, T.K.; Viola, F.; Cipolla, R.; Soga, K. Distortion-free image mosaicing for tunnel inspection based on robust cylindrical surface estimation through structure from motion. *J. Comput. Civ. Eng.* **2015**, *30*, doi:10.1061/(ASCE)CP.1943-5487.0000516
4. Khaloo, A.; Lattanzi, D. Hierarchical dense structure-from-motion reconstructions for infrastructure condition assessment. *J. Comput. Civ. Eng.* **2016**, *31*. [CrossRef]
5. Zhang, D.; Burnham, K.; Mcdonald, L.; Macleod, C.; Dobie, G.; Summan, R.; Pierce, G. Remote inspection of wind turbine blades using UAV with photogrammetry payload. In Proceedings of the 56th Annual British Conference of Non-Destructive Testing-NDT 2017, Telford, UK, 4–7 September 2017.
6. Bemis, S.P.; Micklethwaite, S.; Turner, D.; James, M.R.; Akciz, S.; Thiele, S.T.; Bangash, H.A. Ground-based and UAV-based photogrammetry: A multi-scale, high-resolution mapping tool for structural geology and paleoseismology. *J. Struct. Geol.* **2014**, *69*, 163–178. [CrossRef]
7. Cho, Y.; Clary, R. Application of SfM-MVS Photogrammetry in Geology Virtual Field Trips. In Proceedings of the 81st Annual Meeting of Mississippi Academy Sciences, Hattiesburg, MS, USA, 23–24 February 2017.
8. Bi, H.; Zheng, W.; Ren, Z.; Zeng, J.; Yu, J. Using an unmanned aerial vehicle for topography mapping of the fault zone based on structure from motion photogrammetry. *Int. J. Remote Sens.* **2017**, *38*, 2495–2510. [CrossRef]

9. Kersten, T.P.; Lindstaedt, M. Image-based low-cost systems for automatic 3D recording and modelling of archaeological finds and objects. In *Euro-Mediterranean Conference*; Springer: Berlin/Heidelberg, Germany, 2012; pp. 1–10.
10. Kyriakaki, G.; Doulamis, A.; Doulamis, N.; Ioannides, M.; Makantasis, K.; Protopapadakis, E.; Hadjiprocopis, A.; Wenzel, K.; Fritsch, D.; Klein, M.; et al. 4D reconstruction of tangible cultural heritage objects from web-retrieved images. *Int. J. Herit. Digit. Era* **2014**, *3*, 431–451. [CrossRef]
11. Hixon, S.W.; Lipo, C.P.; Hunt, T.L.; Lee, C. Using Structure from Motion mapping to record and analyze details of the Colossal Hats (Pukao) of monumental statues on Rapa Nui (Easter Island). *Adv. Archaeol. Pract.* **2018**, *6*, 42–57. [CrossRef]
12. Thoeni, K.; Giacomini, A.; Murtagh, R.; Kniest, E. A comparison of multi-view 3D reconstruction of a rock wall using several cameras and a laser scanner. *Int. Arch. Photogramm. Remote Sens.* **2014**, *40*, 573. [CrossRef]
13. Schöning, J.; Heidemann, G. Evaluation of multi-view 3D reconstruction software. In *International Conference on Computer Analysis of Images and Patterns*; Springer: Berlin/Heidelberg, Germany, 2015; pp. 450–461.
14. Nikolov, I.; Madsen, C. Benchmarking close-range structure from motion 3D reconstruction software under varying capturing conditions. In Proceedings of the Euro-Mediterranean Conference Conference, Nicosia, Cyprus, 31 October–5 November 2016.
15. Agisoft. Metashape. 2010. Available online: http://www.agisoft.com/ (accessed on 20 September 2019).
16. Bentley. ContextCapture. 2016. Available online: https://www.bentley.com/ (accessed on 20 September 2019).
17. CapturingReality. Reality Capture. 2016. Available online: https://www.capturingreality.com/ (accessed on 20 September 2019).
18. Kim, B.; Rossignac, J. Geofilter: Geometric selection of mesh filter parameters. In *Computer Graphics Forum*; Wiley Online Library: Hoboken, NJ, USA, 2005; Volume 24, pp. 295–302.
19. Nealen, A.; Igarashi, T.; Sorkine, O.; Alexa, M. Laplacian mesh optimization. In Proceedings of the 4th International Conference on Computer Graphics and Interactive Techniques in Australasia and Southeast Asia 2006, Kuala Lumpur, Malaysia, 29 November–2 December 2006; pp. 381–389.
20. Su, Z.X.; Wang, H.; Cao, J.J. Mesh denoising based on differential coordinates. In Proceedings of the 2009 IEEE International Conference on Shape Modeling and Applications, Beijing, China, 26–28 June 2009; pp. 1–6.
21. Lu, X.; Chen, W.; Schaefer, S. Robust mesh denoising via vertex pre-filtering and l1-median normal filtering. *Comput. Aided Geom. Des.* **2017**, *54*, 49–60. [CrossRef]
22. Wang, P.S.; Liu, Y.; Tong, X. Mesh denoising via cascaded normal regression. *ACM Trans. Graph.* **2016**, *35*, 232. [CrossRef]
23. Lu, X.; Liu, X.; Deng, Z.; Chen, W. An efficient approach for feature-preserving mesh denoising. *Opt. Lasers Eng.* **2017**, *90*, 186–195. [CrossRef]
24. Zheng, Y.; Fu, H.; Au, O.K.C.; Tai, C.L. Bilateral normal filtering for mesh denoising. *IEEE Trans. Vis. Comput. Graph.* **2010**, *17*, 1521–1530. [CrossRef] [PubMed]
25. Wasenmüller, O.; Bleser, G.; Stricker, D. Joint bilateral mesh denoising using color information and local anti-shrinking. *J. Wscg.* **2015**, *23*, 27–34
26. Lavoué, G. A local roughness measure for 3D meshes and its application to visual masking. *ACM Trans. Appl. Percept. (TAP)* **2009**, *5*, 21. [CrossRef]
27. Wang, K.; Torkhani, F.; Montanvert, A. A fast roughness-based approach to the assessment of 3D mesh visual quality. *Comput. Graph.* **2012**, *36*, 808–818. [CrossRef]
28. Song, R.; Liu, Y.; Martin, R.R.; Rosin, P.L. Mesh saliency via spectral processing. *ACM Trans. Graph. (TOG)* **2014**, *33*, 6. [CrossRef]
29. Guo, J.; Vidal, V.; Baskurt, A.; Lavoué, G. Evaluating the local visibility of geometric artifacts. In Proceedings of the ACM SIGGRAPH Symposium on Applied Perception, Tübingen, Germany, 13–14 September 2015; pp. 91–98.
30. Ioannou, Y.; Taati, B.; Harrap, R.; Greenspan, M. Difference of normals as a multi-scale operator in unorganized point clouds. In Proceedings of the 2012 Second International Conference on 3D Imaging, Modeling, Processing, Visualization & Transmission, Zurich, Switzerland, 13–15 October 2012; pp. 501–508.
31. Rabbani, T.; Van Den Heuvel, F.; Vosselmann, G. Segmentation of point clouds using smoothness constraint. *Int. Arch. Photogramm. Remote. Sens. Spat. Inf. Sci.* **2006**, *36*, 248–253.

32. Harwin, S.; Lucieer, A. Assessing the accuracy of georeferenced point clouds produced via multi-view stereopsis from unmanned aerial vehicle (UAV) imagery. *Remote. Sens.* **2012**, *4*, 1573–1599. [CrossRef]
33. Huhle, B.; Schairer, T.; Jenke, P.; Straßer, W. Robust non-local denoising of colored depth data. In Proceedings of the 2008 IEEE Computer Society Conference on Computer Vision and Pattern Recognition Workshops, Anchorage, AK, USA, 23–28 June 2008; pp. 1–7.
34. Shiozaki, A. Edge extraction using entropy operator. *Comput. Vision Graph. Image Process.* **1986**, *36*, 1–9. [CrossRef]
35. Favalli, M.; Fornaciai, A.; Isola, I.; Tarquini, S.; Nannipieri, L. Multiview 3D reconstruction in geosciences. *Comput. Geosci.* **2012**, *44*, 168–176. [CrossRef]
36. D'Amico, N.; Yu, T. Accuracy analysis of point cloud modeling for evaluating concrete specimens. Nondestructive Characterization and Monitoring of Advanced Materials, Aerospace, and Civil Infrastructure 2017. In Proceedings of the International Society for Optics and Photonics, San Diego, CA, USA, 25–29 March 2017; Volume 10169, p. 101691D.
37. Özyeşil, O.; Voroninski, V.; Basri, R.; Singer, A. A survey of structure from motion*. *Acta Numer.* **2017**, *26*, 305–364. [CrossRef]
38. Bay, H.; Tuytelaars, T.; Van Gool, L. Surf: Speeded up robust features. In *European Conference on Computer Vision*; Springer: Berlin/Heidelberg, Germany, 2006; pp. 404–417.
39. Rosten, E.; Drummond, T. Machine learning for high-speed corner detection. In *European Conference on Computer Vision*; Springer: Berlin/Heidelberg, Germany, 2006; pp. 430–443.
40. Rublee, E.; Rabaud, V.; Konolige, K.; Bradski, G. ORB: An efficient alternative to SIFT or SURF. In Proceedings of the 2011 International Conference on Computer Vision, Barcelona, Spain, 6–13 November 2011; pp. 2564–2571.
41. Kodak. *Optical Formulas and Their Applications*; Kodak: Rochester, NY, USA, 1969.
42. Marčiš, M. Quality of 3D models generated by SFM technology. *Slovak J.Civ. Eng.* **2013**, *21*, 13–24. [CrossRef]
43. Zhou, Q.Y.; Park, J.; Koltun, V. Open3D: A Modern Library for 3D Data Processing. **2018**, arXiv:1801.09847.
44. Bradski, G. The OpenCV Library. *J. Softw. Tools* **2000**, *25*, 120–125.
45. Technologies, U. Unity. 2005. Available online: unity.com (accessed on 20 July 2020).
46. Benesty, J.; Chen, J.; Huang, Y.; Cohen, I. Pearson correlation coefficient. In *Noise Reduction in Speech Processing*; Springer:Berlin/Heidelberg, Germany, 2009; pp. 1–4.
47. Pedregosa, F.; Varoquaux, G.; Gramfort, A.; Michel, V.; Thirion, B.; Grisel, O.; Blondel, M.; Prettenhofer, P.; Weiss, R.; Dubourg, V.; et al. Scikit-learn: Machine Learning in Python. *J. Mach. Learn. Res.* **2011**, *12*, 2825–2830.
48. Chawla, N.V.; Bowyer, K.W.; Hall, L.O.; Kegelmeyer, W.P. SMOTE: synthetic minority over-sampling technique. *J. Artif. Intell. Res.* **2002**, *16*, 321–357. [CrossRef]
49. Zhang, D.; Watson, R.; Dobie, G.; MacLeod, C.; Khan, A.; Pierce, G. Quantifying impacts on remote photogrammetric inspection using unmanned aerial vehicles. *Eng. Struct.* **2020**, *209*, 109940. [CrossRef]
50. Saito, T.; Rehmsmeier, M. The precision-recall plot is more informative than the ROC plot when evaluating binary classifiers on imbalanced datasets. *PLoS ONE* **2015**, *10*, e0118432. [CrossRef]

© 2020 by the authors. Licensee MDPI, Basel, Switzerland. This article is an open access article distributed under the terms and conditions of the Creative Commons Attribution (CC BY) license (http://creativecommons.org/licenses/by/4.0/).

Article

Histogram Adjustment of Images for Improving Photogrammetric Reconstruction

Piotr Łabędź , Krzysztof Skabek , Paweł Ozimek * and Mateusz Nytko

Faculty of Computer Science and Telecommunications, Cracow University of Technology, Warszawska 24, 31-155 Kraków, Poland; piotr.labedz@pk.edu.pl (P.Ł.); krzysztof.skabek@pk.edu.pl (K.S.); mateusz.nytko@pk.edu.pl (M.N.)
* Correspondence: pawel.ozimek@pk.edu.pl

Abstract: The accuracy of photogrammetric reconstruction depends largely on the acquisition conditions and on the quality of input photographs. This paper proposes methods of improving raster images that increase photogrammetric reconstruction accuracy. These methods are based on modifying color image histograms. Special emphasis was placed on the selection of channels of the RGB and CIE L*a*b* color models for further improvement of the reconstruction process. A methodology was proposed for assessing the quality of reconstruction based on premade reference models using positional statistics. The analysis of the influence of image enhancement on reconstruction was carried out for various types of objects. The proposed methods can significantly improve the quality of reconstruction. The superiority of methods based on the luminance channel of the L*a*b* model was demonstrated. Our studies indicated high efficiency of the histogram equalization method (HE), although these results were not highly distinctive for all performed tests.

Keywords: photogrammetry; preprocessing; enhancement; point cloud; 3D reconstruction; image processing; image histogram

Citation: Łabędź, P.; Skabek, K.; Ozimek, P.; Nytko, M. Histogram Adjustment of Images for Improving Photogrammetric Reconstruction. *Sensors* **2021**, *21*, 4654. https://doi.org/10.3390/s21144654

Academic Editor: Francesco Pirotti

Received: 2 June 2021
Accepted: 5 July 2021
Published: 7 July 2021

Publisher's Note: MDPI stays neutral with regard to jurisdictional claims in published maps and institutional affiliations.

Copyright: © 2021 by the authors. Licensee MDPI, Basel, Switzerland. This article is an open access article distributed under the terms and conditions of the Creative Commons Attribution (CC BY) license (https://creativecommons.org/licenses/by/4.0/).

1. Introduction

Until recently, photogrammetric reconstruction was available to a narrow community of remote sensing specialists [1,2]. It has become widely available thanks to the development of mobile applications using cameras installed in smartphones [3]. However, reconstructions made with these applications do not provide sufficient reliability for engineering applications [4]. Its potential use is building surveying, cost estimation, modeling space for virtual and augmented reality [5–8]. Unfortunately, it is very difficult to extract precise data allowing for sufficiently precise modeling from point clouds generated by these technologies [9–11]. This led to a search for methods to improve reconstruction quality and reliability [12–14]. The trust of engineers in this reconstruction is crucial [15].

Photographs are not always taken correctly. This depends on the lightning conditions [16], the availability of a given property, the amount of time one can spend on registration, the quality of the photo equipment and many other factors [17,18]. The image file format should also be taken into account when capturing images [19]. The quality of the photographs is especially important for the automated photogrammetric reconstruction process [20,21]. Its impact is particularly noticeable when processing large image datasets [22], or using the markerless method [23]. Therefore, there is a need to correct and improve the quality of these photos. Image preprocessing is used to enhance the quality of images [24] and it is a very significant issue from the point of view of 3D modeling and photogrammetric reconstruction [25]. It includes many different methods such as histogram enhancement, color balancing, denoising, deblurring or filtering [26]. Some methods have already been used to improve photo quality in the photogrammetric reconstruction pipeline. Many of these are associated with the deblurring of images [27–29]. The literature also includes studies on the use of using polarizing filters or High Dynamic

Range (HDR) images, [25] decolorization [30] or novel techniques for the conversion of the color signal into a grayscale [31]. Significant research within this area has been reported by Gaiani et al. [32]. They developed an advanced pipeline consists of color balancing, image denoising, color-to-gray conversion and image content enrichment, including new methods for image denoising and grayscale reduction. Another result of evaluating the impact of image processing for photogrammetry were presented by Feng et al. [33]. Presently, neural networks, especially the ones based on deep models, are used to improve photogrammetric quality [29,34–36].

On the contrary to create the advanced photogrammetric pipeline [32] or use sophisticated tools, such as neural networks we looked for a simple way to improve photos to obtain the best material for photogrammetric reconstruction. We assumed that selected methods of improving images, based on histogram operations, should give the desired results. Our methodology is designed to work directly on photographs, as a preprocessing step in the photogrammetric reconstruction pipeline. It can be used independently of reconstruction software, which required testing with black-box type tool. We verified the correctness of our method by comparing the reconstructions obtained from the preprocessed materials with the most reliable geometric model at our disposal. In addition to methods of improving image quality, finding a good way to compare the reconstruction to a reliable model is not an obvious task.

In our previous work [37] we explored the improvement of input images for photogrammetric reconstruction. Our conclusion was that reconstruction using unprocessed photos does not always give satisfactory results. We tested several methods of image preprocessing based on histogram analysis however, we focused only on grayscale images in that paper. Reconstruction based on such images is correct in terms of structure, but it is also somewhat limited as it is not possible to use colored textures. To correct this inconvenience, we tested the following image processing methods: histogram stretching (HS), histogram equalizing (HE), adaptive histogram equalizing (AHE) and exact histogram matching (EHM) on color images with the use of different color spaces.

2. Materials and Methods

2.1. Research Method

Several methods of processing photographic material were selected for our study. The processing results were used to generate photogrammetric reconstructions of photographed objects. The reconstructions were in the form of point clouds. Each reconstruction was compared with a suitable reference model. The reference models featured a representation of polygonal meshes. The aim of our study was to characterize individual material processing methods in terms of their suitability for generating geometrically correct point clouds. Therefore, the selection of reliable reference models and reliable methods for comparing reconstructions to reference models were an important element of our study.

Two real-world objects were selected as case studies. The first was a porcelain swan figurine. The second was a large historical building. Analyses carried out on objects of such different scales resulted in a wide range of recommendations regarding the methods of processing photographic material that had been tested. At the same time, they required other methods of obtaining reference models. Unfortunately, different modeling methods give models with varying degrees of fidelity. When selecting the appropriate measures for such different objects, the same methods of comparing the reconstruction with the models were used. The results of these comparisons were also characterized in relation to the scale of the object.

2.2. Photo Improvement Methods

In this paper, we focused on improving the quality of images, mainly in terms of contrast and brightness. We did not explore the influence of other factors (such as sharpness), which shall be a part of further research. We focused only on color photographs (Figure 1), as we already published research on monochrome images [37]. Two different color spaces are used in the imaging processing—RGB and L*a*b*. The RGB space image was divided into three separate channels—red, green and blue. Each channel was treated as a single monochrome image, therefore histogram corrections were performed on each channel separately (Figure 2), and the RGB image was recombined afterwards. In the L*a*b* space image was also divided into three separate channels – L*, a* and b*. The L* channel represents brightness (luminance), a* channel color position between red and green and b* channel color position between yellow and blue. In order to preserve the original colors of an image, modifications were introduced into the L* channel only. For this channel to be considered a monochrome image, its values were previously transformed to the [0 1] range.

Figure 1. Example of an underexposed image.

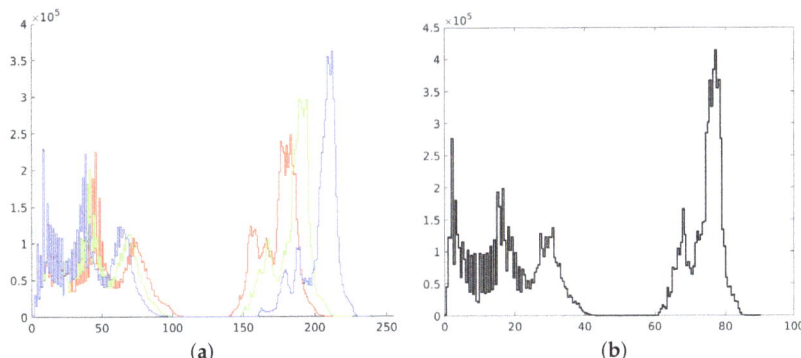

Figure 2. Histograms of RGB channels (**a**) and L channel of L*a*b* space (**b**).

As described in Algorithm 1 each image needs to be converted back into RGB space to perform a photogrammetric reconstruction.

2.2.1. Histogram Stretching (HS)

The first image enhancement method that we used was common histogram stretching. This process maps the intensity values of the image into new values to redistribute the information of the histogram toward the extremes of a gray-level range. This increases the intensity range, although some of the brightness values are not represented in the processed image [38]. Transformed images with their histograms are presented in Figures 3 and 4.

Algorithm 1: Histogram modification in L*a*b* space

Input: I_{RGB}—image in RGB space
1. convert I_{RGB} into I_{L*a*b*}
2. divide I_{L*a*b*} into separate channels
3. leave $a*$ and $b*$ channels without any changes
4. transform $L*$ channel to the [0 1] range
5. perform certain histogram operation on $L*$ channel
6. transform back $L*$ channel to the [0 100] range
7. combine $I_{L*a*b*'}$ from separate channels
8. convert $I_{L*a*b*'}$ into $I_{RGB'}$
9. **return** $I_{RGB'}$—modified image in RGB space

Figure 3. Image transformed with the use of histogram stretching in RGB (**a**) and L*a*b* space (**b**).

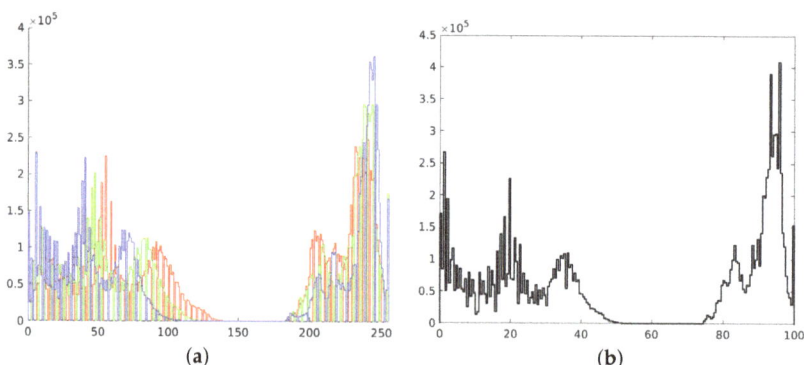

Figure 4. Histograms of RGB channels (**a**) and L channel of L*a*b* space (**b**).

2.2.2. Histogram Equalizing (HE)

The second way to improve image contrast that we tested was the histogram equalization. This technique effectively stretches the most common intensity values—it extends the intensity range of an image. Afterwards, intensities can be distributed better—low-contrast areas will have higher contrast and the cumulative histogram would increase linearly [24]. Performing histogram equalization on separate RGB channels often leads to unrealistic effects. Therefore, this operation is more suitable for images in different color spaces, such as L*a*b* (Figure 5).

The resulting histograms are as flat as possible, without noticeable peaks. However, it should be noticed that many intensities are not represented in obtained images. This is represented as white gaps between the histogram bars (Figure 6).

Figure 5. Image transformed with the use of histogram equalizing in RGB (**a**) and L*a*b* space (**b**).

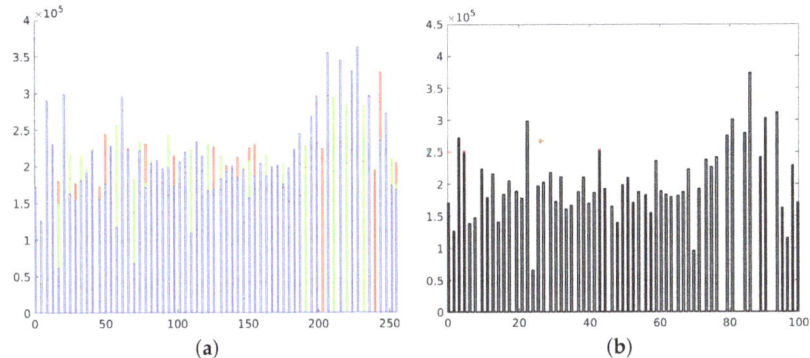

Figure 6. Histograms of RGB channels (**a**) and L channel of L*a*b* space (**b**).

2.2.3. Adaptive Histogram Equalizing (AHE)

The third method of improving images that we tried was adaptive histogram equalization. This method differs from ordinary histogram equalization in the way that the adaptive method (AHE) calculates histograms in separate parts of the image (tiles) instead of the entire image [39]. In each tile, a transformation function is calculated for each pixel based on neighboring values. The classic AHE approach tends to overamplify contrast and noise. Therefore, contrast-limited adaptive histogram equalization (CLAHE) was applied in place of AHE. In this algorithm, the histogram is first truncated at a predefined value and the transformation function is calculated afterwards. This situation takes place especially in homogeneous areas [40].

After performing the equalization, an algorithm combines neighboring tiles using bilinear interpolation to eliminate artificially induced boundaries. The resulting image and its histogram differ significantly from the image obtained with the use of the standard histogram equalization technique (Figures 7 and 8).

(a) (b)

Figure 7. Image transformed with the use of adaptive histogram equalizing in RGB (**a**) and L*a*b* space (**b**).

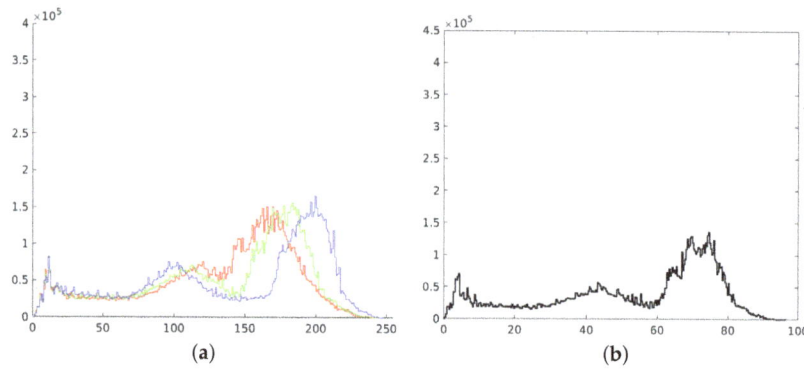

Figure 8. Histograms of RGB channels (**a**) and L channel of L*a*b* space (**b**).

2.2.4. Exact Histogram Matching (EHM)

The last image enhancement process consists of two main stages. In the first stage, a mean histogram of the whole set of images is calculated. This is performed by adding histograms of each image (separately for each channel) and dividing the result by the number of images in the set. In the second step, an exact histogram matching operation [41,42] is used to adapt the histogram of each image to match the obtained average histogram (Figures 9 and 10).

Figure 9. Image transformed with the use of exact histogram matching in RGB (**a**) and L*a*b* space (**b**).

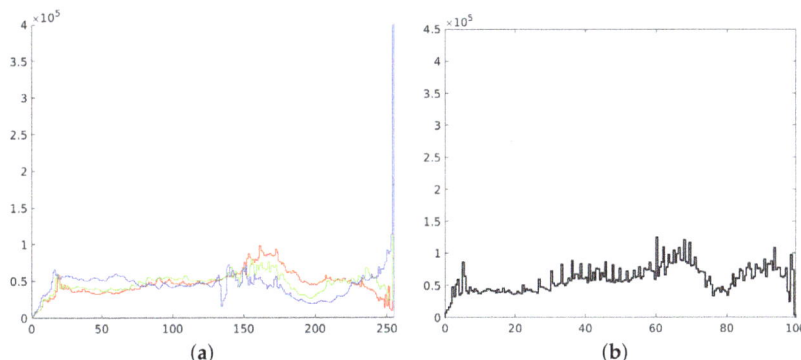

Figure 10. Histograms of RGB channels (**a**) and L channel of L*a*b* space (**b**).

2.3. Photogrammetric Reconstruction

For photogrammetric reconstruction, we used Agisoft Metashape, which is used extensively worldwide. During the reconstruction phase, the same parameters were used for each step to keep the test method consistent. The reconstructions were created independently for each set of photos, i.e., the original photos and those transformed following the algorithms described. We chose two types of objects to validate the method's correctness regardless of the scale of the object subjected to photogrammetric reconstruction. Comparisons between individual reconstructions were performed using Cloud Compare software.

Agisoft Metashape provides a built-in tool for assessing the quality of photos, but this is based on sharpness only. For this purpose, it analyzes contrast between pixels and determines the quality factor which takes a value from 0 to 1. According to the manufacturer's recommendation, images with a factor lower than 0.5 should be excluded from the reconstruction process. This criterion is not always reliable, because in the case of directional blur, which is most common when taking pictures, sharp areas can still be detected by algorithm, and qualify the picture as good quality [43]. There is no information about any image preprocessing techniques implemented in Metashape, therefore it should be treated as a black-box tool.

A DSLR camera was used in the registration of photographs for the photogrammetric reconstruction. The sensor was not calibrated in any of the described case studies, in addition a markerless method was used. This was due to the intent to present a method that can be used in a wide range of cases without having to meet any special requirements.

2.3.1. Reference Model of the Porcelain Figurine

The first testing datum was a little porcelain figurine, about 12 cm high. Its reference model was acquired with the use of a 3D scanner that operates in the field of structural light (Figure 11a). The scanner, unfortunately, was not able to reach all the covered parts of the object. The fragments of the figurine that did not have a correct digital representation were filtered out in the process of analysis. They were also the same fragments for which obtaining a correct representation during the photogrammetric reconstruction met with failure.

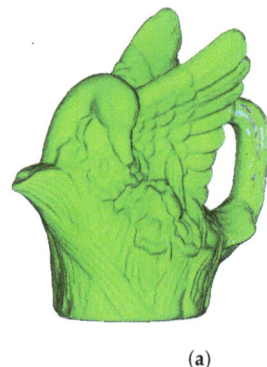 

(a)   (b)

Figure 11. Porcelain figurine models: (**a**) mesh acquired by the 3D scanner, (**b**) point cloud obtained with the use of photogrammetric reconstruction.

2.3.2. Reference Model of the Castle in Nowy Wiśnicz

The castle in Nowy Wiśnicz is in the north-eastern part of the town, less than 400 m from the Market Square. The origins of the castle date back to the fourteenth century, but its present shape is from the seventeenth century. It is a typical example of a palazzo in fortezza. Its quadrangular shape is accentuated by four corner towers. There is an additional segment to the south. The entire layout is surrounded by bastion fortifications in the shape of a pentagon with the longest dimension of 190 m. Together with bastions and curtain walls, it covers approx. 19,500 m^2 (Figure 12b). The height of the castle itself, measured from the level of the courtyard inside the fortifications, is about 36 m (Figure 12a) [44].

Obtaining a reference model of such a large object was not possible using a 3D scanner. It was created using photogrammetric methods and verified via comparison with a point cloud obtained using airborne LiDAR. It is a combination of several partial models covering various parts of the castle. The individual partial models were obtained from separate sets of photos taken to ensure high quality. The photos were taken from human eye-level and from different heights, for which an UAV (Unmanned Aerial Vehicle) was used. There were 2300 photos in total. Partial reconstructions of the castle's buildings were generated in Agisoft Metashape. They were then compiled using orthogonal versor matrix multiplication in Cloud Compare and combined into a single point cloud and then converted into a mesh model.

The point cloud used to validate the model represented a digital terrain and cover model. It was obtained from the state surveying repository. In Poland, by virtue of the law [45], the numerical terrain model and land cover data are currently available free of charge. They can be downloaded from the governmental servers of National Geoportal [46] as a grid of points with x, y, z coordinates, deployed at 1 m intervals. There are also LAS standard point cloud data available [47], acquired as a part of the ISOK project (National Land Cover IT System) [48]. As a result of this project, 98% of the territory of Poland was scanned, with a density of between 6–12 points/m^2. The point cloud is available in the form of LAS files, where each point is represented by X, Y, Z, coordinates, RGB color (Figure 13a) and assigned to one of four classes: ground, structure, water and vegetation (Figure 13b) [47]. This enables the individual segmentation layers to be compared separately [49]. These data are reliable from a geolocation point of view, but insufficient for many engineering applications. Especially in the case of building walls, which, as elements with a mostly vertical geometry, are very poorly covered with points, as they are recorded via airborne LiDAR flyovers [50]. The fixed measurement interval does not provide the coordinates of distinctive elements of building geometry (corners, ridges, tops of towers). However, this is a feature that can be used for verification. The ISOK model points were used in this work for the systematic sampling of model correctness. The sparse ISOK point cloud precisely defines the space into which the dense cloud of the reference model must fit. After appropriate scaling and fitting of the reference model, the correctness was assessed by comparing the sparse ISOK cloud to the dense cloud of the reference model by examining the root mean square distance (RMS) between the superimposed structures (Figure 14). The RMS error estimation method allows the determination of a fitting error's statistical values, which are expressed in spatial distance, which is well understood by engineers potentially using such models [51].

Figure 12. Polygonal models of the castle in Nowy Wiśnicz: (**a**) building only (4,133,352 faces), (**b**) the building with fortifications (1,966,487 faces).

Figure 13. ISOK point cloud: scanning colors (**a**); class colors (**b**).

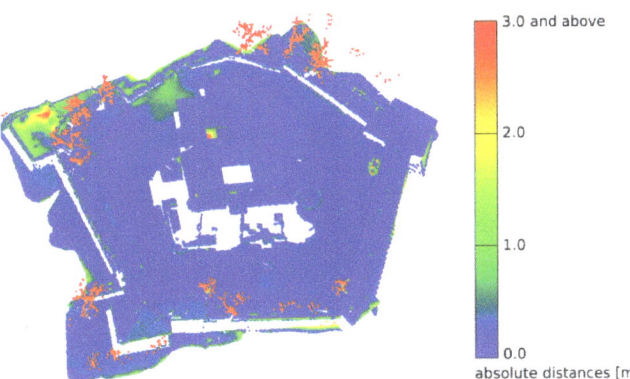

Figure 14. Distance error (RMS) for reconstructed model and reference LAS ISOK: LAS ISOK to reconstructed model—point-to-mesh mapping.

3. Results

The subject of our research was to determine whether the preprocessing of photos used to perform photogrammetric reconstruction affects its quality. Our research on grayscale images has shown that uniformity of pictures in terms of brightness and contrast can significantly improve the quality of a model obtained via reconstruction. However, using monochrome images has the disadvantage that the resulting texture does not precisely reflect the object's appearance. For this reason, the photographs of the objects mentioned in the previous section were processed in color spaces in accordance with the methods presented.

The general scheme of the tests carried out on each set of photos is as follows:

1. Photo correction using one of the methods mentioned.
2. Creation of a sparse and then dense point cloud by photogrammetric reconstruction.
3. Registration using the ICP method to match the received data sets [52].
4. Comparison of individual reconstructions using the distance calculation: reconstructed model cloud to reference model cloud.
5. Statistical and visual analysis of the obtained results.

Classical statistical measures such as mean or standard deviation can be used when a variable is quantitative in nature. However, when the variable is ordinal, it is better to use positional measures such as median or interquartile range. In the case under study, the variable is the distance of a point derived from the photogrammetric reconstruction from the reference model obtained with the scanner. Therefore, the variable can be considered ordinal – the smaller the distance, the more correct the result. The analyzed statistical quantities are quartiles and in the interquartile range. Subsequent quartiles provide information about how far away from the original model the 25% (Q_1), 50% (Q_2—median), and 75% points (Q_3) are. These values are calculated from distances given in absolute (unsigned) quantities, so a distance of 0.10 cm is treated the same as a distance of −0.10 cm. However, in the case at hand, this is irrelevant because the objective is to determine the number of points that are within a given range of distances, without taking into account whether they are outside or inside the reference model. The smaller the value of a given quartile, the better the result because it means that a given percentage of points lies at a closer distance to the reference model. For example, in the case of the swan figure and the HE method in the L*a*b* space, the third quartile is 0.17 cm, which means that 75% of the points are closer to the reference model than this value. For the same method in RGB space, these 75% of the points are closer than 0.20 cm, which is a larger value, and therefore the points are further away from the reference model (Table 1).

Another measure that is calculated is the interquartile range (IQR), which is one of the values that determines dispersion. This measure is determined using a signed number and determines the degree of diversity—the higher the interquartile range value, the greater the variety of a feature. The value of this measure is calculated based on the difference between the third and first quartiles: $IQR = Q_3 - Q_1$. In other words: the 50% of points are within the interval defined by the value of IQR. A narrower interval means a higher concentration of points in closer proximity to the reference model, so in the case under analysis, a lower value of IQR indicates a better result.

Based on quartile values obtained, it was also possible to perform an analysis of the length of the "tails", i.e., the number of outlying points beyond a specific value of the distance from the reference model. The average of the values of Q_3 for all the measurements taken was assumed to be this particular value (σ). The values obtained determined how many points were at a significant distance from the reference model. These points can therefore be treated as incorrectly reconstructed. Due to the different number of points in the cloud for each reconstruction, what is important here is the value in percentage terms—the smaller it is, the fewer points are at a significant distance from the reference model, and thus it can be considered more accurate.

3.1. Case Study 1—The Porcelain Figurine

The set of images used for photogrammetry reconstruction consisted of 33 photographs. Twelve of them were underexposed, and eleven of them were overexposed. The dense clouds obtained in the figurine reconstruction process consisted of several hundred thousand points. These were then reviewed for duplicates, leading to the removal of approximately 10% of points from each cloud. The final number of obtained points is shown in Table 1. The obtained reconstructions are quite difficult to compare in terms of quality visually (Figure 15). However, the differences in contrast and color saturation of the textures are well demonstrated. Particularly notable is the low contrast in the reconstruction created using the original images (Figure 15a). As described in Section 2.2, the modification of the image in L*a*b* space involves transforming the histogram on the luminance channel without modifying the color channels, which can easily be seen in Figure 15f–i. On the other hand, modifying individual channels in the RGB color space results in changes in the saturation of individual colors (Figure 15b–e) and can also lead to errors in texture colors.

Table 1. Number of points and statistical measures for the reconstruction of the porcelain figurine.

	Method	Aligned Cams	Points	Q1	Median	Q3	IQR
	original	24	350,232	0.05	0.11	0.19	0.23
L*a*b*	AHE	23	365,903	0.05	0.12	0.20	0.23
	EHM	23	300,097	0.05	0.10	0.19	0.19
	HE	24	313,622	0.05	0.10	0.17	0.20
	HS	27	338,858	0.05	0.11	0.19	0.21
RGB	AHE	23	337,649	0.05	0.11	0.19	0.21
	EHM	23	280,334	0.05	0.11	0.20	0.21
	HE	23	316,222	0.05	0.11	0.20	0.20
	HS	27	373,312	0.05	0.12	0.22	0.23
						avg:	0.20

Upon a statistical analysis of the results (presented in Table 1), one can see very similar results for all investigated methods. The value of the first quartile in each case was 0.05 cm. The median values also differed very little. Only values for the third quartile showed a slightly bigger variation, where the best value was attained for the HE method in the L*a*b* space. However, it should be noted that the differences in individual values were minuscule and approximate the reconstruction obtained using unprocessed photograph.

The situation was slightly different when tail analysis was performed (Table 2). The σ value for the figurine was taken as 0.2 cm. This value was exceeded by 23.40% of points when reconstructed from images without enhancement, and only 17.84% of points when corrected using the HE method in L*a*b* space. In general, methods operating in L*a*b* space performed better in this case, with an average of 22.16% of points above the σ value, compared to 25.48% for methods operating in RGB space. A similar relationship can also be observed by analyzing the distribution of values for 2σ and 3σ. It should also be noted that the number of points located at a distance above 2σ, i.e., above 4 mm from the reference model, was relatively small—for the previously mentioned HE method in L*a*b* space, it is 3.62% of the total number of points in the cloud.

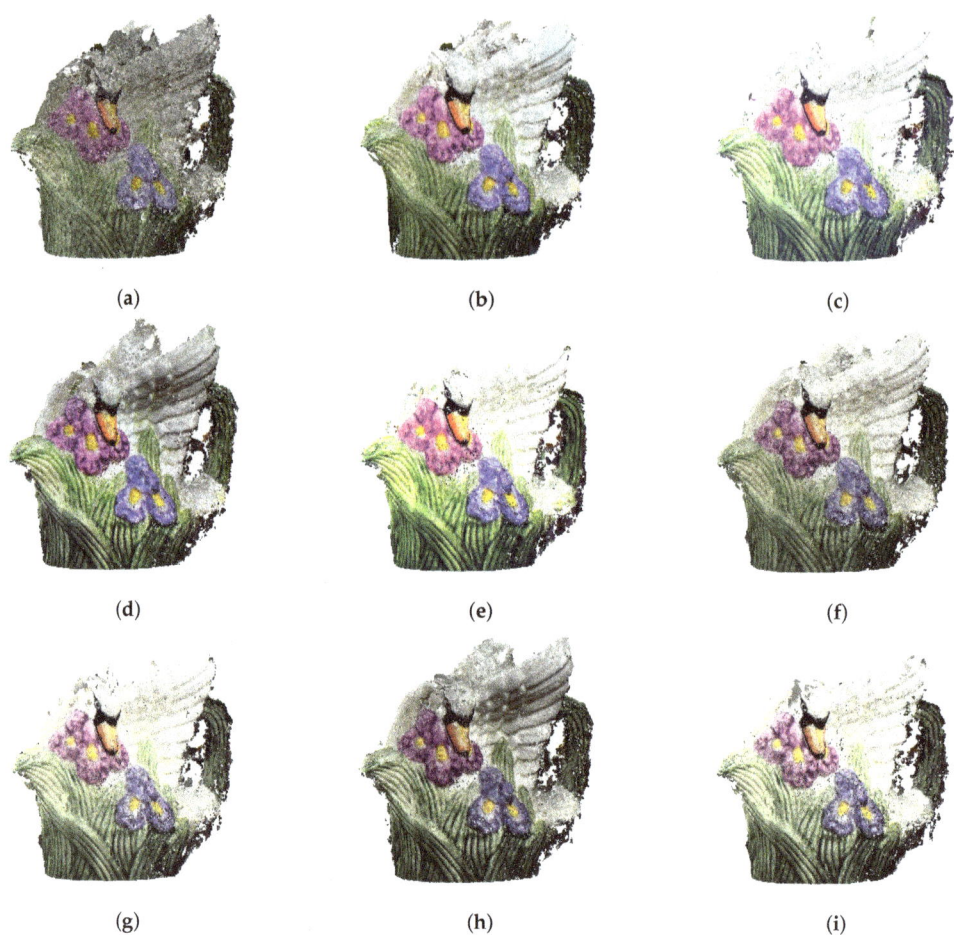

Figure 15. Dense point clouds of the porcelain figurine. Reconstructions based on photographs: (**a**) original; preprocessed in the RGB space with the use of: (**b**) histogram stretching, (**c**) histogram equalizing, (**d**) adaptive equalizing, (**e**) exact histogram matching; preprocessed in the L*a*b* space with the use of: (**f**) histogram stretching, (**g**) histogram equalizing, (**h**) adaptive equalizing, (**i**) exact histogram matching.

The distribution of deviations is not uniform over the entire surface of the model. The largest number is found in the ear region of the figure, as this section was the most difficult to reconstruct for the photogrammetric reconstruction algorithms (Figure 16). The different reconstructions were similar to each other in terms of error deviations. Visually, the largest number of points, located at distances above 3σ, was found for the HE method in RGB space (Figure 16c), which is confirmed by the numerical values (Table 2). It is also notable that for the method based on the EHM algorithm, the upper part of the figure was not correctly reconstructed for both RGB and L*a*b* space processed photographs (Figure 16e,i).

Table 2. Tail analysis for swan figure ($\sigma = Q3 = 0.2$).

	Method	σ	[%]	2σ	[%]	3σ	[%]
	original	81,949	23.40	16,829	4.81	4232	1.21
L*a*b*	AHE	92,521	25.29	12,898	3.52	4756	1.30
	EHM	69,061	23.01	14,381	4.79	3695	1.23
	HE	55,938	17.84	11,367	3.62	3165	1.01
	HS	76,199	22.49	13,910	4.10	3846	1.13
RGB	AHE	78,988	23.39	15,156	4.49	4268	1.26
	EHM	69,099	24.65	13,064	4.66	4015	1.43
	HE	79,759	25.22	18,559	5.87	6565	2.08
	HS	106,967	28.65	26,207	7.02	4564	1.22

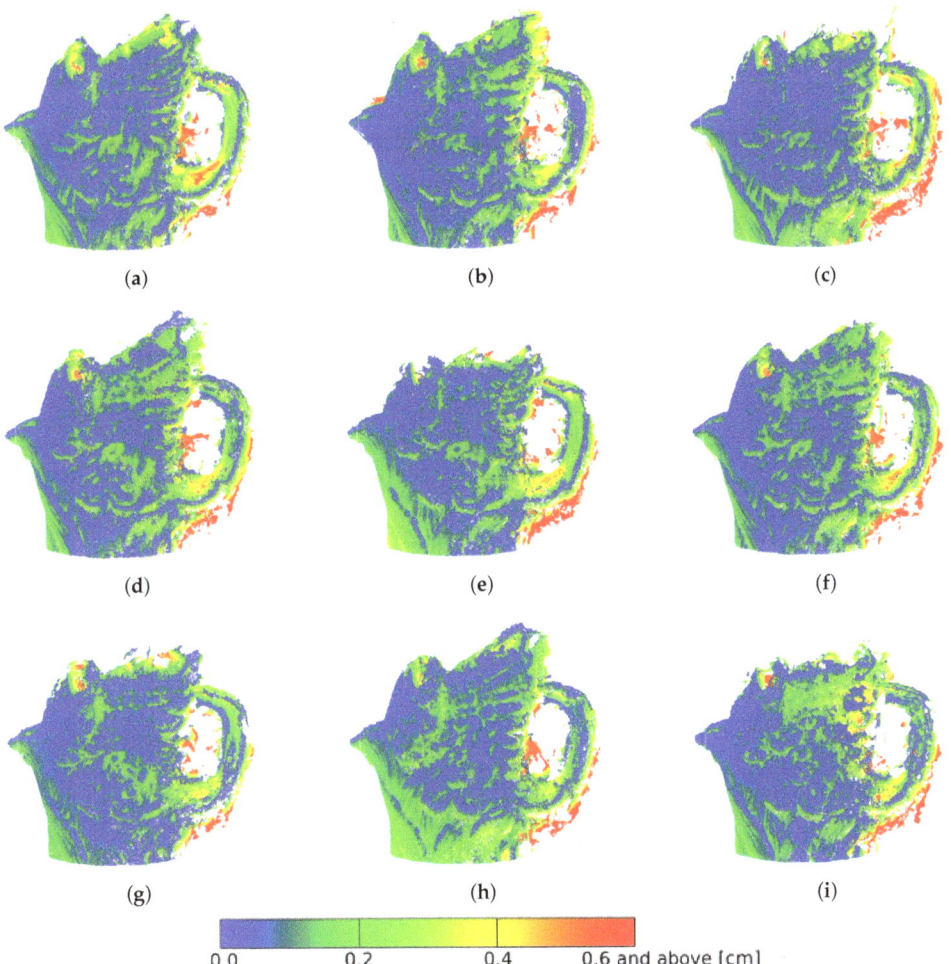

Figure 16. Dense point clouds of the porcelain figure. Distances to the model up to $3\sigma = 0.6$ cm: (**a**) original; preprocessed in the RGB space with the use of: (**b**) histogram stretching, (**c**) histogram equalizing, (**d**) adaptive equalizing, (**e**) exact histogram matching; preprocessed in the L*a*b* space with the use of: (**f**) histogram stretching, (**g**) histogram equalizing, (**h**) adaptive equalizing, (**i**) exact histogram matching.

As mentioned, both original and preprocessed reconstructions are quite similar. This may suggest that in the case of easily accessible objects of small size, repeating the acquisition of a series of photographs with more attention to lighting conditions and correct camera settings would prove a better reconstruction improvement method than source image processing. To verify this hypothesis, an additional analysis was performed with photographs that met the proper acquisition conditions.

Upon visual comparison of the reconstruction (Figure 17), it can be seen that the point cloud contained more detail than the others shown in Figure 15. Numerical data also confirmed this. The cloud consisted of 521,404 points, which was an increase of 39.7% relative to the reconstruction with the highest number of points (retrieved using the HS method in RGB space) and 48.9% more than the reconstruction made from the original images. However, the number of points in the cloud itself was not conclusive, as the points may not have been reconstructed correctly. Nevertheless, the repeated reconstruction gained some advantage here as well. Both quartiles and IQR values were 0.01 lower on average than the best reconstructions obtained by the histogram improvement method. The value of IQR, which is a measure of dispersion, was also 0.05 (21.7%) lower than the value obtained for the reconstructions made from the originally acquired images. The tail analysis also confirmed the desirability of re-taking the image sequence. The number of points above σ equaled 17.15%, with 23.40% for the original reconstruction and 17.84% for the best reconstruction obtained using the HE method in L*a*b* space. The number of points above 3σ represented only 0.46% of the total number of points, which was a decrease by 54.5% from the best reconstruction and 61.9% from the original one.

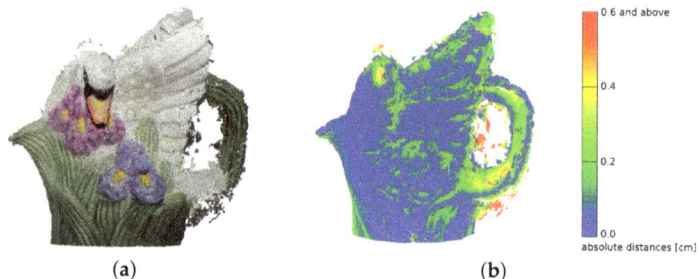

Figure 17. Improved reconstruction of the porcelain figurine: (**a**) RGB dense cloud, (**b**) absolute distance mapping.

3.2. Case Study 2—The Castle in Nowy Wiśnicz

The castle in Nowy Wiśnicz is an object several times larger than the presented model of the figurine. The number of images used in its reconstruction process was about ten times greater, and the number of points of the dense cloud about 100 times greater (about 30 million points—Table 3). The whole set of selected images used for photogrammetric reconstruction consisted of 338 photographs, and about 20% of them were underexposed. The first look at the resulting models allowed us to notice that the reconstruction made with the use of original photographs were incorrectly made (Figure 18a). In fact, this error inspired the research towards improving the quality of the reconstruction without repeating the photographic registration process. The defects could also be seen with the use of the AHE method in RGB space. The deviations were more clearly visible when analyzing images where the distance from the reference model was marked with the corresponding color (Figure 19). Similar to the reconstruction of the porcelain figurine, differences in contrast and color saturation of the textures were also apparent here. For reconstructions created using images processed in RGB space, color distortion was noticeable (Figure 18c–e). On the other hand, when using the AHE method in L*a*b* space, considerable texture brightening was visible.

Table 3. Number of points and statistical measures for the reconstruction of the castle in Nowy Wisnicz.

	Method	Aligned Cams	Points	Q1	Median	Q3	IQR
	original	338	41,833,893	11.78	47.73	87.30	75.90
L*a*b*	AHE	338	30,328,057	0.05	0.10	0.17	0.19
	EHM	338	31,144,627	0.04	0.10	0.19	0.20
	HE	338	28,241,790	0.05	0.10	0.17	0.19
	HS	338	29,980,758	0.04	0.10	0.16	0.18
RGB	AHE	338	30,774,070	0.06	0.16	0.46	0.31
	EHM	338	30,223,244	0.05	0.10	0.17	0.19
	HE	338	30,163,377	0.05	0.11	0.17	0.19
	HS	338	31,055,419	0.05	0.11	0.19	0.20
					avg:	0.21	

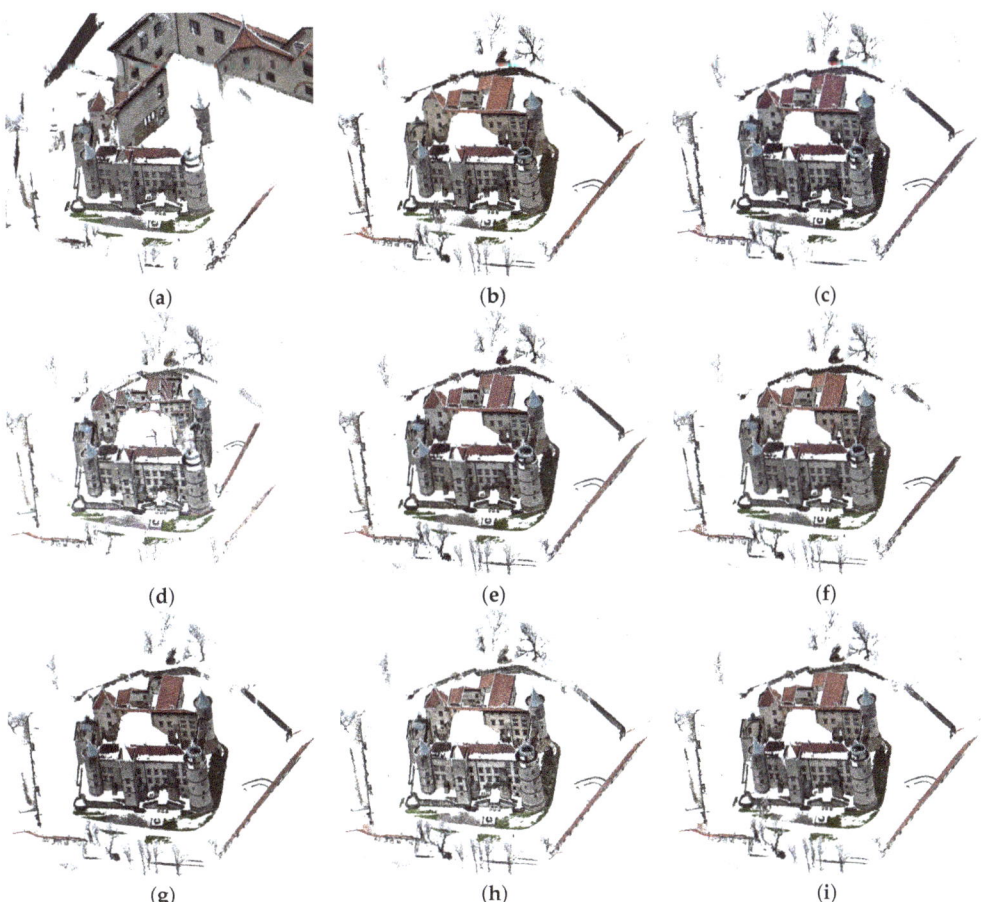

Figure 18. Dense point clouds of the castle. Reconstructions based on photographs: (**a**) original; preprocessed in the RGB space with the use of: (**b**) histogram stretching, (**c**) histogram equalizing, (**d**) adaptive equalizing, (**e**) exact histogram matching; preprocessed in the L*a*b* space with the use of: (**f**) histogram stretching, (**g**) histogram equalizing, (**h**) adaptive equalizing, (**i**) exact histogram matching.

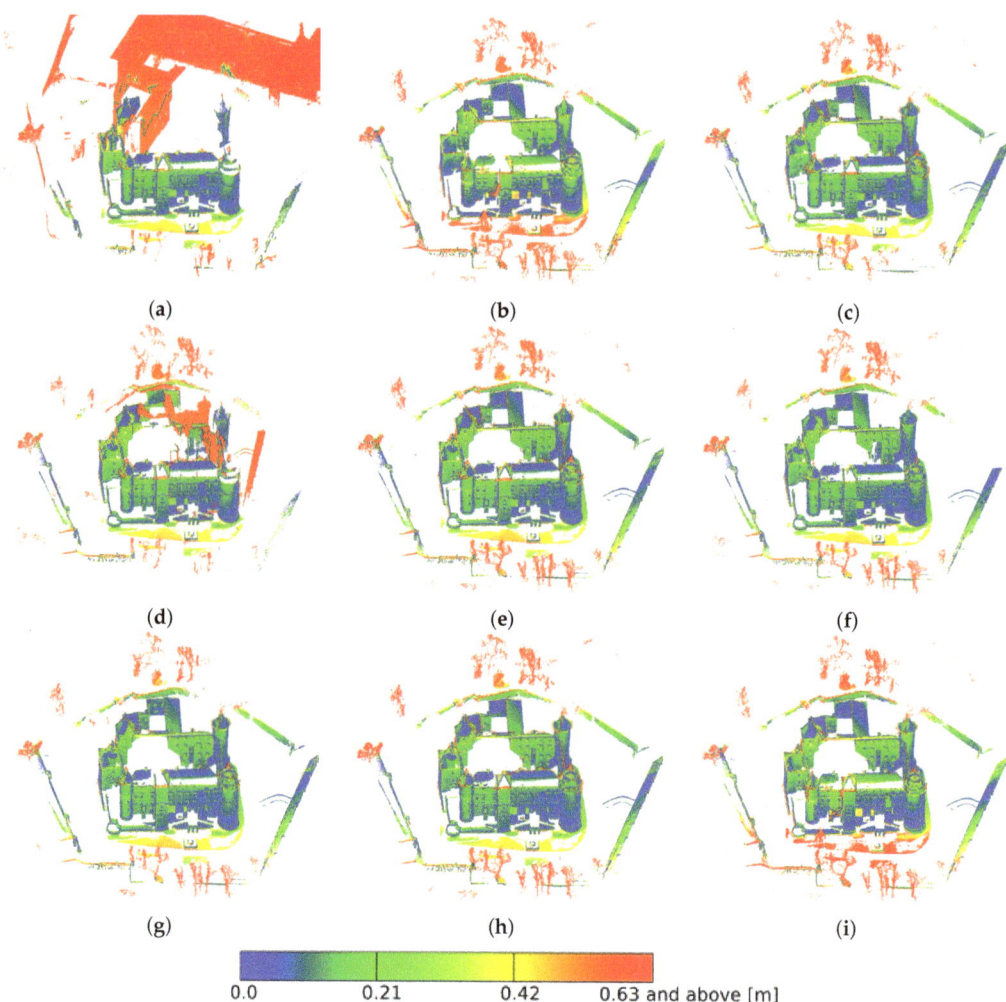

Figure 19. Dense point clouds of the castle. Distances to the model up to $3\sigma = 0.63$ m: (**a**) original; preprocessed in the RGB space with the use of: (**b**) histogram stretching, (**c**) histogram equalizing, (**d**) adaptive equalizing, (**e**) exact histogram matching; preprocessed in the L*a*b* space with the use of: (**f**) histogram stretching, (**g**) histogram equalizing, (**h**) adaptive equalizing, (**i**) exact histogram matching.

Analysis of the statistical values strongly highlighted the defect in the original reconstruction. The value of Q_1 reached 11.78 m and IQR, which is a measure of dispersion as high as 75.90 m (Table 3). For reconstructions based on modified images, these values were several orders of magnitude smaller. The value of the first quartile oscillated around 5 cm, while the third quartile for almost all reconstructions was less than 20 cm. It was larger only in the previously mentioned AHE method in RGB space and amounted to 46 cm. This numerically confirmed the visually observed abnormality in this reconstruction. However, it should be noted that the Q_3 value was a small fraction (about 0.4%) of the castle dimensions, which demonstrates the correctness of the reconstruction since 75% of the points lied within this dimensional deviation from the reference model. It is interesting that each of the 338 photographs, was aligned by the software. This indicates that the reconstruction error occurs already at the keypoint extraction stage. This observation should be deeply investigated in subsequent studies.

The above conclusions are also confirmed by tail analysis (Table 4). The σ value was assumed to be 0.21 cm in this case. Almost all (89.22%) points for the reconstruction performed from unenhanced images were above this value. The lowest and therefore best values were achieved for the HS and HE enhancement methods in L*a*b* space—14.58% and 14.96%, respectively. Again, methods operating in L*a*b* space tended to perform better—on average, 16.47% of points were above the σ value, compared to 23.32% for the methods operating in RGB space. Of course, this was affected by the aforementioned reconstruction incorrectness for the AHE method in RGB, for which the σ value was 39.47%. The number of significant deviations from the reference model was the smallest for the methods operating in L*a*b* space. A deviation of more than 63 cm (3σ) was observed for less than 1% of the points for the three methods operating in this space. Among the methods operating in RGB, only in two cases, a deviation above 3σ was noted for about 1% of the points, while for the AHE method in this space, it was more than 23% of the points.

Table 4. Tail analysis ($\sigma = Q3 = 0.21$) for the reconstruction of the castle in Nowy Wisnicz.

	Method	σ	[%]	2σ	[%]	3σ	[%]
	original	37,325,857	89.22	36,496,517	87.24	36,157,816	86.43
L*a*b*	AHE	4,706,784	15.52	759,082	2.50	254,641	0.84
	EHM	6,477,239	20.80	1,889,706	6.07	736,932	2.37
	HE	4,225,890	14.96	661,988	2.34	159,952	0.57
	HS	4,372,294	14.58	764,709	2.55	258,619	0.86
RGB	AHE	12,145,976	39.47	8,087,475	26,28	7,209,241	23.43
	EHM	4,945,613	16.36	975,302	3.23	323,573	1.07
	HE	4,659,034	15.45	799,797	2.65	261,054	0.87
	HS	6,829,555	21.99	2,295,160	7.39	987,314	3.18

The fragments of the reconstruction with the largest deviation from the reference model were marked in Figure 19 in red. The reconstruction defect for the original images is clearly visible (Figure 19a) as is the one for the AHE method in RGB (Figure 19d), where additional elements of the castle towers and walls, positioned at the wrong angle, are visible. In all cases, the deviations occurred within the courtyard in front of the main entrance to the castle. In two cases, RGB HS and L*a*b* EHM, they were slightly larger (Figure 19b,i). An indication of this was also reflected in the statistics featured in Table 4. It is also noteworthy that a significant amount of the points that diverged the most from the reference model were associated with vegetation that was not present on the reference. Removing this vegetation would require manual manipulation of the reconstruction, which the authors wanted to avoid in order not to distort the results of the comparisons. Other inaccuracies included roof edges that were difficult to reconstruct. The least amount of such inaccuracies was found in the HS improvement in L*a*b* space (Figure 19f).

4. Discussion and Conclusions

The research presented in this paper was intended to verify the effectiveness of image enhancement methods for photogrammetric reconstruction. It is an extension of earlier studies performed on grayscale reduced images, which demonstrated that modifying the histogram of individual images can significantly improve reconstruction quality. However, monochromatic images do not fully represent the reconstructed object correctly due to the lack of realistic color reproduction. For this reason, we explored the performance of image modifications using histogram enhancement methods (HS, HE, AHE, EHM) in color spaces (RGB and L*a*b*). As demonstrated, histogram modifications affected the final shape of the reconstruction. The clearest example of this is the case study presented in Section 3.2, regarding the castle in Nowy Wiśnicz. Although the reconstruction based on the original photographs was incorrect, the other reconstructions, obtained using the modified photographs, were already accurate, which should be considered a success of the presented method. This is evident both from the visual side (Figure 18) and from the statistical analyses (Tables 3 and 4). Each of the reconstructions looked slightly different, which also proves that modifications of the source photographs impacted the final result.

Two case studies were selected so as to present the method's application on objects of different sizes. For a small object, the influence on the quality of the reconstruction was more visually noticeable. Statistically, the deviations of the reconstruction from the reference model were rather small, and it is difficult to unambiguously indicate which of the point clouds obtained represent better quality. However, this is influenced by the fact that the reconstruction performed based on original images was roughly correct, i.e., it reflected the object's original shape. The reconstruction obtained from photographs taken with attention to appropriate lighting conditions indicates that proper image acquisition is advantageous over image preprocessing methods. However, registration cannot always be repeated for various reasons, such as the temporary availability of the object to be reconstructed. In such cases, improving the quality of the photographs using appropriate methods can give the expected results.

Architectural objects, such as the castle in Nowy Wiśnicz presented in Section 2.3.2, are significantly more challenging in terms of obtaining the correct exposure of the collected material. It is impossible to choose the optimal lighting by oneself, as it is directly related to the current weather conditions. Additionally, in the case of such an object, the number of photos that must be taken is much greater, which results in longer recording and data post-processing times. In most situations, it is not possible to verify the correctness of registration at the place of acquisition, e.g., by performing preliminary reconstruction. In such cases, it is reasonable to use image preprocessing, as demonstrated in this paper.

Based on the histogram correction methods analyzed, several conclusions can be drawn. Visual analyses of the obtained reconstructions (Figures 15 and 18) indicated the superiority of methods that operate in L*a*b* space. In the RGB space, all the color channels are modified, which leads to color distortions in the textures. Similarly, from a statistical point of view, methods that operate in the L*a*b* space are more efficient. It is possible that this is related to the aforementioned falsification of the colors of the photographs, but such a conclusion requires further research. Our experiments did not result in any unequivocal recommendations regarding the superiority of specific histogram methods; however promising results were obtained using histogram equalization and histogram stretching in L*a*b* space. This is important since these methods are well-known and considered basic, are mathematically simple and do not require significant computation power. However, further research in this direction is required.

Author Contributions: Conceptualization , P.Ł., P.O. and K.S.; methodology, P.Ł., P.O. and K.S.; validation, P.Ł., P.O. and K.S.; formal analysis, P.Ł., P.O. and K.S.; investigation, P.Ł., P.O., K.S. and M.N.; resources, P.Ł., P.O. and K.S.; writing—original draft preparation, P.Ł., P.O., K.S. and M.N.; writing—review and editing, P.Ł., P.O., K.S. and M.N.; visualization, P.Ł., P.O., K.S. and M.N. All authors have read and agreed to the published version of the manuscript.

Funding: This research received no external funding.

Institutional Review Board Statement: Not applicable.

Informed Consent Statement: Not applicable.

Data Availability Statement: The study used data in LAS format, which is freely available at https://mapy.geoportal.gov.pl/imap/Imgp_2.html (accessed on 25 March 2021).

Acknowledgments: We would like to thank the management of the castle in Nowy Wiśnicz for allowing us to access the building for scientific research.

Conflicts of Interest: The authors declare no conflict of interest.

Abbreviations

The following abbreviations are used in this manuscript:

HS	histogram stretching
HE	histogram equalizing
AHE	adaptive histogram equalizing
EHM	exact histogram matching
CLAHE	contrast-limited adaptive histogram equalization
UAV	Unmanned Aerial Vehicle
LiDAR	Light Detection and Ranging
ISOK	IT System of the Country Protection
RMS	Root Mean Square
ICP	Iterative Closest Point
IQR	interquartile range

References

1. Di Angelo, L.; Di Stefano, P.; Guardiani, E.; Morabito, A.E. A 3D Informational Database for Automatic Archiving of Archaeological Pottery Finds. *Sensors* **2021**, *21*, 978. [CrossRef]
2. Apollonio, F.I.; Fantini, F.; Garagnani, S.; Gaiani, M. A Photogrammetry-Based Workflow for the Accurate 3D Construction and Visualization of Museums Assets. *Remote Sens.* **2021**, *13*, 486. [CrossRef]
3. Nocerino, E.; Poiesi, F.; Locher, A.; Tefera, Y.T.; Remondino, F.; Chippendale, P.; Gool, L.V. 3D Reconstruction with a Collaborative Approach Based on Smartphones and a Cloud-Based Server. *Int. Arch. Photogramm. Remote Sens. Spat. Inf. Sci.* **2017**, *42*, 187–194. [CrossRef]
4. Javadnejad, F.; Slocum, R.K.; Gillins, D.T.; Olsen, M.J.; Parrish, C.E. Dense Point Cloud Quality Factor as Proxy for Accuracy Assessment of Image-Based 3D Reconstruction. *J. Surv. Eng.* **2021**, *147*, 04020021. [CrossRef]
5. Osello, A.; Lucibello, G.; Morgagni, F. HBIM and Virtual Tools: A New Chance to Preserve Architectural Heritage. *Buildings* **2018**, *8*, 12. [CrossRef]
6. Carnevali, L.; Lanfranchi, F.; Russo, M. Built Information Modeling for the 3D Reconstruction of Modern Railway Stations. *Heritage* **2019**, *2*, 2298–2310. [CrossRef]
7. Croce, V.; Caroti, G.; De Luca, L.; Jacquot, K.; Piemonte, A.; Véron, P. From the Semantic Point Cloud to Heritage-Building Information Modeling: A Semiautomatic Approach Exploiting Machine Learning. *Remote Sens.* **2021**, *13*, 461. [CrossRef]
8. Chan, T.O.; Xia, L.; Chen, Y.; Lang, W.; Chen, T.; Sun, Y.; Wang, J.; Li, Q.; Du, R. Symmetry Analysis of Oriental Polygonal Pagodas Using 3D Point Clouds for Cultural Heritage. *Sensors* **2021**, *21*, 1228. [CrossRef]
9. Moyano, J.; Nieto-Julián, J.E.; Bienvenido-Huertas, D.; Marín-García, D. Validation of Close-Range Photogrammetry for Architectural and Archaeological Heritage: Analysis of Point Density and 3D Mesh Geometry. *Remote Sens.* **2020**, *12*, 3571. [CrossRef]
10. Surový, P.; Yoshimoto, A.; Panagiotidis, D. Accuracy of Reconstruction of the Tree Stem Surface Using Terrestrial Close-Range Photogrammetry. *Remote Sens.* **2016**, *8*, 123. [CrossRef]
11. Klein, L.; Li, N.; Becerik-Gerber, B. Imaged-based verification of as-built documentation of operational buildings. *Autom. Constr.* **2012**, *21*, 161–171. [CrossRef]
12. Li, Y.; Wu, B. Relation-Constrained 3D Reconstruction of Buildings in Metropolitan Areas from Photogrammetric Point Clouds. *Remote Sens.* **2021**, *13*, 129. [CrossRef]
13. Aldeeb, N.H. Analyzing and Improving Image-Based 3D Surface Reconstruction Challenged by Weak Texture or Low Illumination. Ph.D. Thesis, Technical University of Berlin, Berlin, Germany, 2020.
14. Yang, J.; Liu, L.; Xu, J.; Wang, Y.; Deng, F. Efficient global color correction for large-scale multiple-view images in three-dimensional reconstruction. *ISPRS J. Photogramm. Remote Sens.* **2021**, *173*, 209–220. [CrossRef]
15. Xiao, Z.; Liang, J.; Yu, D.; Asundi, A. Large field-of-view deformation measurement for transmission tower based on close-range photogrammetry. *Measurement* **2011**, *44*, 1705–1712. [CrossRef]
16. Remondino, F.; Spera, M.G.; Nocerino, E.; Menna, F.; Nex, F. State of the art in high density image matching. *Photogramm. Rec.* **2014**, *29*, 144–166. [CrossRef]
17. Skabek, K.; Tomaka A. Comparison of photogrammetric techniques for surface reconstruction from images to reconstruction from laser scanning. *Theor. Appl. Inform.* **2014**, *26*, 161–178.
18. Dikovski, B.; Lameski, P.; Zdravevski, E.; Kulakov, A. Structure from motion obtained from low qualityimages in indoor environment, Conference Paper. In *Proceedings of the 10th Conference for Informatics and Information Technology (CIIT 2013)*; Faculty of Computer Science and Engineering (FCSE): Mumbai, India, 2013
19. Alfio, V.S.; Costantino, D.; Pepe, M. Influence of Image TIFF Format and JPEG Compression Level in the Accuracy of the 3D Model and Quality of the Orthophoto in UAV Photogrammetry. *J. Imaging* **2020**, *6*, 30. [CrossRef]
20. Pierrot-Deseilligny, M.; de Luca, L.; Remondino, F. Automated image-based procedures for accurate artifacts 3D modeling and orthoimage generation. *Geoinform. FCE CTU J.* **2011**, *6*, 291–299. [CrossRef]

21. Remondino, F.; Del Pizzo, S.; Kersten, T.P.; Troisi, S. Low-cost and open-source solutions for automated image orientation—A critical overview. In Progress in Cultural Heritage Preservation. In Proceedings of the 4th International Conference, EuroMed 2012, Lemessos, Cyprus, 29 October–3 November 2012; pp. 40–54.
22. Snavely, N.; Seitz, S.M.; Szeliski, R. Modeling the world from internet photo collections. *Int. J. Comput. Vis.* **2008**, *80*, 189–210. [CrossRef]
23. Barazzetti, L.; Scaioni, M.; Remondino, F. Orientation and 3D modeling from markerless terrestrial images: Combining accuracy with automation. *Photogramm. Rec.* **2010**, *25*, 356–381. [CrossRef]
24. Gonzalez, R.F.; Woods, R. *Digital Image Preprocessing*; Prentice Hall: Upper Saddle River, NJ, USA, 2007.
25. Guidi, G.; Gonizzi, S.; Micoli, L.L. Image pre-processing for optimizing automated photogrammetry performances. *ISPRS Int. Ann. Photogramm. Remote Sens. Spat. Inf. Sci.* **2014**, *II-5*, 145–152. [CrossRef]
26. Maini, R.; Aggarwal, H. A comprehensive review of image enhancement techniques. *J. Comput.* **2010**, *2*, 8–13.
27. Klein, G.; Murray, D. Improving the agility of keyframe-based SLAM. In Proceedings of the 10th ECCV Conference, Marseille, France, 12–18 October 2008; pp. 802–815.
28. Lee, H.S.; Kwon, J.; Lee, K.M. Simultaneous localization, mapping and deblurring. In Proceedings of the IEEE ICCV Conference, Barcelona, Spain, 6–13 November 2011; pp. 1203–1210.
29. Burdziakowski, P. A Novel Method for the Deblurring of Photogrammetric Images Using Conditional Generative Adversarial Networks. *Remote Sens.* **2020**, *12*, 2586 [CrossRef]
30. Verhoeven, G.; Karel, W.; Štuhec, S.; Doneus, M.; Trinks, I.; Pfeifer, N. Mind your gray tones—Examining the influence of decolourization methods on interest point extraction and matching for architectural image-based modelling. *Int. Arch. Photogramm. Remote Sens. Spat. Inf. Sci.* **2015**, *XL-5/W4*, 307–314. [CrossRef]
31. Ballabeni, A.; Gaiani, M. Intensity histogram equalisation, a colour-to-grey conversion strategy improving photogrammetric reconstruction of urban architectural heritage. *J. Int. Colour Assoc.* **2016**, *16*, 2–23.
32. Gaiani, M.; Remondino, F.; Apollonio, F.I.; Ballabeni, A. An Advanced Pre-Processing Pipeline to Improve Automated Photogrammetric Reconstructions of Architectural Scenes. *Remote Sens.* **2016**, *8*, 178. [CrossRef]
33. Feng, C.; Yu, D.; Liang, Y.; Guo, D.; Wang, Q.; Cui, X. Assessment of Influence of Image Processing On Fully Automatic Uav Photogrammetry. *ISPRS Int. Arch. Photogramm. Remote Sens. Spat. Inf. Sci.* **2019**, *XLII-2/W13*, 269–275. [CrossRef]
34. Pashaei, M.; Starek, M.J.; Kamangir, H.; Berryhill, J. Deep Learning-Based Single Image Super-Resolution: An Investigation for Dense Scene Reconstruction with UAS Photogrammetry. *Remote Sens.* **2020**, *12*, 1757. [CrossRef]
35. Eastwood, J.; Zhang, H.; Isa, M.; Sims-Waterhouse, D.; Leach, R.K.; Piano, S. Smart photogrammetry for three-dimensional shape measurement. In Proceedings of the Optics and Photonics for Advanced Dimensional Metrology, Online Only, France, 6–10 April 2020; SPIE: Bellingham, WA, USA, 2020; Volume 11352, p. 113520A.
36. Alidoost, F.; Arefi, H.; Tombari, F. 2D Image-To-3D Model: Knowledge-Based 3D Building Reconstruction (3DBR) Using Single Aerial Images and Convolutional Neural Networks (CNNs). *Remote Sens.* **2019**, *11*, 2219. [CrossRef]
37. Skabek, K.; Łabędź, P.; Ozimek, P. Improvement and unification of input images for photogrammetric reconstruction. *Comput. Assist. Methods Eng. Sci.* **2019**, *26*, 153–162
38. Petrou M.; Petrou, C. *Image Processing*; The Fundamentals: Wiley, UK, 2010.
39. Pizer, S.M.; Amburn, E.P.; Austin, J.D.; Cromartie, R.; Geselowitz, A.; Greer, T.; Romeny, B.H.; Zimmerman, J.B.; Zuiderveld, K. Adaptive Histogram Equalization and Its Variations. *Comput. Vis. Graph. Image Process.* **1987**, *39*, 355–368. [CrossRef]
40. Zuiderveld, K. Contrast limited adaptive histogram equalization. In *Graphics Gems IV*; Heckbert, P., Ed.; Academic Press: Cambridge, MA, USA, 1994; pp. 474–485.
41. Coltuc, D.; Bolon, P.; Chassery, J.-M. Exact histogram specification. *IEEE Trans. Image Process.* **2006**, *15*, 1143–1152 [CrossRef] [PubMed]
42. Semechko, A. Exact Histogram Equalization and Specification. Available online: https://www.github.com/AntonSemechko/exact_histogram (accessed on 7 July 2020).
43. Agisoft LLC. *Agisoft Metashape (Version 1.6.3)*; Agisoft LLC: Saint Petersburg, Russia, 2020.
44. The Castle in Wisnicz. Available online: http://zamekwisnicz.pl/zamek-w-wisniczu-2/?lang=en (accessed on 4 March 2021)
45. Prawo geodezyjne i kartograficzne z dnia 17 maja 1989 r., Dz. U. 1989 Nr 30 poz. 163, art. 40a ust. 2 pkt.1. Available online: https://isap.sejm.gov.pl/isap.nsf/DocDetails.xsp?id=WDU19890300163 (accessed on 25 March 2021).
46. Geoportal Krajowy. Available online: https://mapy.geoportal.gov.pl/imap/Imgp_2.html (accessed on 25 March 2021).
47. LAS Specification 1.4—R14. *The American Society for Photogrammetry & Remote Sensing*. Available online: http://www.asprs.org/wp-content/uploads/2019/03/LAS_1_4_r14.pdf (accessed on 26 March 2021).
48. Informatyczny System Osłony Kraju. 2012. Available online: http://www.isok.gov.pl/en/about-the-project (accessed on 25 March 2021)
49. Orlof, J.; Ozimek, P.; Łabędź, P.; Widłak, A.; Nytko, M. Determination of Radial Segmentation of Point Clouds Using K-D Trees with the Algorithm Rejecting Subtrees. *Symmetry* **2019**, *39*, 1451. [CrossRef]
50. Chen, J.; Yi, J.S.K.; Kahoush, M.; Cho, E.S.; Cho, Y.K. Point Cloud Scene Completion of Obstructed Building Facades with Generative Adversarial Inpainting. *Sensors* **2020**, *20*, 5029. [CrossRef] [PubMed]
51. Ozimek A.; Ozimek P.; Skabek K.; Łabędź, P. Digital Modelling and Accuracy Verification of a ComplexArchitectural Object Based on Photogrammetric Reconstruction. *Buildings* **2021**, *11*, 206. [CrossRef]
52. McKay, N.D.; Besl, J. A method for registration of 3-D shapes. *IEEE Trans. Pattern Anal. Mach. Intell.* **1992**, *14*, 239–256.

UAV Block Geometry Design and Camera Calibration: A Simulation Study

Riccardo Roncella * and Gianfranco Forlani

Department of Engineering and Architecture, University of Parma, 43124 Parma, Italy; gianfranco.forlani@unipr.it
* Correspondence: riccardo.roncella@unipr.it; Tel.: +39-05-2190-5972

Abstract: Acknowledged guidelines and standards such as those formerly governing project planning in analogue aerial photogrammetry are still missing in UAV photogrammetry. The reasons are many, from a great variety of projects goals to the number of parameters involved: camera features, flight plan design, block control and georeferencing options, Structure from Motion settings, etc. Above all, perhaps, stands camera calibration with the alternative between pre- and on-the-job approaches. In this paper we present a Monte Carlo simulation study where the accuracy estimation of camera parameters and tie points' ground coordinates is evaluated as a function of various project parameters. A set of UAV (Unmanned Aerial Vehicle) synthetic photogrammetric blocks, built by varying terrain shape, surveyed area shape, block control (ground and aerial), strip type (longitudinal, cross and oblique), image observation and control data precision has been synthetically generated, overall considering 144 combinations in on-the-job self-calibration. Bias in ground coordinates (dome effect) due to inaccurate pre-calibration has also been investigated. Under the test scenario, the accuracy gap between different block configurations can be close to an order of magnitude. Oblique imaging is confirmed as key requisite in flat terrain, while ground control density is not. Aerial control by accurate camera station positions is overall more accurate and efficient than GCP in flat terrain.

Keywords: UAV; photogrammetry; camera calibration; GNSS-assisted block orientation; dome effect; Monte Carlo simulation

1. Introduction

Accurate knowledge of camera interior orientation elements and proper mathematical modelling of the image formation process are key elements for image metrology. UAV photogrammetry is no exception in this respect [1]. Camera calibration, the process leading to the estimation of such model parameters, has long been (and still is) one of the most researched topics in close range photogrammetry [2] as well as in computer vision [3,4]. At least in the former area, there is general agreement on conditions providing optimal results [1,5]: camera parameters should be estimated in a Least Squares Bundle Block Adjustment (BBA) of a highly redundant camera network with strong geometry (highly convergent images, orthogonal roll angles, more than six rays per point, and large scale variations in images), a testfield with appropriate targets, highly accurate image matching of targets, image points covering full frame format, and significance tests to avoid overparametrization [6–8]. Not all conditions need to be satisfied nor are Ground Control Points (GCP) generally necessary.

In the context of UAV camera calibration, assessing the accuracy of calibration parameters computed in various image block configurations by on-the-job self-calibration is still a disputed argument. Current technology also allows, besides the traditional case of block control by GCP, GNSS-assisted self-calibration. Evaluating the effects of residual calibration errors on tie point accuracy, in the case of pre-calibration as well as of on-the-job self-calibration, on the other hand, is of relevant interest, especially from a practical point of view.

In this paper a set of UAV synthetic photogrammetric blocks, built by varying terrain shape, surveyed area shape, block control (ground and aerial), strip type (longitudinal, cross and oblique), image observation, and control data precision has been synthetically generated. Through a set of Monte Carlo simulations the actual performance of each single configuration has been investigated. From an operational standpoint, analytical camera calibration comes in two versions: pre-calibration or on-the-job calibration. Both use a BBA with additional parameters; the former is normally executed in a laboratory test field under optimal camera network geometry, with estimated parameters kept fixed later in actual surveys; the latter estimates camera parameters as a by-product of the BBA of the actual survey block [9].

How to transfer close-range expertise on camera calibration, with its strong roots in industrial and metrology applications, to UAV photogrammetry is still an investigated topic. In a way, UAV photogrammetry is indeed a mix of close-range and aerial photogrammetry, as it inherits consumer cameras from the former and block geometry features from the latter (e.g., a basic flight plan made of nadir imagery along parallel strips). To complicate matters, UAV platforms come in two versions, fixed-wing and rotary-wing, with marked differences in flight management and camera pointing flexibility. The wealth of ongoing research devoted to UAV camera calibration witnesses a not-yet-settled issue, with many questions still open and even "old" certainties put under scrutiny [10].

Pre-calibration is well suited when the camera is mechanically stable and repeatable in focusing operations [11]; a further constraint is that it should be operated in the field under similar conditions (image scale, scene depth, etc.) to that of calibration. Most software packages provide specific camera calibration tools, with calibration patterns and automatic target detection to speed up operations. With fixed-wing platforms, cameras are easily removed from the drone body and so pre-calibration can take place in laboratory settings. With rotary-wing platforms both indoor and outdoor options are generally feasible. It should be noted, however, that if similarity of image scale between calibration and survey block is sought, indoor or laboratory calibration can be troublesome, especially with longer focal length optics.

As far as the alternative between pre-calibration and on-the-job calibration is concerned, the outcomes of the many study cases on UAV camera calibration are not all consistent, and the situation looks poised to remain so. The results of [12] found that, with dense ground control, differences between on-the-job and pre-calibration were not substantial. In [13], proper distortion modelling is the goal to pursue to avoid systematic errors; pre-calibration is recommended together with an after-flight calibration check based on k1-k2 parameters' equifinality. Oblique imaging in the range of 20° to 45° with respect to nadir amounting to at least 10% of block images should be included to reduce doming. The authors of [14] recommend robust pre-calibration (longitudinal and double cross with a few oblique ones) and claim that an on-site block as small as 20 images, with four oblique images at block corners, in a scene with sufficient height variations, might be enough to achieve this aim. Additionally, using pre-calibrated parameters, they found virtually the same residuals on GCP for two flights executed at a three-day distance over the same test field, implying a good short-term stability of camera parameters. In a rectangular block with high-overlap nadir imagery, [15] found pre-calibration to be more accurate than on-the-job calibration, though the main improvement came from accurate camera distortion modelling. On the other hand, it has been found in empirical tests [16–18] that Interior Orientation (IO) elements are not stable or that the reliability of the pre-computed parameters is questionable, due perhaps to poor repeatability of focusing, shocks in landing or different ambient temperatures. According to [1], pre-calibration remains the best option in the case that basic conditions for self-calibration cannot be met on site. However, in practice, on-the-job calibration is the method of choice, perhaps optimizing flight parameters to meet both survey requirements and safe conditions for self-calibration.

The progress in feature-based matching, with tens of thousands of tie points extracted and often matched across more than a dozen images, makes self-calibration without targets possible [19,20], on condition of a reasonably textured scene. Therefore, tie points' distribution over the full frame format and accurate image matching can be taken for granted in most survey flights. In his analysis, [1] highlights the importance of scale changes within images as a key factor allowing, even with limited geometric block strength, full or partial recovery of IO and distortion parameters. Flight planning software for UAVs commonly incorporates the so-called double-grid option, with cross strips providing the orthogonal roll angles to reduce projective coupling between Exterior Orientation (EO) and IO parameters. Multi-scale self-calibration, with scale changes between images arising from blocks flown at different altitudes, has been shown [14] being less effective, at least unless GCP are introduced [10,21].

Simulations, as well as empirical studies, showed that large systematic elevation errors (the so-called doming effect) could arise from inaccurate estimations of calibration parameters [22]. The addition of oblique imaging to nadir imagery along parallel strips has been proposed and shown to be beneficial [13,23–25]. Rather than adding another flight layer, even flying the longitudinal strips with moderate-to-strong (30° to 45°) camera axis inclination along flight direction [12,21,26] proved effective in eschewing systematic errors in elevation. The effectiveness of the gently oblique (20° camera pitch) double grid proposed by [26] has also been confirmed by [27]. More radically, the very advantage of using nadir images at all, as well as of the large overlaps of UAV blocks, has been questioned: from homologous ray intersection analysis, [10] suggests switching, whenever feasible, to a simple or double grid image acquisition mode where the UAV camera always points towards the center of the area of interest at ground level. On the other hand, a simulation study [28] showed that, with only gently inclined camera axes, otherwise negligible correlations among decentring and radial distortion parameters may arise and affect calibration results as well as reduce the doming effect mitigation of oblique imaging.

Most flight planning software allows for simple and double grid schemes and, for multi rotors, for Point Of Interest (POI) mode, where the UAV takes a circular path around a ground target that is always kept centred in the camera frame. It should be noted, however, that (to the best of authors' knowledge) all experimental studies with oblique imaging have been performed with multi-rotor platforms, where the camera is normally mounted on a gimbal. Oblique imaging with fixed wings, though an option available in some platforms, is more difficult to achieve in practice, so meeting optimal conditions for on-the-job self-calibration with these platforms may be harder; [24] suggest including gently banked turns in the flight plan to this aim.

In aerial blocks, the basic camera network geometry is determined by image overlap (side and forward), as the area of interest is typically covered by nadir imagery along parallel strips. Increasing overlap to a much higher degree than necessary for stereo coverage is common in UAV blocks; due to high repeatability of extracted key points, it increases ray multiplicity and so network strength. How effective this larger overlap is in improving self-calibration is, however, questioned [10], as the average ray intersection angle decreases with increasing overlap.

In aerial and UAV photogrammetry, block georeferencing and block control by GCP are intertwined and enforced in the BBA. Finding rules for determining the most efficient density and distribution of GCP in a UAV survey is not a trivial task, given the number of parameters involved. Indeed, the topic is still a debated subject of investigation [29,30] and is further complicated if accurate camera station positions are employed. Using Camera Stations (CS) determined by on-board Global Navigation Satellite System (GNSS) receivers to georeference and control the block is indeed a more than 30-year-old technique [31], known as GPS-supported or GPS-assisted aerial triangulation [32,33]. In many of today's papers this technique is (improperly, in the author's opinion) referred to as Direct Georeferencing (DG), a term that should be restricted to blocks where camera E.O. data are all determined by GNSS-assisted inertial navigation, and in principle there is no need for tie points. The availability on the market of both fixed-wing and multi-rotor platforms equipped with dual frequency GNSS receivers with Real Time Kinematic (RTK) technology enables GNSS-assisted block georeferencing and control, minimising the need for control at ground level [34,35]. As this technology becomes less expensive and satellite constellations improve their coverage, ensuring cm-level accuracy, it can be expected that it will gain ground, especially whenever site conditions make GCP survey difficult [36,37]. Notice that RTK is not strictly necessary, though it allows quick, on-site checking of the positioning quality. Indeed, the GNSS observations might as well be recorded on board and elaborated later in Post Processing Kinematic (PPK) mode, exploiting more sophisticated processing options and possibly improving positioning accuracy [35,38].

Agreeing with the Computer Vision approach, [13] believe that GCP or GNSS-determined CS need not to be involved in the BBA but instead used to compute an Helmert transformation from the BBA arbitrary reference frame and the mapping reference frame. However, it is also acknowledged in the paper that GCP or GNSS-determined CS help to refine calibration or limit block deformations that may arise from un-modelled systematic errors (such as residual calibration errors) and, to some extent, might also improve calibration parameter estimation. It is therefore worth investigating whether moving the control from ground points to CS changes the accuracy of the calibration parameters in a self-calibrating BBA. A few experiences [11] as well as previous simulation studies [13,39] suggest camera calibration with UAV blocks flown with GNSS-assisted block georeferencing and control deserves a more systematic investigation. In particular, in early tests [34,40] and later ones [35] it has consistently been found that in nadir-only imagery blocks a bias in elevation could arise using self-calibration and that a way to cope with this problem is to use at least one GCP. Lately, however, no need for such single GCPs has been found if oblique images are added [27].

In the context of UAV camera calibration, this paper therefore has two objectives. The main one is to assess the accuracy of calibration parameters computed in various image block configurations by on-the-job self-calibration under realistic conditions, representative of two widespread operating scenarios in UAV surveys. Besides the traditional case of block control by GCP, a well-searched topic, of special interest in authors' view is the performance of a GNSS-assisted self-calibrating BBA as a function of the number of GCP; more precisely, just one at block centre or none at all.

The second paper goal is to assess the effects of residual calibration errors on tie point accuracy in case of pre-calibration as well as of on-the-job self-calibration, again as a function of different block configurations.

Compared to other papers on the subject, the experiments herein try for a more systematic approach through simulations, to gain insight on the influence of several factors affecting UAV camera calibration. To this aim, a set of synthetic UAV photogrammetric blocks has been generated that encompasses overall 144 different combinations of landform, surveyed area shape, block control type (ground and aerial), number and type of strip layers, precision of image coordinates and control data. In a Monte Carlo (MC) scheme, each simulated block combination has been adjusted by a self-calibrating BBA where the simulated, true values of image and control data have been corrupted with random errors, executing 1000 runs for each combination. A similar approach, here applied in a more comprehensive test setup, has been already proposed by [41,42], applied to GNSS-assisted block orientation by [39] and also adopted by [35] to generate precision maps. Another example of a Monte Carlo simulation study focused on the dome effect is also presented in [43].

Of course, the problem dimensionality is so large that many other factors could have been considered in the simulations (first of all image overlap, instead kept fixed to values frequently adopted in today's UAV surveys). A choice was made to limit computing time and memory storage.

2. Materials and Methods

For the simulated blocks to be as realistic as possible, it has been decided to build them from the BBA output of two real blocks, each flown over a different landform according to the same flight plan. The motivation for this choice is to avoid an unrealistic distribution of the tie points over a regular grid and, most of all, an artificially high and fairly homogeneous distribution of the tie point ray multiplicity i.e., of the number of images an object point is observed on. As the two sites present rather different characteristics, the tie point distribution and their multiplicity can be expected to differ as well; this should help in clarifying whether and how these two factors affect the calibration accuracy. In the following, first the characteristics of the real blocks are described, then the procedure to build the synthetic blocks is illustrated.

2.1. Characteristics of the Two Real Survey Flights

The first block (Flat) images the Torrente Baganza riverbed (44°43'3" N, 10°14'18" E) made of bare terrain (gravel and sand) with bushes and a few rows of high trees. The second block (Hilly) images a steep ravine located in the Appenines, about 20 km South-West of Parma (Italy) (44°40'29" N, 10°8'57" E) with bare terrain, boulders as well as trees, grass and bushes. Images have been acquired with a DJI (Shenzhen, China) Phantom 3 equipped with a FC300X camera with a resolution of 4000 × 3000 pixels, a pixel size of 1.56 micrometres and a 3.61 mm nominal focal length (21 mm equivalent 35 mm format focal length). The flight plan (see Figures 1 and 2) is made of three different strip types, all flown at constant elevation above sea level (a.s.l.), i.e., with nominal camera station positions all in the same horizontal plane:

- 7 nadir-imaging longitudinal strips with 80% forward overlap and 70% sidelap;
- 12 nadir-imaging cross strips with 80% forward overlap and 70% sidelap matching the longitudinal strips;
- 2 rings of 36 oblique images, regularly spaced along a horizontal circle, with camera axes pointing downwards at the circle centre ground projection (POI mode), with an angle from nadir close to 49 degrees. As the longitudinal strips length is designed to be twice the block width, the centre of each ring has been designed to be close to the (square) half-block centre while the circle radius is slightly larger than half the half-block diagonal.

Pix4D capture flight planning software has been used with the double grid option for shooting the longitudinal and cross strips as well as the ring of oblique images. Two separate flights have been executed in each site, one for the double grid, the other for the rings. The flight elevation above ground level (a.g.l.) is computed with respect to the lowest terrain point in Flat block and to the highest in Hilly block. Both full blocks (i.e., including all images of all strip types) have been oriented with Agisoft's Metashape v. 1.5.3 and georeferenced on navigation data only. Figure 1 shows the orthophotos (top) and the DEMs (bottom) of both areas. The camera stations are shown, color-coded according to strip type, superimposed to the orthophotos. Table 1 summarizes the main characteristics of the two flights, that show different average GSD, number of extracted tie points, average image overlap and reprojection error.

Table 1. Summary of the real blocks' characteristics used as a basis for simulated data generation.

Description	Block "Flat"	Block "Hilly"
Area size (Width × Height) (m)	480 × 290	450 × 300
Terrain type	Flat	Hilly
DSM hmin–hmax a.s.l. (m)	136–160 (tree tops)	218–305
# images	396	
# long. strips/# images per strip	7/24	
# cross strips/# images per strip	13/12	
# oblique images rings	2 × 36	
Flight abs. elevation a.s.l. (m)	190	355
Flight elevation a.g.l. (m)	49	49

Table 1. *Cont.*

Description	Block "Flat"	Block "Hilly"
GSD min–max (cm)	1.3–2.4	2.2–6.0
baselenght (long. & cross) (m)	13	13
# tie points, effective overlap	111,000, 7	85,000, 15
Reprojection error (pix)	0.97	1.59

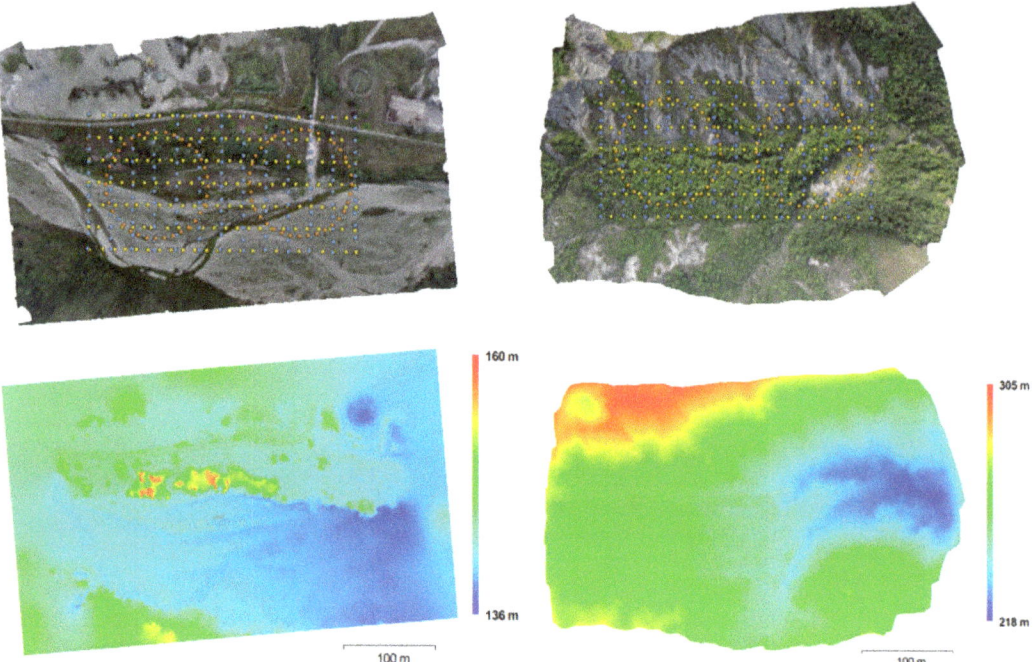

Figure 1. The top row shows the orthophoto of the Flat (**left**) and Hilly (**right**) areas, with superimposed the camera stations locations, colour-coded per strip type: longitudinal (yellow), cross (blue), oblique (orange). The bottom row shows the DEM generated from the sparse tie points.

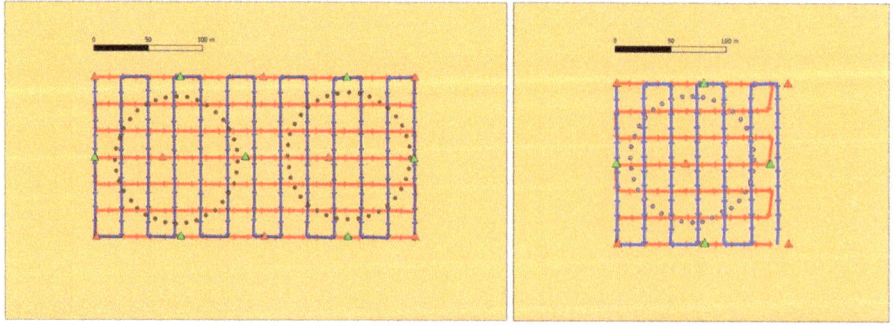

Figure 2. The surveyed area (orange background) with the trajectories of the three strip types: Longitudinal (blue); Cross (red); POI (brown). The figure depicts a Control type GCP with Enhanced control tightness (see Table 3) where GCP are represented by triangles. Different colours of triangles refer to control tightness: Basic (red) or Enhanced (red + green). (**Left**): LCO block configuration. (**Right**): HLCO block configuration.

On-the-job self-calibration has been executed in the BBA, enabling the estimation of the camera parameters listed in Table 2. Notice that the camera mount of the Phantom 3 is such that the largest side of the sensor is perpendicular to flight direction. As such, the Y axis is "along strip" and the X axis is "across strip": the image coordinate system is oriented with the Y axis in flight direction and the X axis 90° clockwise with respect to Y.

Table 2. Camera parameters estimated in the BBA.

	Principal Distance	Principal Point	Radial Distortion	Decentring Distortion	Scaling and Non-Orthogonality
Acronym	c	PPx, PPy	k1, k2, k3	P1, P2	B1, B2

No particular refinement of the BBA adjustment results has been carried out, as the goal of the operation was simply to provide data for the simulations.

2.2. Generation of Block Configurations for the Simulation

Different block configurations are generated by selectively removing from the full block one of the strip types (namely: the cross strips, the oblique images or both) in the original projects. Each block configuration is labeled according to the strip type it contains using the letters L, C and O to label longitudinal strips, cross strips and POI images, respectively. For instance, an LCO configuration corresponds to the original full block, while an LC configuration represents a block made of longitudinal and cross strips only, and so on. A block configuration is therefore made of one, two or three different strip types.

In order to account for the effect of different area shapes on calibration, exploiting the 1:2 width-to-heigth ratio of the rectangular original block, the second half of the original full block (LCO) has been cut out, allowing the generation of square block configurations also from the (original) first half-block. Configurations derived from this square block recieve the prefix H. As such, the HLO configuration is made of a longitudinal square block complemented with a ring of oblique images; HO is a POI (single ring) over the square area, and so on. For each block configuration, camera stations and pose, tie point ground coordinates and camera calibration parameters are exported to act as true data for the simulation.

Overall, four block configurations (LCO, LC, LO and L) have been considered for the rectangular area and five (HLCO, HLC, HLO, HL and HO) for the square area. Each configuration has been generated for both the Flat and Hilly areas.

As far as block control is concerned, both ground-only (GCP case) and GNSS-assisted (GNSS case) have been tested with both Basic and Enhanced tightness (see Table 3). In GCP Basic (see red triangles in Figure 2), 8 and 5 GCP are placed at the corners and in the middle of the block square(s), respectively, for the rectangular and square area. In GCP Enhanced, 15 and 9 GCP are arranged in three rows along the longitudinal flight lines, respectively, for the rectangular and square blocks (see red and green triangles in Figure 2). In GNSS Basic as well as in GNSS Enhanced, all camera stations' positions are used as control information; however, in the former no additional GCP are used, while in the latter a single GCP, located at the block centre, is fixed.

Table 3. The two block control cases each with two different tightness levels, depending on the number of GCP fixed.

Block Control Case	Tightness
GCP (Ground)	Basic: 5–8 GCP Enhanced: 9–15 GCP
GNSS (Aerial)	Basic: no GCP Enhanced: 1 GCP

Two sets of observation precisions have been considered to simulate both a medium as well as a high-precision data set (see Table 4). Though "average" users cannot do much to improve the precision of the control data, in principle a better quality of positioning data can be foreseen. As far as the GNSS case is concerned, a better hardware (especially a better antenna), a good satellite configuration and expertise in PPK GNSS data might do the job. As far as the GCP case is concerned, using a Total Station shifts the accuracy range below the cm level [30]. Tie point image coordinate precisions depend on image quality and object texture characteristics so, once the image block is acquired, the user has a limited ability to intervene. Some differences can be expected in point identification performances, if different Structure from Motion algorithms (i.e., different software packages) are used, or if different processing parameters are chosen (e.g., the Orientation quality parameter in Metashape). A matching precision of 1 pixel and of 1/3rd of a pixel has been considered.

Table 4. Measurement precisions levels considered in the simulations.

Precision Level	CS Coord. (cm)	GCP Coord. (cm)	Tie Point Img. Coord. (pix)
Medium	3.0	0.5	1
High	1.0	0.17	0.33

Table 5 summarizes the combinations of BBA configurations tested in the MC simulation. The combination of Area shape and Strip types yields nine different configurations. Combining them in all possible ways (16) with the parameters Landform, Measurement precision, Block control type and Block control tightness, a total of 144 different cases were investigated in the MC simulations.

Table 5. Parameters accounted for in the simulated BBA.

Parameter	
Area shape and Block type	rectangular: L, LC, LCO, LO square: HL, HLC, HLCO, HLO, HO
Landform	Flat, Hilly
Measurement precision	Medium, High
Block control case	GNSS, GCP
Block control tightness	Basic, Enhanced

2.3. Generation of True Values and True Errors for the Synthetic Data

The true values of the exterior and interior orientation parameters (including the camera optical and sensor distortion parameters) and of the tie points' ground coordinates of the simulated blocks are taken from the real blocks, i.e., from the estimated parameters values after a free-net self-calibrating BBA executed with Agisoft MetaShape (Agisoft, St. Petersburg, Russia) on the two real blocks. The tie points' distribution and their multiplicity in the simulated block is also taken from the real blocks. To this aim, the list of tie points in each image has been exported from MetaShape and the tie point image coordinates' true values have been generated by projecting the ground point coordinates with the collinearity equations, according to the estimated exterior orientation parameters and camera parameters. The synthetic image coordinates so obtained incorporate the optical and sensor frame distortion estimated for the real block. Therefore, though the same camera has been used in both real flights, the synthetic block's IO parameters are slightly different. For instance, the focal length true value for the Flat terrain block is 2335 pixels (21.01 mm equivalent 35 mm format focal length) while for the Hilly terrain block it is 2320 pixels (20.88 mm equivalent 35 mm format focal length). Normally distributed errors with standard deviations according to Table 4 have been generated in each run of the MC simulation and added to the true observations.

Running the BBA in the MC simulations, the standard deviations assigned to the observations should be the same as those reported in Table 4. This is true for the tie points' image coordinates; for the GCP and the CS coordinates, however, in a real block it might be advisable to reduce their standard deviations (i.e., by a factor three/four) to reduce unmodeled errors, as the much larger number of image observations with respect to the other observation types needs to be counterbalanced by increasing the weights of the latter [1,40,42,44]. In this case, however, being the observations affected by zero mean gaussian errors, the effect of varying to some extent the weights of the observations have negligible effects on the final results.

2.4. Accuracy Evaluation of Camera Calibration Parameters

The calibration parameters' accuracy will be investigated at the single parameter level as well as at a global (image) level. The former analysis will focus primarily on the three IO parameters, the latter on the largest residual distortion over the whole image frame.

Correlations between parameters play a key role in estimation errors, not affected by the MC simulations, and will also be considered in the evaluation of the nine configurations. Given this paper's objectives, special attention will be given to comparison between GNSS-assisted and traditional GCP block control.

To present the results, the nine block configurations have been (albeit arbitrarily) ranked according to a decreasing block "strength score" (see Table 6).

Table 6. Block configurations ranked by decreasing "strength".

1	2	3	4	5	6	7	8	9
LCO	LO	HLCO	HLO	LC	HLC	L	HL	HO

The overall calibration accuracy of each block configuration will be measured by the largest residual distortion. To this aim, a grid of 20 by 15 points has been set over the image frame. At each iteration of the MC scheme, the maximum residual distortion error on such grid points (i.e., the distance between the true image distortion correction and the one computed with the estimated distortion parameters) is recorded. Upon completion of MC simulations, the average and standard deviation of the maxima per iteration are computed. To weigh the alternative between GNSS and GCP block control in the calibration, the percentage gain in modelling distortion (reducing the average max residual distortion) will be computed for identical block configurations and similar block control tightness.

2.5. Accuracy Evaluation of Ground Coordinates

The accuracy of the ground coordinates in each of the 144 cases is evaluated by comparing, for each tie point coordinate, the true values of the coordinates against the estimated value in each block adjustment. For each check point coordinate, the mean error, the error standard deviation and the RMSE obtained in the 1000 MC iterations is computed and averaged over all the tie points common to all block configurations. As the number of tie points depends on the block configuration, in order for the comparisons to be made on an equal basis, only points common to all configurations have been used in computing the error statistics. As such points are fairly distributed over the survey area, restricting the analysis to the common set doesn't affect the statistics' significance. However, this common tie point set has been built excluding the HO configuration (POI case with oblique images only), as very few tie points turned out to be common to other blocks. Using such a small amount, in our opinion, would have affected the significance of the results for all configurations too much. As such, the sample size of the error statistics for HO configuration is not homogeneous with the other configurations [10]. As a matter of fact, in our test the HO case turned out to yield in most cases quite singular results, hardly in agreement with a trend that could be spotted in the other configurations and mainly quite poor [10]. Additionally, in our experiment design we did not think of the

POI as a real standalone configuration, but rather as a complement of nadir imagery. A reason for such disappointing results might be that the POI is not that far from an "orbital motion" critical configuration [45]. We present them anyway, with this caveat and without any further comment.

2.6. Dome Effect and Pre-Calibration

Our test has not been specifically designed to study the so-called dome effect [22] that may show up when residual calibration errors and weak block geometry produce systematic errors in tie point coordinates, mostly apparent in elevation. Indeed, except in one case (GNSS control case with noGCP), the block control applied in the simulations (see Figure 2) always foresees at least one GCP in the block centre, therefore limiting the magnitude of the Z coordinate error at the block center. However, taking advantage of the 144,000 camera calibration parameter sets estimated in the BBA of the MC simulations in the nine block configurations of the experiment, we investigated the 3D tie point coordinates sensitivity to (inaccurate) pre-calibrated camera parameters, i.e., the dependence of the dome size on the pre-calibration block configuration, through a second MC simulation.

To set "better" conditions for the dome effect to show up, a slightly modified L (ongitudinal) image block configuration has been extracted from the Flat block. In addition to the original tie points, in this block a set of more than 1600 check points has also been generated as follows. The horizontal coordinates of each check point are taken from the nodes of a regular 5×5 m grid set over the area, while their elevation is set equal to the average elevation of all the tie points in the original Flat block. The synthetic image coordinates of tie points and check points are then generated error-free, i.e., by projection on the images according to true values of camera parameters, ground coordinates and EO parameters. Finally, only four GCP located at the block corners are used as control in the modified L block.

In the new MC simulation, consisting of 144,000 runs over the modified L block, random errors with a standard deviation of 1 pixel (Medium precision in Table 4) are applied to tie points' image coordinates only, while check point image coordinates are left unperturbed. The modified L block observations are then adjusted, fixing the four GCP at the corners, in pre-calibration mode, i.e., using fixed camera calibration parameters. Such camera parameters are taken, in each run of the new MC simulation, from one of the 144,000 calibration parameter sets estimated in the first MC simulation. After the BBA, the ground coordinates of the check points are computed by forward intersection (i.e., keeping fixed the estimated EO parameters). In this way, only the effect of the tie point image errors and of the pre-calibrated parameters is transferred via the EO parameters to the check points ground coordinates, as the check points image coordinates are error-free. On completion of the MC simulation then we get 144,000 *Z error* sets for the check points, each set representing the dome effect generated in the flat area by application to the modified L block of a pre-calibrated camera parameter set coming from one of the 144 originally tested configurations.

We divide the 144,000 error sets in 144 groups, according to the block configuration the camera parameters have been estimated on. To summarize the results, for each group (block configuration) the average *Z error*, calculated as the average over all check points of the mean *Z error* of the 1000 MC runs in each check point, is computed. Moreover, the error range due to each pre-calibration configuration is computed as follows. Out of the 1000 runs over each check point, the largest positive, largest negative and standard deviation (being the mean approximately zero for all the points) of the *Z error* are recorded. Finally, the average of all check points differences between the largest positive and negative error (i.e., the maximum range) is computed, hereafter named the *Z error* range. The analysis of both errors should highlight the influence of the pre-calibration block configuration on a dome effect-prone block such as the modified L block.

3. Results

In the following, all figures and tables, unless explicitly stated, refer to simulations with random errors generated under Medium precision (see Table 4): 3 cm for CS, 0.5 cm for GCP and 1 pixel for image coordinates.

3.1. Camera Calibration Parameters

3.1.1. Principal Distance

Figure 3 presents the RMSE in pixels of the estimated principal distance (as a function of the nine block configurations, of the terrain type (Flat, Hilly) and of the control tightness (Basic or Enhanced).

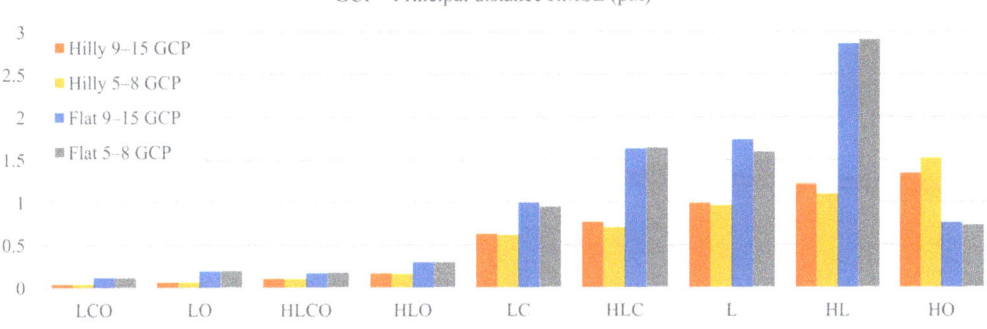

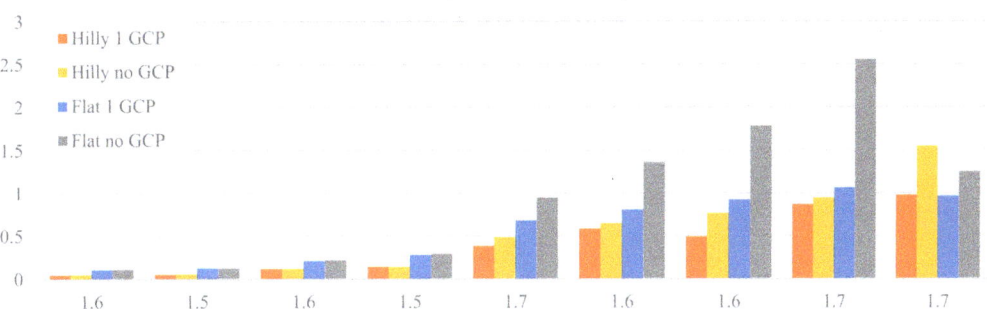

Figure 3. RMSE of the principal distance as a function of block type, terrain type and control tightness: (**top**): GCP case; (**bottom**): GNSS case.

The plots show an accuracy deterioration from strong to weak block configurations which is comparatively larger in flat terrain, especially in the GCP case. Oblique images are necessary for accurate estimation of the principal distance: if they are included, the accuracy range is from 0.05 to 0.29 pixel irrespective of control type and tightness as well as terrain type. If they are missing, a sharp decrease in accuracy may occur, from 0.4 to 2.9 pixels. The importance of oblique images becomes apparent when computing the ratio between the principal distance average RMSE of the four LC, HLC, L and HL configurations without oblique images and the corresponding average RMSE of the configurations LCO, LO, HLCO and HLO with oblique images (see Table 7).

Table 7. Ratio between average principal distance RMSE for LC, HLC, L and HL configurations and average principal distance RMSE for LCO, LO, HLCO and HLO configurations for different terrain types, block control type and tightness.

Control	Enhanced		Basic	
	Hilly	Flat	Hilly	Flat
GNSS	6.1	4.7	7.5	8.7
GCP	9.0	9.1	8.7	9.0

As can be seen, the accuracy gap in principal distance determination without and with oblique images ranges from a factor 5 to 9. In the GCP case the gap is largest and almost the same, irrespective of terrain type and control tightness. In the GNSS case, if no GCP is fixed (Basic) the gap is quite significant, while it is the lowest if 1 GCP is fixed (Enhanced), especially in flat terrain. This on the one hand means that, in flat terrain, oblique images are even more necessary than in hilly ones; on the other hand, that GNSS control with 1 GCP partly compensates for a less geometrically strong block configuration.

Without oblique images, in the GCP case it is the terrain type that ensures (Hilly) or prevents (Flat) accurate determination of the principal distance, while control tightness plays only a minor role; flat terrain is critical also in the GNSS case as, unless a single GCP is employed, the estimation error raises quickly well above 1 pixel. In both GCP and GNSS cases, the same block configurations in hilly terrain provides better results than in a flat one (on average about two times better in our test settings). The HO case (a single ring of oblique images) stands out: it is the only case where, with GCP control, Flat is more precise than Hilly and in the GNSS case Flat Basic (noGCP) is not markedly worse than Flat Enhanced (1 GCP).

3.1.2. Principal Point Location

Figures 4 and 5 represent the RMSE in pixels of the Principal Point (PP) coordinates PPx and PPy as a function of the nine block configurations, of the terrain type (Flat, Hilly) and of the control tightness (Basic or Enhanced).

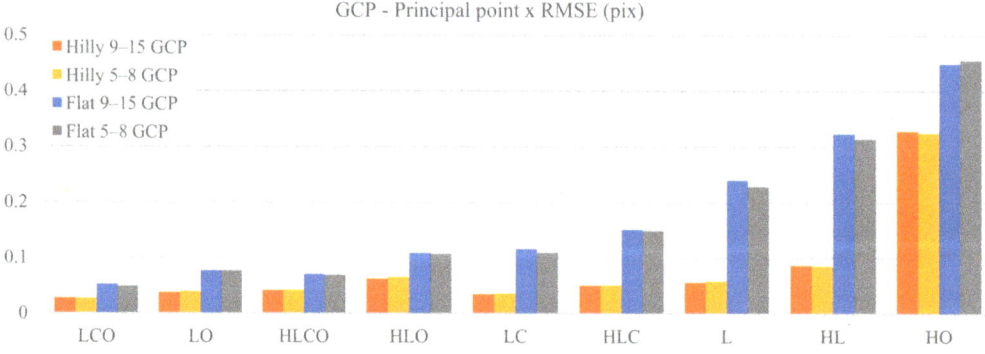

Figure 4. *Cont.*

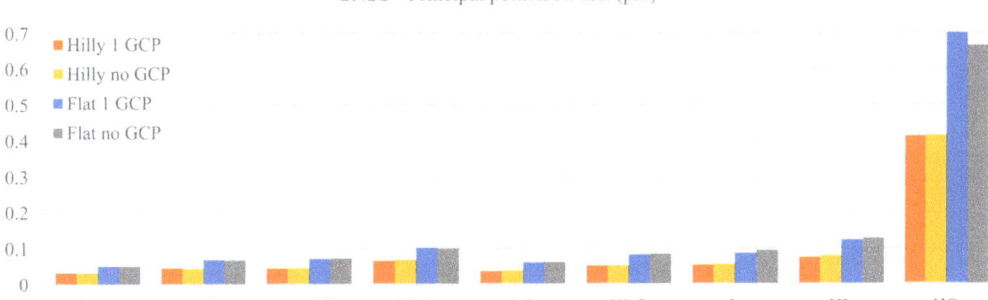

Figure 4. RMSE of the principal point x coordinate as a function of block configuration, terrain type (Flat, Hilly) and control tightness (Basic or Enhanced): (**top**): GCP case; (**bottom**): GNSS case.

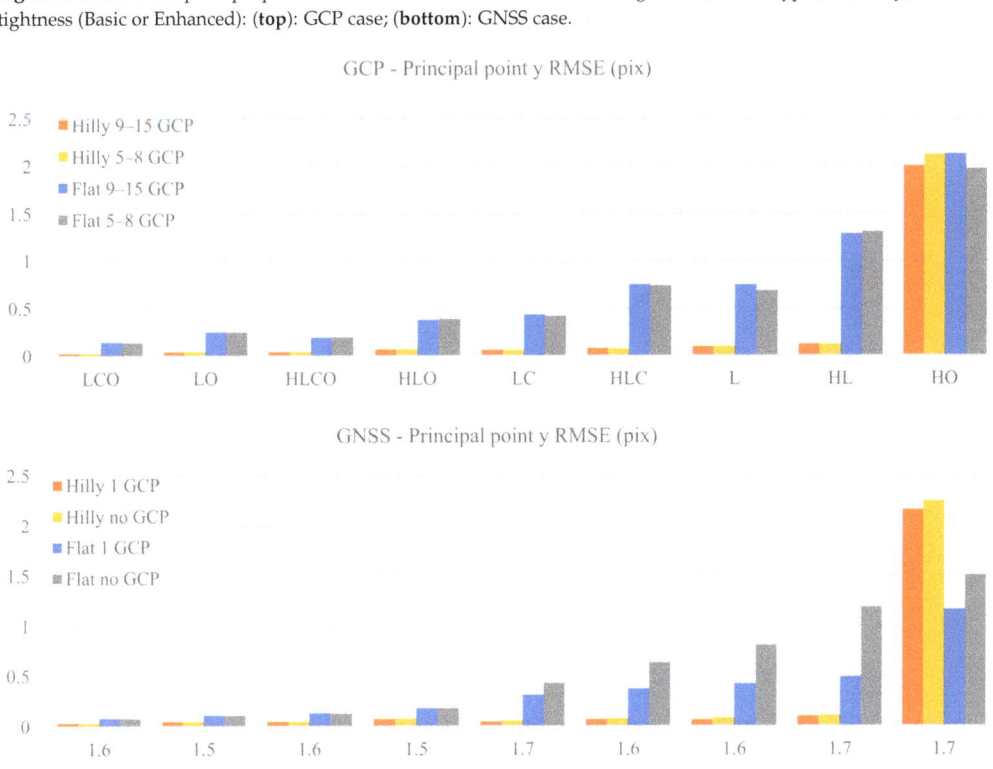

Figure 5. RMSE of the principal point y coordinate as a function of block configuration, terrain type (Flat, Hilly) and control tightness (Basic or Enhanced): (**top**): GCP case; (**bottom**): GNSS case.

With both GCP and GNSS block control, control tightness is not critical for PPx estimation, as the values for Basic and Enhanced cases are very similar. In the GCP case, the accuracy gap between hilly and flat terrain deteriorates markedly moving from strong to weak block configurations, up to a factor 3.8 in HL. To the contrary, in the GNSS case, both for flat and hilly terrain, the accuracy level is weakly dependent on the block configuration and control tightness: indeed, the accuracy gap between the two terrain types is quite stable and never exceeds a factor of 1.7. Finally, the HO case appears again as a singular and critical one, both with GCP or GNSS-assisted block control, and particularly so in the latter case, with an eightfold decrease in accuracy.

PPy accuracy is overall substantially worse than PPx, at least in weak block configurations. In hilly terrain the accuracy gap with respect to PPx is limited for both GCP and GNSS cases: the ratio RMSE_PPx/RMSE_PPy ranges from 1 to 1.7; in flat terrain, to the contrary, the PPy RMSE is worse by a factor ranging from 1.5 (LCO Enhanced) up to 9 (HL Basic).

The plot of the GCP case shows that ground control tightness is not critical for PPy estimation in both terrain types. In the GNSS case, however, this is true only with hilly terrain, while in flat terrain and block configurations lacking oblique images the 1 GCP case is significantly more accurate. In both GCP and GNSS cases and hilly terrain the PPy accuracy is very stable with respect to block configuration and always better than in flat terrain under the same block configuration. In the GNSS case and flat terrain, moreover, the error increases when moving from strong to weak block configurations with a marked jump and at a higher rate when oblique images are removed; a growing gap also opens between Basic (no GCP) and Enhanced (1 GCP) control tightness. The overall relative accuracy gap between the strongest and the weakest block configurations is significant: for the GCP case the error increases by a factor of 3 in hilly terrain to a factor of 9 in flat terrain, while the respective figures for the GNSS case are from 2.4 to 15. Finally, also for PPy, the HO case is critical.

3.1.3. Calibration Overall Accuracy

Figure 6 shows the average of the maximum distortion error value registered over the image frame as a function of the nine block configurations, of the terrain type (Flat, Hilly) and of the control tightness (Basic or Enhanced), respectively, in the GCP and GNSS cases.

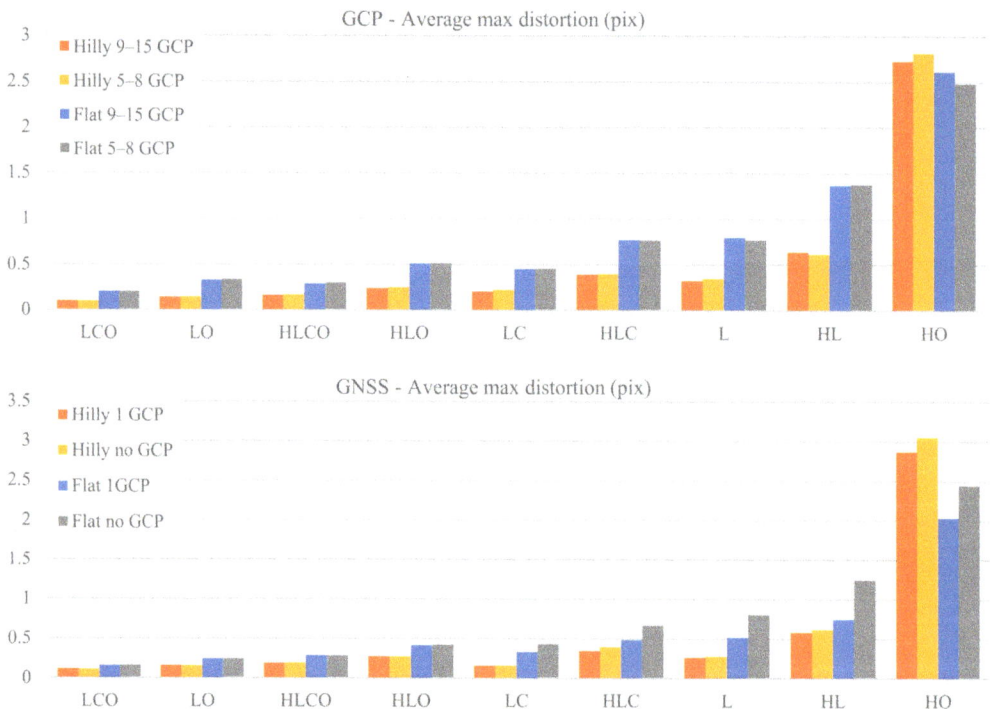

Figure 6. Average value of the maximum distortion error (in pixels) registered over the image frame as a function of the nine cases, of the terrain type (Flat, Hilly) and of the control tightness (Basic or Enhanced), respectively, in the GCP and GNSS cases.

Both the GCP and GNSS cases show similar trends, with a slight degradation of accuracy for decreasing block configuration strength. A hilly terrain yields more accurate distortion modelling than a flat one: by a factor of 1.8 to 2.3 in the GCP case and from 1.4 to 2.9 in the GNSS case. The HO case is somehow apart, with the largest values about four times worse than the worst result of the other eight block configurations. As far as the block control type is concerned, while in the GCP case the control tightness has little or no influence on distortion accuracy, in the GNSS case with flat terrain the Enhanced control (1 GCP) is clearly more effective when oblique images are missing.

To measure, if any, the overall calibration accuracy gap between the GCP and GNSS case, Figure 7 plots the percentage accuracy gain of performing camera calibration in a GCP or GNSS case, for the nine block configurations. More precisely, for each pair of GCP and GNSS identical configurations, the difference of the average max distortions is computed and expressed as percentage.

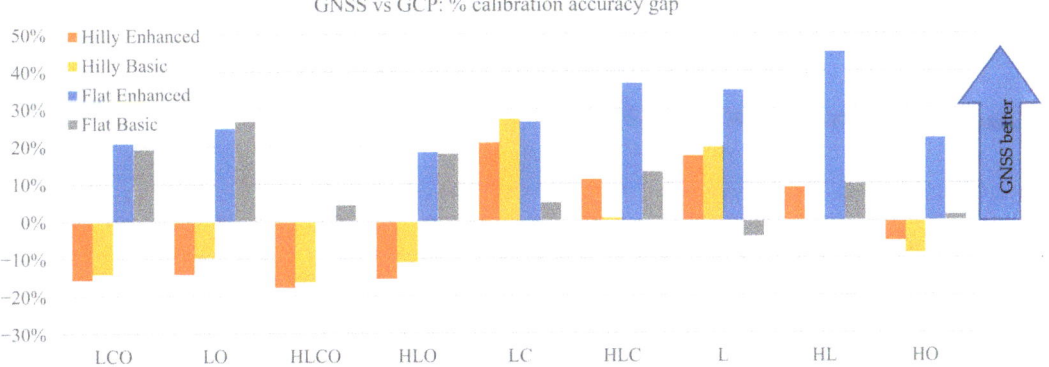

Figure 7. Percentage accuracy gap in camera distortion modelling between GCP and the GNSS case for each of the nine block configurations as a function of the terrain type (Flat, Hilly) and of the control tightness (Basic or Enhanced).

Δ_{maxD} of the distortion in the GCP case for each of the nine block configurations, two terrain types and two control cases:

$$\Delta_{maxD} = \frac{maxD(\text{GCP}) - maxD(\text{GNSS})}{maxD(\text{GCP})} \qquad (1)$$

where: $maxD(\text{GCP})$ = average value of maximum distortion error over the image frame in the 1000 MC runs when the block configuration is adjusted with the GCP control type. $maxD(\text{GNSS})$ = average value of maximum distortion error over the image frame in the 1000 MC runs when the block configuration is adjusted with the GNSS control type.

In Figure 7 a positive value means the GNSS case is more accurate in modelling the overall image distortion than the GCP case, and vice versa for negative values. Overall, the GNSS delivers a better calibration in most cases, sometimes with quite a significant improvement (up to 45%). In the four strongest block configurations (all with oblique images) GNSS performs markedly better in flat terrain (+23% on average), while GCP is better in hilly terrain (+14% on average). In weaker blocks GNSS performs almost always better than GCP (+20% on average). The largest gains are in flat terrain if at least 1 GCP is used (Enhanced tightness case), with three cases exceeding a 30% gain.

3.2. Ground Point Coordinate Accuracy

The ground coordinates accuracy is evaluated by comparing the true against the estimated coordinates for a set of tie points common to all block configurations (see Section 2.5). Such coordinates are influenced by the estimated interior orientation and distortion parameters, whose accuracy, as shown in the previous sections, can vary strongly with the block configuration and control type. At the same time, different block configurations (e.g., LCO vs. HO) have different tie point projections redundancy, projecting ray intersection angles and image multiplicity which affect the accuracy of the tie points as well.

Rather than the magnitude (absolute values) of the coordinates' RMSE, it seems more appropriate here to present a relative comparison among the different block configurations, as this provides a measure of the accuracy gain when flying according to one or another block configuration. More precisely, the relative accuracy loss Δ_{RMSE} has been computed as:

$$\Delta_{RMSE} = \frac{RMSE(LCO) - RMSE(CFG_i)}{RMSE(LCO)} \qquad (2)$$

where $RMSE(CFG_i)$: average RMSE on tie points in the CFG_i configuration, with CFG_i = LCO, LO, ..., HL and HO; and $RMSE(LCO)$: average RMSE on tie points in the LCO configuration.

Figure 8 shows the percentage loss Δ_{RMSE} of the ground coordinates RMSE of every block configuration with respect to the reference configuration (LCO) as a function of terrain type and control tightness in the GCP case and in the GNSS case. The top figures refer to horizontal coordinates and the bottom ones to elevation. Please note that the previously used sequence order of the block configuration labels in the graphs has been modified in such a way as to have a monotonic decreasing accuracy.

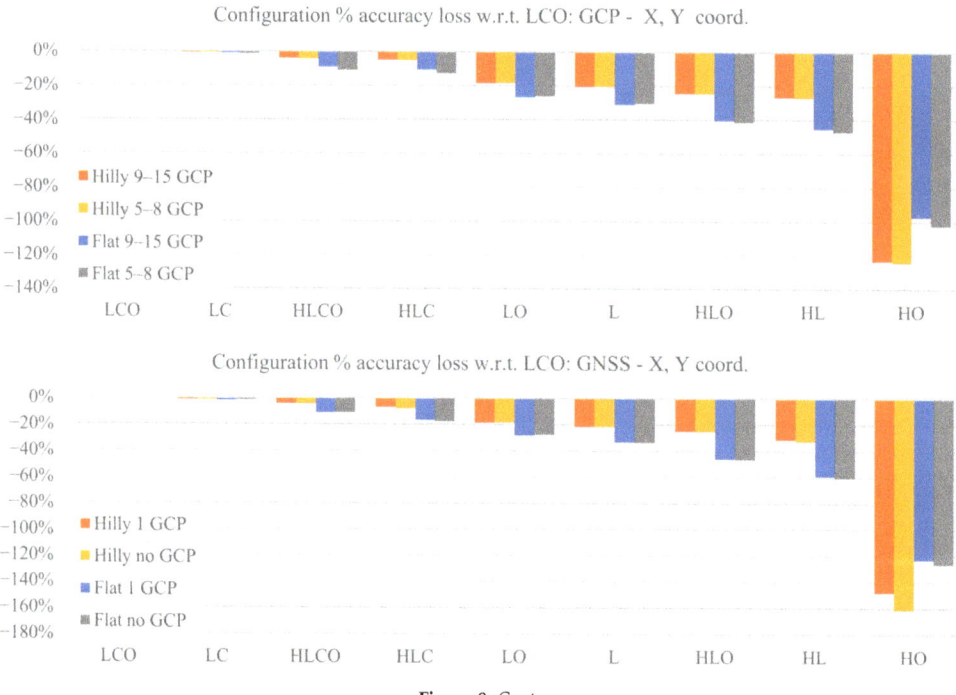

Figure 8. *Cont.*

Figure 8. X, Y and Z tie points' coordinates in GCP and GNSS cases: percentage RMSE loss of each of the nine block configurations w.r.t. the reference block configuration (LCO), as a function of terrain type and control tightness; (**top**): X, Y coordinates; (**bottom**): Z coordinate.

In both the GCP and the GNSS case, the block configurations split in three groups of similar accuracy: (1) LCO, HLCO, LC and HLC; (2) LO, HLO, L and HL; and (3) HO, which is a singular case. This suggests that cross strips look more important than oblique images to ensure accurate ground coordinates, while the opposite is true for camera calibration parameter estimation accuracy (see Table 7).

For the horizontal coordinates, in the GCP case and hilly terrain, group (1) blocks are roughly equally accurate (differences below 10%); group (2) blocks are 20% to 30% less accurate than LCO; and block HO is 120% less accurate than LCO. In flat terrain the accuracy gap range in group (2) is larger (30% to 50%). In the GNSS case the accuracy gap pattern is basically the same as the GCP case, with a larger group (2) gap (from 35% to 60%).

As far as elevations are concerned, in the GCP case the accuracy gaps in group (2) range from 24% to 35% in hilly terrain and from 35% to 70% in flat terrain. Moreover, in flat terrain a noticeable dependence on control tightness is apparent. A smaller accuracy gap is found in the HO case (from 60% to 80%) with respect to horizontal coordinates. In the GNSS case the picture is more complex. In group (2) the rate of accuracy decrease in flat terrain is larger than in hilly terrain, and even more so between Basic and Enhanced control tightness (the accuracy gap reaches 180%). In HO configuration the accuracy gap goes from 60% (Hilly Dense) to 150% (Flat Sparse).

For a comparison between GCP and GNSS case, Figure 9 reports for the tie point coordinates RMSE the percentage gain (or loss) relative to the GCP case. More precisely, the relative accuracy gaps Δ_{RMSE_CT} between GCP and GNSS RMSE for the same configuration have been computed as:

$$\Delta_{RMSE\_CT} = \frac{RMSE_{GCP}(CFG_i) - RMSE_{GNSS}(CFG_i)}{RMSE_{GCP}(CFG_i)} \qquad (3)$$

where: $RMSE_{GCP}(CFG_i)$: average RMSE on tie points in CFG_i configuration with GCP block control type; $RMSE_{GNSS}(CFG_i)$: average RMSE on tie points in CFG_i configuration with GNSS block control type.

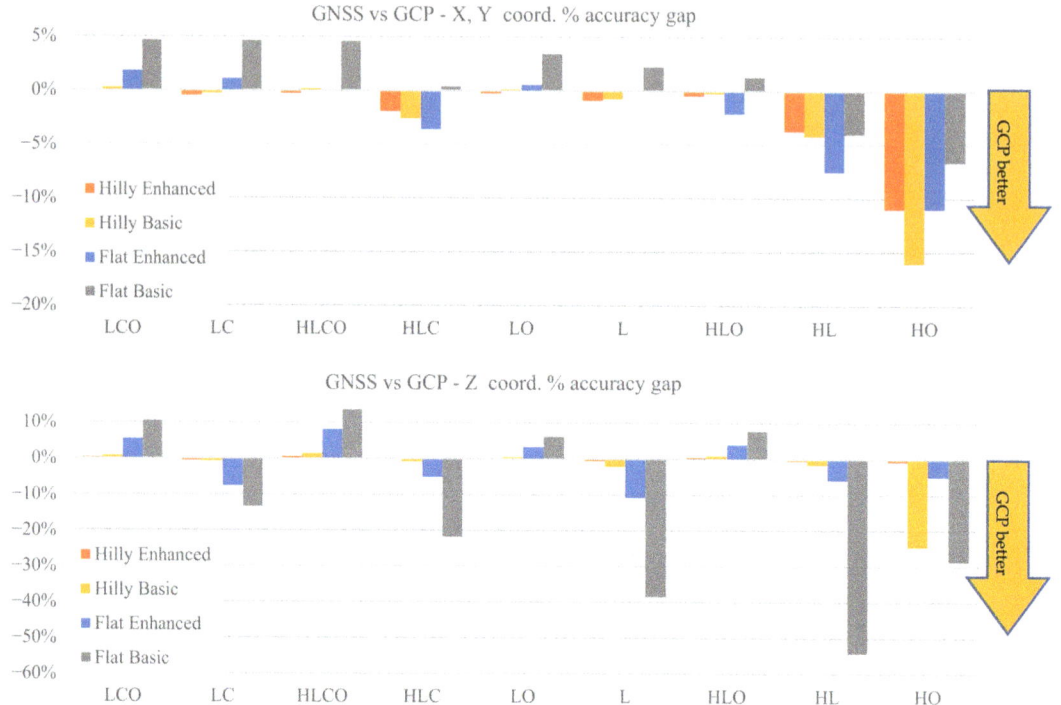

Figure 9. Percentage accuracy gap between the GCP and the GNSS cases, relative to the GCP case, for tie point coordinates' RMSE as a function of the nine configurations, of the terrain type (Flat, Hilly) and of the control tightness (Basic or Enhanced). Positive values mark a better RMSE for GNSS compared to GCP, and vice versa for negative values. (**Top**): horizontal coordinates; (**bottom**): elevations.

A positive value means the GNSS case is more accurate than the GCP case and vice versa for negative values.

From Figure 9 it can be seen that the horizontal coordinates' accuracy does not show significant differences between the GCP and GNSS cases: the largest for all configurations (less than 5%) can be expected in flat terrain with basic control; in hilly terrain the differences are below 1%. The HL and HO configurations are (partial) exceptions, with differences up to 8% and 16%, respectively. In elevation the pattern is somehow similar, with differences even more insignificant in hilly terrain. However, in flat terrain there is a clear distinction for block with and without oblique images. In the former case the GNSS case is better (up to 14%) while in the latter the GCP case is markedly better unless the single GCP (Enhanced control case) is fixed: the gap grows from 14% (LC) to almost 55% (HL).

Comparison of GNSS-controlled blocks vs. GCP-controlled ones is strongly influenced by the instruments' actual precision in Camera Station and GCP coordinate determination and by the weights assigned to such information in the BBA (see Section 2.3). Such precisions, in author's opinion, are representative of the current state-of-the-art of most UAV surveys. In our test context the two solutions (GCP control network vs. GNSS-assisted orientation) are largely balanced and provide similar tie point accuracy results. Should this not be the case (e.g., should a less-precise on-board receiver be used) one solution would provide significantly better performance than the other.

3.3. Effect on Tie Points RMSE of Increased Block Control Precision

As pointed out at the beginning of this section, all the above-presented results refer to errors in GNSS-determined camera stations, GCP coordinates and image coordinates generated according to Medium precision in Table 4. With error magnitudes three times smaller i.e., generated according to High precision observations as reported in Table 4, the tie points RMSE patterns as a function of block configuration are broadly similar to those shown in the previous paragraph. To measure the improvement (if any) brought by the increased measurement precision, Figure 10 shows the percentage accuracy gain for the tie points ground coordinates achievable with High precision measurements as opposed to Medium precision measurements. More precisely, the accuracy gain Δ_i has been computed separately for horizontal (X, Y) and vertical (Z) tie point coordinates as:

$$\Delta_i = \frac{\text{RMSE}_i(\text{Medium}) - \text{RMSE}_i(\text{High})}{\text{RMSE}_i(\text{Medium})} \quad i = (X, Y), Z \qquad (4)$$

where $\text{RMSE}_i(\text{Medium})$ = average value of the tie points' coordinate i RMSE over the 1000 MC runs with observation errors of image coordinates, camera stations and GCP, respectively, of 1 pixel, 3 cm and 0.5 cm. $\text{RMSE}_i(\text{High})$ = average value of the tie points' coordinate i RMSE over the 1000 MC runs with observation errors of image coordinates, camera stations and GCP, respectively, of 0.33 pixel, 1 cm and 0.17 cm.

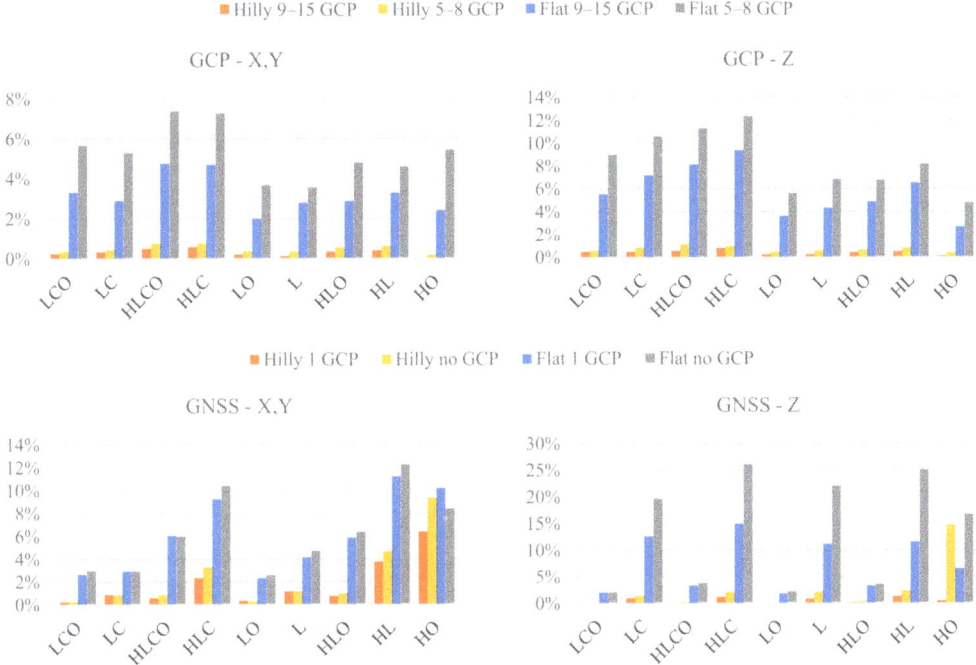

Figure 10. Percentage accuracy gain of the tie points' ground coordinates when measurement error standard deviations in the MC simulations decrease from 3 cm for CS, 0.5 cm for GCP and 1 pixel for image coordinates (Medium precision in Table 4) to 1 cm, 0.17 cm and 0.33 pixel, respectively (High precision in Table 4). (**Top**): GCP control case; (**bottom**): GNSS control case.

Overall, the ground coordinates' accuracy increase is very limited in hilly terrain in both the GNSS and GCP cases, where it is lower than 2% in almost all cases, and even less for the horizontal coordinates. In flat areas, accuracy gains, though in absolute terms mostly small, are larger than in hilly terrain in both control cases and, again in both control cases, are more significant for elevations. The gains pattern as a function of the block configuration is, however, different. In the GCP case, perhaps surprisingly, the largest gains (from 5 to 7% in horizontal coordinates and from 6 to 12% in elevation) are found for the stronger configurations with cross strips (LCO, LC, HLCO and HLC). In the GNSS case the only noticeable gains in horizontal coordinates are for the square block (from 6 to 12%). Still in the GNSS case, the largest accuracy gains (from 20% to 26%) are registered for the elevations, in flat terrain and Basic control tightness (no GCP) in block configurations without oblique images.

3.4. Dome Effect

As anticipated in Section 2, a second MC simulation has been carried out to evaluate whether and to what extent applying pre-calibrated parameters may still cause the occurrence of the dome effect in the current block being adjusted. In particular, the influence of the pre-calibration block characteristics is investigated. From this new simulation, 144,000 Z error sets, each computed on 1600 check points, have been obtained. Every error set represents the dome effect generated on the check points by the application of a pre-calibrated camera parameter set, obtained in one of the first 144,000 MC simulations, in the adjustment of the simulated image observations of a L (ongitudinal) block configuration flown over a flat area, with four GCPs at the corners (see Section 2.6 for details).

The 144,000 error sets have been divided in 144 groups, according to the configuration type of the pre-calibration block. To summarize the results, for each group the average Z error and the Z error range have been computed. Rather than measuring the magnitude of elevation distortion, here the focus is on the effectiveness of the pre-calibration block configurations in preventing it. Therefore, Figure 11 shows the percentage increase of the average Z error and of the Z error range of each pre-calibration block configuration with respect to the reference configuration LCO. Both values refer to Z errors computed over the 1600 check points in the 1000 adjustments of the modified L (ongitudinal) block configuration with the 1000 camera parameter sets obtained in the former MC simulation. More precisely, the relative percentage differences Δ_Z for the average Z error have been computed as:

$$\Delta_Z = \frac{Z\ error(\text{CFG}_i\ pre\_cal.) - Z\ error\ (\text{LCO}\ pre\_cal.)}{Z\ error\ (\text{LCO}\ pre\_cal.)} \quad (5)$$

where $Z\ error$ (CFG_i *pre-cal.*): average Z error on 1600 check points in a L block in a flat area adjusted with camera parameters from a pre-calibration block in CFG_i configuration, with CFG_i = LCO, LO, ..., HL and HO; and $Z\ error$ (LCO *pre-cal.*): average Z error on 1600 check points in a L block in a flat area adjusted with camera parameters from the reference pre-calibration block. The reference LCO varies according to block control type (GCP or GNSS), control tightness and terrain type.

Likewise, the relative percentage differences Δ_{Z_range} for the Z error range have been computed as:

$$\Delta_{Z\_range} = \frac{eZ\ range(\text{CFG}_i\ pre\_cal.) - eZ\ range\ (\text{LCO}\ pre\_cal.)}{eZ\ range\ (\text{LCO}\ pre\_cal.)} \quad (6)$$

where $eZ\ range$ (CFG_i *pre-cal.*): average Z error range (difference of the largest positive and the largest negative Z error) on 1600 check points in a L block in a flat area adjusted with camera parameters from a pre-calibration block in CFG_i configuration, with CFG_i = LCO, LO, ..., HL and HO; and $eZ\ range$ (LCO *pre-cal.*): average Z error range (difference of the largest positive and the largest negative Z error) on 1600 check points in a L block in a flat area adjusted with camera parameters from the reference pre-calibration block. The reference LCO varies according to block control type (GCP or GNSS), control tightness and terrain type.

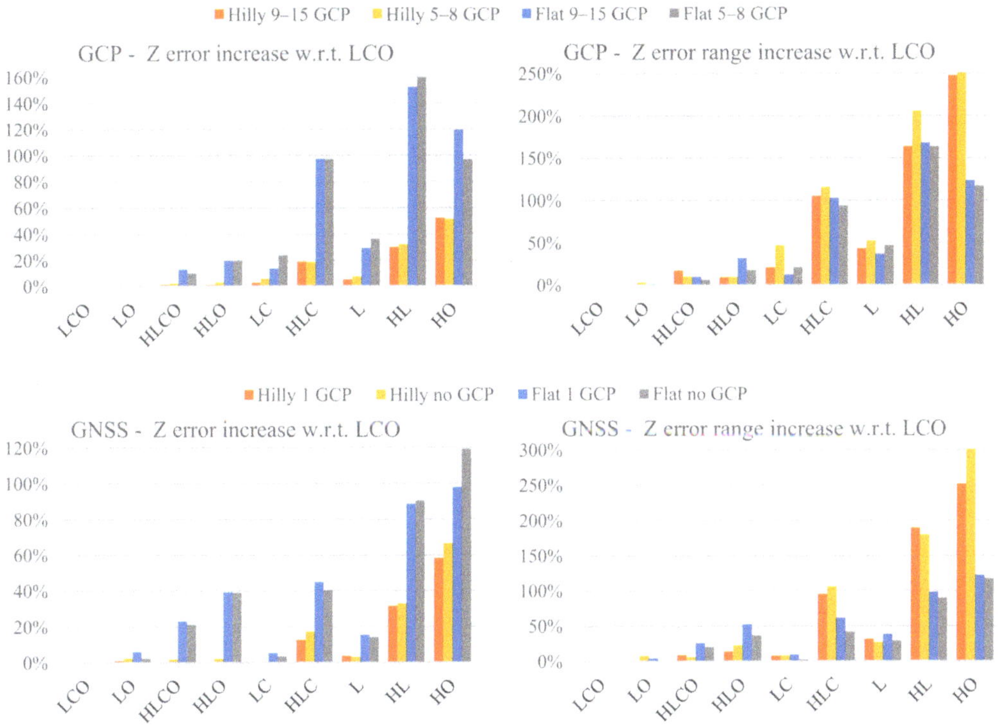

Figure 11. Average *Z error* increment (left) and *Z error* range increment (right) when applying to a L configuration block a pre-calibrated camera parameter set estimated with a given configuration with respect to a pre-calibration set estimated with LCO configuration. The results are presented for both the GCP (**top**) and GNSS (**bottom**) block control cases, as a function of block configuration, block control and terrain type.

From Figure 11 left, it is apparent that the percentage error gap can be dramatic, especially in flat terrain and in square blocks, if oblique images are missing: the worst case is HL, with 150% and 90% increase in GCP and GNSS case, respectively. In all cases pre-calibration parameters estimated on hilly terrain perform better compared to those on a flat terrain: except for the HO case, the percentage increase of the *Z error* is always less than half compared to that from a calibration over flat terrain, and much less so in the strongest block configurations.

In the GCP case, with pre-calibration executed over a hilly terrain, LO, HLCO and HLO configurations are on par with LCO pre-calibration. This applies also to LC and, perhaps surprisingly, to L (only 7% worse than LCO). Square blocks without oblique images (HLC or HL), on the other hand, deliver calibration parameters that produce *Z errors* 20 to 30% worse. Pre-calibration parameters estimated over a flat terrain with square blocks are not as effective even with oblique images (HLCO +13% and HLO +20%), and much worse without (+97% and +160% in HLC and HL, respectively).

In the GNSS case with pre-calibration executed over hilly terrain, all rectangular configurations (LO, LC and L) and the square configurations with oblique imaging (HLCO and HLO) are on a par with LCO. As in the GCP case, HLC (+17%) and HL (+33%) produce instead significantly larger *Z errors*. In flat terrain a pre-calibration with GNSS rectangular blocks perform better than square ones, as in the GCP case, even if they include oblique images (HLCO +23% and HLO +39%). Comparing GNSS and GCP pre-calibration, GNSS is always better in rectangular blocks and in all square blocks except those including oblique images.

In both the GNSS and the GCP case, the pre-calibration block control tightness seems to play a marginal role (i.e., increasing block control does not significantly reduce the gap with respect to the reference case LCO).

The *Z error* range looks rather independent of the terrain type and block control; it is larger for weaker configurations, but without a clear monotonic trend (i.e., matching the decreasing "block strength" emerging from the previous analysis). The ratio between average height of the dome (Volume/Area) and error range (difference between maximum and minimum height of the dome) is almost constant in flat terrain, from 6 to 7; in hilly terrain instead, it increases from 1.9 to 4.9 with decreasing block strength.

A comparison between the effectiveness of camera calibration when taking advantage of GNSS-determined camera stations and when using GCP is among the paper objectives. The average *Z error* obtained using camera parameters from the same calibration block configuration adjusted with GNSS-determined camera stations with respect to the equivalent error obtained from adjustments with camera parameters obtained with GCP control is shown in Figure 12. To compare both pre-calibrations, the percentage difference $\Delta_{pre\text{-}cal}$ has been computed as:

$$\Delta_{pre-cal} = \frac{Z\ error(\text{GCP }pre\_cal.) - Z\ error\ (\text{GNSS }pre\_cal.)}{Z\ error\ (\text{GCP }pre\_cal.)} \qquad (7)$$

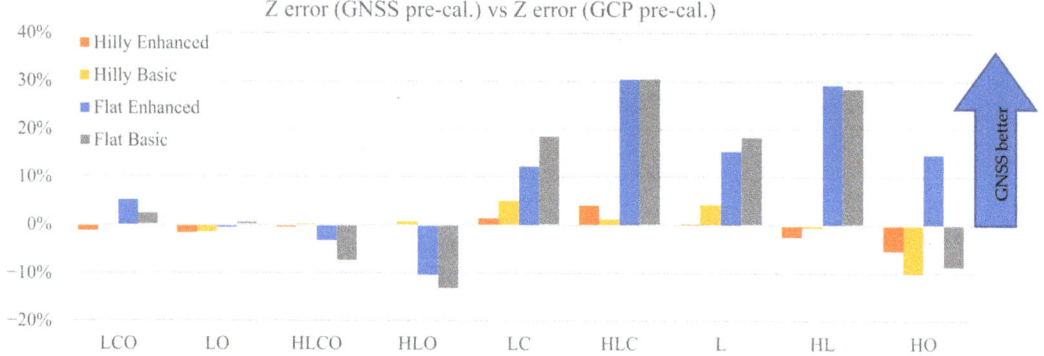

Figure 12. Percentage *Z error* difference on check points for blocks pre-calibrated with camera parameter sets estimated with GNSS-determined camera stations or GCP. Percentages are shown as a function of the nine cases of the terrain type (Flat, Hilly) and of the control tightness (Basic or Enhanced). Positive values mark comparatively smaller *Z errors* for GNSS w.r.t GCP and vice versa for negative values.

Positive $\Delta_{pre\text{-}cal}$ values mark comparatively smaller *Z errors* for GNSS pre-calibration w.r.t. GCP pre-calibration and vice versa for negative values.

It is apparent that in hilly terrain both control types are basically equivalent, as differences are below 5%. Likewise, block control tightness is not very important as differences between Basic and Enhanced are also below 5%. In flat terrain with oblique images GCP performs better; however, just slightly so, with differences ranging from almost insignificant (LO, less than 1%) to small (HLO, 13%). Without oblique images, GNSS pre-calibration is better, with improvements up to 30% for square blocks (HLC and HL) and a bit smaller (up to 19%) in rectangular shaped blocks (L and LC). Interestingly, also with Basic tightness (no GCP) the GNSS case seems to deliver better calibration parameters than GCP when flying over a flat terrain.

4. Discussion

4.1. Camera Calibration Parameters

Overall, as far as the estimation accuracy of the IO parameters is concerned, the GNSS and the GCP cases show similar trends with respect to block configuration, terrain type and block control. Accurate estimation of the principal distance (see Figure 3) is ensured if POI oblique images are included to complement nadir-imaging longitudinal (and possibly cross) strips; if they are missing, the accuracy becomes two-to-five times worse. The HL case in flat terrain is particularly critical for both the GNSS and GCP case (up to ten times worse than the best case). It should also be noted that [10] in a single POI block (HO case) the accuracy is five times worse than the best case. Cross strips, on the other hand, provide only a marginal improvement. Although, at first thought, this might seem surprising, the image block being much more rigid with cross-strips, it is actually in line with findings from [10,18] where cross strips attained less than expected improvements or worse results. It should be noted, as far as principal distance is concerned, that nadir-imaging cross strips do not introduce significant new geometrical constraints (from a projective point of view) for its estimation. On the contrary, having a more significant depth change in the scene pictured by the oblique images (as well as due to the object geometry e.g., as in the hilly study area), drastically increases the accuracy of the estimation.

The accuracy of Principal Point (PP—see Figures 4 and 5) estimation in hilly terrain is very stable with respect to block configuration and control tightness, while in flat terrain the accuracy gets worse with weak block geometries. It should be noted, in this context, that the use of cross strips increases, although not drastically, the determination of the PP location. In fact, it is well known (see for instance [8]) that the use of 90-degree-rolled images in a calibration image block prevents, or at least reduces, the insurgence of unwanted correlations between the parameters and in particular the ones associated to the PP.

The average and standard deviation of the maximum residual distortion affecting the image coordinates after camera calibration parameter estimation (see Figure 6) show trends quite similar to those of PP accuracy estimation. It is worth pointing out that, in this analysis, the distortion error considers both the effects due to a not accurate estimation of the radial and tangential calibration parameters and the ones induced by a not accurate Principal Distance and Principal Point estimation. In other words, the reported errors represent the image coordinates error on image plane due to all the estimated parameters. It is therefore intuitive that this analysis shows similar trends of the ones in Figures 4 and 5. The HO case is a stunning exception in the GNSS case as, even in hilly terrain, the accuracy is more than ten times worse than the best case. This is also true for the maximum average distortion, where HO shows a clear gap compared to other configurations.

To summarize the comparison between block control by GNSS or GCP, there is perhaps no outright winner, but a clear edge for the GNSS case, which performs better especially in weaker block geometry configurations. In agreement with findings from [10,27], accurate determination of all interior orientation parameters is possible with GNSS even without GCP, if oblique images can be included. At first sight, this seems to contradict authors' [40] and others' previous tests [34]. However, it should be noted that in both the cited cases the GNSS-assisted blocks were made of nadir images only, as the flights were performed with fixed-wing platforms. Moreover, the authors of [35], flying only longitudinal strips, found adding 1 GCP necessary and sufficient to recover bias in elevation due to inaccurate determination of principal distance.

The question about the optimal survey block configuration is likely to remain open, as the variety of parameters to explore is really too large. As far as our contribution to this point is concerned, a few basic configurations and their combinations have been taken into account. However, some promising variants in the imaging geometry i.e., flying the longitudinal strips with moderate-to-strong (30° to 45°) camera axis inclination along flight direction [26] that recently received attention [10,27,28] were not considered. Another caveat applies to block size and shape, especially in the GNSS case, as pointed out in [27]: should large blocks be composed by juxtaposing basic, optimized sub-block tiles? Do results found with this and other simulations apply to any block size and shape? Is a complete layer of oblique images necessary to complement a basic longitudinal strip layer or, as suggested in [14], is taking just one at each block corner enough? From the results, longitudinal nadir-only blocks should be limited to hilly terrain (in the presented case the largest image scale was three times bigger than the smallest one), where calibration is still fine and the accuracy loss on ground coordinate (see next section) compared to LCO is negligible in horizontal coordinates and does not exceed 20% in elevation. This agrees with [1]. Adding two flight layers (C and O) to the basic longitudinal one delivers of course the top results. It should be noted that, in most cases (see Figures 4–6) dropping one of the two results in significant worse accuracies (at least as far as the percentage error increment is considered) of the estimated camera model parameters, but does not result in a significant accuracy loss for the ground coordinates (see Figure 8), except for the GNSS control case on flat terrain. If a choice is to be made between cross and oblique, our results are ambiguous. In flat terrain oblique images are necessary for accurate determination of all IO parameters, while cross strips are only effective with PP coordinate estimation. On the other hand, LC and HLC configurations for rectangular and square blocks show significantly better RMSE on tie point coordinates for hilly terrain, and better or comparable ones for flat terrain compared to LO and HLO.

Do GNSS-based and GCP-based image blocks deliver equivalent calibration accuracy? Broadly speaking the answer is negative, as the former performs always better than the latter in flat terrain if 1 GCP is used, with improvements up to 30%, while the latter is 10% to 20% better with strong block configurations in hilly terrain. It should be noted, however, that from a practical standpoint, GNSS-assisted UAV surveys come with significantly fewer operational constraints than traditional GCP-based ones, especially if the area investigated presents accessibility issues and if total time of operation is critical. The simulations seem to confirm what several of the previously cited authors illustrate in their contributions: the current state of the GNSS technologies implemented in most of the modern RTK UAV systems are already precise enough to implement accurate, and maybe also reliable, GCP-free surveys in most of (if not all) operational conditions. In author's opinion, acquiring also some GCP (at least one) remains an important requirement nonetheless: as far as the accuracy of the ground points is concerned, introducing at least one GCP might highlight some RTK solution bias and reduce it to some extent. In author's experience the GNSS UAV navigation solution is sometimes affected by systematic errors, easily masked in a pure GNSS-assisted solution. Additional independent ground control constraints can therefore dramatically increase the survey reliability. At the same time, as the simulations highlighted, including at least one GCP in the GNSS-assisted block might increase significantly (though not drastically) also the quality of the IO and distortion parameters estimation, especially for the weaker image block geometries.

4.2. Check Point Coordinates Accuracy

With the exception of HO case, the accuracy loss of horizontal coordinates as a function of the block configuration grows from just 1% (LC) to 30% (HL) in hilly terrain but reaches 60% (HL) in flat terrain. The double grid configurations show the lower loss (Figure 8). The pattern is similar for the GNSS and GCP cases, though in flat terrain the loss rate is more pronounced for the former.

As far as elevations are concerned, in hilly terrain the pattern is similar to horizontal coordinates, though the loss is higher (38% in HL case) in both the GCP and GNSS cases. In flat terrain, however, the GNSS and the GCP show, to the contrary, clear differences. In the former, without oblique images, the accuracy loss is quite sensitive (up to 175% in HL) to afford the lack of ground control. Adding (at least) a single GCP does not really solve the problem as the overall loss remains very high (75% in HL). To the contrary, with inclusion of oblique images, there is no difference between adding or not the single GCP and the overall loss is below 50% in the worst case (HLO). This suggests that adding the GCP as proposed in [40] is not the best solution to error estimation in the principal distance: using a stronger block configuration is more effective. In the light of [27] results and of authors' findings, a double grid with a moderate pitch angle configuration seems the best trade-off, though perhaps not yet an operational solution for many fixed-wing platforms.

In the GCP case, increasing the control tightness does not bring substantial improvements in horizontal coordinates; in elevations the gains are a bit higher, but not much. Though a meaningful comparison is difficult, this result only partly agrees with findings in [12].

Accuracy gains by increasing by a factor three the control precision (Figure 10) are very limited in hilly terrain, being less than 5% in both horizontal and vertical coordinates. In flat terrain the situation is more complex. In GCP case the improvement is between 5% and 10% in elevation and mostly less than 5% in horizontal coordinates. This point agrees with [12] results. With aerial control, the improvement in horizontal coordinates is still modest, below 10%. In elevation, to the contrary, configurations without oblique images gain from 15% (with a single GCP fixed) to 25% (without GCP) while the remaining are basically not affected.

4.3. Dome Effect

Before discussing the results of Section 3.4, it should be stressed again that they refer to the case of pre-calibration only. In other words, what has been presented is an analysis of the pre-calibration block configuration performance in possibly delivering an effective camera calibration parameter set. All the IO and distortion parameter sets evaluated in the different image block configurations were applied (i.e., were used as pre-calibrated parameters) in a single (always the same) L (ongitudinal) image block. For the results presented in Figure 11, the LCO configuration has been taken as "gold standard" and the results of the other configuration types have been measured relative to that case, in order to measure the calibration accuracy loss when pre-calibrating with a weaker block configuration.

As far as *Z error* increase is concerned (see Figure 11 left), a pre-calibration over hilly terrain with both control types (GCP and GNSS) is always better than one over flat terrain. Moreover, except for some weaker configurations (i.e., HLC and HL) the increase in *Z error* is very limited i.e., up to 7% worse. For HLC the increase is ca. 19%, while for HL is stronger (32%). In flat terrain, on the other hand, if the configuration includes oblique imaging, the accuracy loss is minimal only for rectangular block shapes (LO) while in square blocks (HLCO and HLO) the gap is noticeable (10 and 20% in GCP case and 23 to 40% in the GNSS case). Without oblique imaging, there are again similarities between GCP and GNSS control, but the gap loss with respect to LCO is reversed (now GCP is almost twice worse than GNSS). In other words, the weaker the pre-calibration block configuration, the more that an accurate camera station position helps in camera calibration. Again, square blocks are less effective than rectangular ones: LC and L are about three to four times better (GCP case) or even more (GNSS case) than HLC and HL. Motivations for this behaviour should be further investigated. In fact, the differences between the square vs. rectangular image blocks resides only, in authors' opinion, in a number of observations approximately two times larger, that should not be enough to justify the results. At the same time, analysing the results of camera model parameters estimations and the connected ground point accuracy (see previous sections), even if square (H) blocks provide usually worse results, the differences with rectangular configurations are much smaller.

It is interesting to compare the results presented in Figure 11 with those concerning the actual accuracy in determining the IO and calibration parameters (shown in Figures 3–6) and the associated behaviour of the different image block configurations.

Looking at the increase in the *Z error* range (see Figure 11 right) three points can be stressed: the loss with respect to LCO is generally much larger; pre-calibration on hilly terrain does not rule out the chance of large errors; the gap between flat and hilly terrain is mostly small in the GCP case but, in the GNSS case for HLC and HL configurations, the error range for pre-calibration in hilly terrain is even larger than in flat terrain (a fact yet without a clear explanation).

The comparison between GCP-based and GNSS-assisted (camera-based) block control shows that pre-calibration with the latter is generally a better option, as smaller Z errors compared to GCP control are obtained. Indeed, results shown in Figure 12 indicate that, as long as oblique imaging is included in the block, it makes little difference in terms of Z error whether block control is achieved with GCP or GNSS, as all LCO, LO, HLCO and HLO configurations obtain similar errors with both control types. This can be seen as in agreement with the claim of [13] that calibration is first and foremost a matter of block imaging geometry and camera modelling and that oblique imaging is an essential element of such imaging geometry in blocks flown over flat terrain as well as, generally, with all previous works on optimal imaging for camera calibration. On the other hand, looking at weaker configurations, when imaging geometry is less robust (LC, HLC, L and HL), camera projection centres are more helpful than GCP in flat terrain. There also seems to be a dependence of the improvement amount on the block shape, while cross strips seem less important. Indeed, with our test settings the gain is limited (from 12 to 19%) in rectangular blocks (LC and L), while it is larger (up to 30%) with square blocks (HLC and HL). In short, if, for whatever reason, oblique imaging is not applicable in a survey over flat terrain, using GNSS-assisted orientation is more advisable than using GCP; the remarkable indication is that this applies also when no GCP are available on ground i.e., in the Basic tightness control case, which is in agreement with results shown in Figures 6 and 7.

5. Conclusions

Drawing conclusions in a topic as complex as UAV camera calibration with reasonable confidence on their scope and validity is never easy, as the results always come out of given experiment settings, never exhaustive of the multi-dimensional space of the process relevant parameters. As such, keeping in mind the test characteristics depicted in Section 2, a few conclusions are presented in the following.

As far as accuracy of interior orientation parameters is concerned, though trivial to say, the calibration block configuration matters a lot: the accuracy decrease could be as high as 30 times in the worst case for the principal distance, though less (nine times) for the PP coordinates. Oblique images help a lot (LO is almost as good as LCO), though a POI-only (HO) calibration is not recommendable: in our findings nadir looking images are also necessary. The comparison between ground (GCP) and aerial (GNSS-assisted) block control configurations shows that over flat terrain the latter deliver 20% to 60% more accurate calibration parameters than the former, in almost all configurations and for all IO parameters. In hilly terrain GCP control is generally better, though no more than 20%. Unless oblique images are included, estimation of principal distance in the GNSS case over flat terrain might result in large errors.

Estimation errors of the calibration parameters in a pre-calibration block, when applied as fixed parameters in a subsequent BBA, affect ground point coordinates. In this respect our conclusions are that configuration of the pre-calibration block matters in general, and particularly when flying over flat terrain. The average Z error increase for weaker configurations compared to LCO can be as large as 150% with GCP control; less so, but still up to 90% for GNSS control. GNSS-assisted block control is in most cases a better option than GCP control in pre-calibration (only with oblique imaging included the difference is minimal). In weaker configurations over flat terrain, camera station positions constrain the block more than GCP.

As far as tie point ground coordinates RMSEs are concerned, weakening the calibration block configuration leads to sizeable but limited percentage accuracy losses in hilly terrain (below 35%) while losses reach 70% in elevation in flat terrain. Block control by GNSS or by GCP are in practice equally accurate in horizontal coordinates, while in elevation GNSS without oblique imaging and no GCP might perform up to 50% worse.

Simulations also confirm that, as many practical experiments have shown, in the GNSS case GCP are generally not necessary for both horizontal coordinates and elevations; however, in flat terrain oblique imaging is necessary to avoid errors in the latter.

Author Contributions: Conceptualization, R.R., G.F.; methodology and software, R.R.; data curation, R.R., G.F.; validation and investigation, R.R., G.F.; writing—original draft preparation, G.F.; writing—review and editing, R.R., G.F. Both authors have read and agreed to the published version of the manuscript.

Funding: This research received no external funding.

Institutional Review Board Statement: Not applicable.

Informed Consent Statement: Not applicable.

Data Availability Statement: Data supporting the findings of this study could be available from the authors on request.

Conflicts of Interest: The authors declare no conflict of interest.

References

1. Fraser, P.C. Camera Calibration Considerations for UAV Photogrammetry Cameras for Drones/UAS/UAVs. In Proceedings of the ISPRS Technical Commission II Symposium, Riva del Garda, Italy, 3–7 June 2018.
2. Brown, D.C. Close-Range Camera Calibration. *Photogramm. Eng.* **1971**, *37*, 855–866.
3. Tsai, R.Y. A Versatile Camera Calibration Technique for High-Accuracy 3D Machine Vision Metrology Using Off-the-Shelf TV Cameras and Lenses. *IEEE J. Robot. Autom.* **1987**, *3*, 323–344. [CrossRef]
4. Zhang, Z. A Flexible New Technique for Camera Calibration. *IEEE Trans. Pattern Anal. Mach. Intell.* **2000**, *22*, 1330–1334. [CrossRef]
5. Remondino, F.; Fraser, C.; Remondino, F. Digital Camera Calibration Methods Considerations and Comparisons digital camera calibration methods: Considerations and comparisons. *Int. Arch. Photogramm. Remote Sens. Spat. Inf. Sci.* **2006**, *36*, 266–272. [CrossRef]
6. Fraser, C.S. On the Use of Nonmetric Cameras in Analytical Close-Range Photogrammetry. *Can. Surv.* **1982**, *36*, 259–279. [CrossRef]
7. Gruen, A.; Beyer, H.A. System Calibration Through Self-Calibration. In *Calibration and Orientation of Cameras in Computer Vision 2001*; Springer: Berlin/Heidelberg, Germany, 2001; pp. 163–193.
8. Luhmann, T.; Robson, S.; Kyle, S.; Boehm, J. *Close-Range Photogrammetry and 3D Imaging*; De Gruyter: Berlin, Germany, 2019.
9. Clarke, T.A.; Fryer, J.G. The Development of Camera Calibration Methods and Models. *Photogramm. Rec.* **1998**, *16*, 51–66. [CrossRef]
10. Sanz-Ablanedo, E.; Chandler, J.H.; Ballesteros-Pérez, P.; Rodríguez-Pérez, J.R. Reducing Systematic Dome Errors in Digital Elevation Models through Better UAV Flight Design. *Earth Surf. Process. Landf.* **2020**, *45*, 2134–2147. [CrossRef]
11. Cramer, M.; Przybilla, H.J.; Zurhorst, A. UAV Cameras: Overview and Geometric Calibration Benchmark. In Proceedings of the International Archives of the Photogrammetry, Remote Sensing and Spatial Information Sciences—ISPRS Archives, International Conference on Unmanned Aerial Vehicles in Geomatics, Bonn, Germany, 4–7 September 2017; Volume 42, pp. 85–92.
12. Harwin, S.; Lucieer, A.; Osborn, J. The Impact of the Calibration Method on the Accuracy of Point Clouds Derived Using Unmanned Aerial Vehicle Multi-View Stereopsis. *Remote Sens.* **2015**, *7*, 11933–11953. [CrossRef]
13. Carbonneau, P.E.; Dietrich, J.T. Cost-Effective Non-Metric Photogrammetry from Consumer-Grade SUAS: Implications for Direct Georeferencing of Structure from Motion Photogrammetry. *Earth Surf. Process. Landf.* **2017**, *42*, 473–486. [CrossRef]
14. Radford, C.R.; Bevan, G. A Calibration Workflow for "Prosumer" Uav Cameras. In Proceedings of the International Archives of the Photogrammetry, Remote Sensing and Spatial Information Sciences—ISPRS Archives, ISPRS Geospatial Week 2019, Enschede, The Netherland, 10–14 June 2019; Volume 42, pp. 553–558.
15. Griffiths, D.; Burningham, H. Comparison of Pre- and Self-Calibrated Camera Calibration Models for UAS-Derived Nadir Imagery for a SfM Application. *Prog. Phys. Geogr.* **2019**, *43*, 215–235. [CrossRef]
16. Rosnell, T.; Honkavaara, E. Point Cloud Generation from Aerial Image Data Acquired by a Quadrocopter Type Micro Unmanned Aerial Vehicle and a Digital Still Camera. *Sensors* **2012**, *12*, 453–480. [CrossRef]
17. Forlani, G.; Diotri, F.; Morra Di Cella, U.; Roncella, R. UAV block georeferencing and control by on-board gnss data. In Proceedings of the International Archives of the Photogrammetry, Remote Sensing and Spatial Information Sciences—ISPRS Archives, XXIV ISPRS Congress (2020 Edition), 31 August–2 September 2020; Volume 43, pp. 9–16. Available online: https://www.proquest.com/openview/37db9d7ecadbebcaae8192398a690c8f/1?pq-origsite=gscholar&cbl=2037674 (accessed on 6 September 2021).
18. Cramer, M.; Zhang, S. Quality Assessment of High-Resolution UAV Imagery and Products. In Proceedings of the 40. Wissenschaftlich-Technische Jahrestagung der DGPF in Stuttgart, Stuttgart, Germany, 4–6 March 2020; pp. 33–46.

19. Barazzetti, L.; Mussio, L.; Remondino, F.; Scaioni, M. Targetless camera calibration. In Proceedings of the ISPRS—International Archives of the Photogrammetry, Trento 2011 Workshop, Trento, Italy, 2–4 March 2011; pp. 335–342. [CrossRef]
20. Fraser, C.S. Automatic Camera Calibration in Close Range Photogrammetry. *Photogramm. Eng. Remote Sens.* **2013**, *79*, 381–388. [CrossRef]
21. Nesbit, P.R.; Hugenholtz, C.H. Enhancing UAV-SfM 3D Model Accuracy in High-Relief Landscapes by Incorporating Oblique Images. *Remote Sens.* **2019**, *11*, 239. [CrossRef]
22. Wackrow, R.; Chandler, J.H. A Convergent Image Configuration for DEM Extraction That Minimises the Systematic Effects Caused by an Inaccurate Lens Model. *Photogramm. Rec.* **2008**, *23*, 6–18. [CrossRef]
23. Wackrow, R.; Chandler, J.H. Minimising Systematic Error Surfaces in Digital Elevation Models Using Oblique Convergent Imagery. *Photogramm. Rec.* **2011**, *26*, 16–31. [CrossRef]
24. James, M.R.; Robson, S. Mitigating Systematic Error in Topographic Models Derived from UAV and Ground-Based Image Networks. *Earth Surf. Process. Landf.* **2014**, *39*, 1413–1420. [CrossRef]
25. Zhou, Y.; Rupnik, E.; Meynard, C.; Thom, C.; Pierrot-Deseilligny, M. Simulation and Analysis of Photogrammetric UAV Image Blocks-Influence of Camera Calibration Error. *Remote Sens.* **2020**, *12*, 22. [CrossRef]
26. Taddia, Y.; Stecchi, F.; Pellegrinelli, A. Coastal Mapping Using Dji Phantom 4 RTK in Post-Processing Kinematic Mode. *Drones* **2020**, *4*, 9. [CrossRef]
27. Stott, E.; Williams, R.D.; Hoey, T.B. Ground Control Point Distribution for Accurate Kilometre-Scale Topographic Mapping Using an Rtk-Gnss Unmanned Aerial Vehicle and Sfm Photogrammetry. *Drones* **2020**, *4*, 55. [CrossRef]
28. James, M.R.; Antoniazza, G.; Robson, S.; Lane, S.N. Mitigating Systematic Error in Topographic Models for Geomorphic Change Detection: Accuracy, Precision and Considerations beyond off-Nadir Imagery. *Earth Surf. Process. Landf.* **2020**, *45*, 2251–2271. [CrossRef]
29. Sanz-Ablanedo, E.; Chandler, J.H.; Rodríguez-Pérez, J.R.; Ordóñez, C. Accuracy of Unmanned Aerial Vehicle (UAV) and SfM Photogrammetry Survey as a Function of the Number and Location of Ground Control Points Used. *Remote Sens.* **2018**, *10*, 1606. [CrossRef]
30. Tonkin, T.N.; Midgley, N.G. Ground-Control Networks for Image Based Surface Reconstruction: An Investigation of Optimum Survey Designs Using UAV Derived Imagery and Structure-from-Motion Photogrammetry. *Remote Sens.* **2016**, *8*, 786. [CrossRef]
31. Friess, P. Empirical Accuracy of Positions Computed from Airborne GPS Data. In *High Precision Navigation*; Springer: Berlin/Heidelberg, Germany, 1989; pp. 163–175.
32. Ackermann, F.; Schade, H. Application of GPS for Aerial Triangulation. *Photogramm. Eng. Remote Sens.* **1993**, *59*, 1625–1632.
33. Mirjam, B.; Eija, H.; Juha, J. GPS supported aerial triangulation using untargeted ground control. *Int. Arch. Photogramm. Remote Sens.* **1998**, *32*, 2–9.
34. Hugenholtz, C.; Brown, O.; Walker, J.; Barchyn, T.; Nesbit, P.; Kucharczyk, M.; Myshak, S. Spatial Accuracy of UAV-Derived Orthoimagery and Topography: Comparing Photogrammetric Models Processed with Direct Geo-Referencing and Ground Control Points. *Geomatica* **2016**, *70*, 21–30. [CrossRef]
35. Zhang, H.; Aldana-Jague, E.; Clapuyt, F.; Wilken, F.; Vanacker, V.; van Oost, K. Evaluating the Potential of Post-Processing Kinematic (PPK) Georeferencing for UAV-Based Structure-from-Motion (SfM) Photogrammetry and Surface Change Detection. *Earth Surf. Dyn.* **2019**, *7*, 807–827. [CrossRef]
36. Zhou, Y.; Rupnik, E.; Faure, P.H.; Pierrot-Deseilligny, M. GNSS-Assisted Integrated Sensor Orientation with Sensor Pre-Calibration for Accurate Corridor Mapping. *Sensors* **2018**, *18*, 2783. [CrossRef]
37. Forlani, G.; Diotri, F.; di Cella, U.M.; Roncella, R. Indirect UAV Strip Georeferencing by On-Board GNSS Data under Poor Satellite Coverage. *Remote Sens.* **2019**, *11*, 1765. [CrossRef]
38. Peppa, M.V.; Hall, J.; Goodyear, J.; Mills, J.P. Photogrammetric Assessment and Comparison of Dji Phantom 4 pro and Phantom 4 Rtk Small Unmanned Aircraft Systems. In Proceedings of the International Archives of the Photogrammetry, Remote Sensing and Spatial Information Sciences—ISPRS Archives, ISPRS Geospatial Week 2019, Enschede, The Netherlands, 10–14 June 2019; Volume 42, pp. 503–509.
39. James, M.R.; Robson, S.; Smith, M.W. 3-D Uncertainty-Based Topographic Change Detection with Structure-from-Motion Photogrammetry: Precision Maps for Ground Control and Directly Georeferenced Surveys. *Earth Surf. Process. Landf.* **2017**, *42*, 1769–1788. [CrossRef]
40. Benassi, F.; Dall'Asta, E.; Diotri, F.; Forlani, G.; Cella, U.M.; Roncella, R.; Santise, M. Testing Accuracy and Repeatability of UAV Blocks Oriented with Gnss-Supported Aerial Triangulation. *Remote Sens.* **2017**, *9*, 172. [CrossRef]
41. James, M.R.; Robson, S. Straightforward Reconstruction of 3D Surfaces and Topography with a Camera: Accuracy and Geoscience Application. *J. Geophys. Res. Earth Surf.* **2012**, *117*, 3017. [CrossRef]
42. Dall'Asta, E.; Delaloye, R.; Diotri, F.; Forlani, G.; Fornari, M.; di Cella, U.M.; Pogliotti, P.; Roncella, R.; Santise, M. Use of Uas in a High Mountain Landscape: The Case of Gran Sommetta Rock Glacier (AO). In Proceedings of the International Archives of the Photogrammetry, Remote Sensing and Spatial Information Sciences—ISPRS Archives, ISPRS Geospatial Week 2015, La Grande Motte, France, 28 September–3 October 2015; Volume 40.
43. Roncella, R.; Forlani, G.; Diotri, F. A monte carlo simulation study on the dome effect. *Int. Arch. Photogramm. Remote Sens. Spat. Inf. Sci.* **2021**, *43*, 53–60. [CrossRef]

44. James, M.R.; Robson, S.; d'Oleire-Oltmanns, S.; Niethammer, U. Optimising UAV Topographic Surveys Processed with Structure-from-Motion: Ground Control Quality, Quantity and Bundle Adjustment. *Geomorphology* **2017**, *280*, 51–66. [CrossRef]
45. Sturm, P. Critical Motion Sequences for Monocular Self-Calibration and Uncalibrated Euclidean Reconstruction. In Proceedings of the IEEE Computer Society Conference on Computer Vision and Pattern Recognition, San Juan, PR, USA, 17–19 June 1997; pp. 1100–1105. [CrossRef]

MDPI
St. Alban-Anlage 66
4052 Basel
Switzerland
Tel. +41 61 683 77 34
Fax +41 61 302 89 18
www.mdpi.com

Sensors Editorial Office
E-mail: sensors@mdpi.com
www.mdpi.com/journal/sensors

www.ingramcontent.com/pod-product-compliance
Lightning Source LLC
LaVergne TN
LVHW070157120526
838202LV00013BA/1333